Taking SIDES

Clashing Views on Controversial Issues in Science, Technology, and Society

Third Edition

Edited, Selected, and with Introductions by

Thomas A. Easton

Thomas Colle

Dushkin/McGraw-Hill
A Division of The McGraw-Hill Companies

Photo Acknowledgments

Cover image: © 1998 by PhotoDisc, Inc.

Cover Art Acknowledgment

Charles Vitelli

Manufactured in the United States of America

Third Edition

10 9 8 7 6 5 4 3 2

Library of Congress Cataloging-in-Publication Data

Main entry under title:
Taking sides: clashing views on controversial issues in science, technology, and society/edited, selected, and with introductions by Thomas A. Easton.—3rd ed.
Includes bibliographical references and index.
1. Science—Social aspects. 2. Technology—Social aspects. I. Easton, Thomas A., *comp.*
306.45

0-07-292815-8

PREFACE

Those who must deal with scientific and technological issues—scientists, politicians, sociologists, business managers, and anyone who is concerned about a neighborhood dump or power plant, government intrusiveness, expensive space programs, or the morality of medical research, among many other issues—must be able to consider, evaluate, and choose among alternatives. Making choices is an essential aspect of the scientific method. It is also an inescapable feature of every public debate over a scientific or technological issue, for there can be no debate if there are no alternatives.

The ability to evaluate and to select among alternatives—as well as to know when the data do not permit selection—is called critical thinking. It is essential not only in science and technology but in every other aspect of life as well. *Taking Sides: Clashing Views on Controversial Issues in Science, Technology, and Society* is designed to stimulate and cultivate this ability by holding up for consideration 17 issues that have provoked substantial debate. Each of these issues has at least two sides, usually more. However, each issue is expressed in terms of a single question in order to draw the lines of debate more clearly. The ideas and answers that emerge from the clash of opposing points of view should be more complex than those offered by the students before the reading assignment.

The issues in this book were chosen because they are currently of particular concern to both science and society. They touch on the nature of science and research, the relationship between science and society, the uses of technology, and the potential threats that technological advances can pose to human survival. And they come from a variety of fields, including computer and space science, biology, environmentalism, law enforcement, and public health.

Organization of the book For each issue, I have provided an *issue introduction,* which provides some historical background and discusses why the issue is important. I then present two selections, one pro and one con, in which the authors make their cases. Each issue concludes with a *postscript* that brings the issue up to date and adds other voices and viewpoints. At the back of the book is a listing of all the *contributors to this volume,* which gives information on the scientists, technicians, professors, and social critics whose views are debated here.

Which answer to the issue question—yes or no—is the correct answer? Perhaps neither. Perhaps both. Students should read, think about, and discuss the readings and then come to their own conclusions without letting my or their instructor's opinions (which perhaps show at least some of the time!) dictate theirs. The additional readings mentioned in both the introductions and the postscripts should prove helpful. It is worth stressing that the issues covered in this book are all *live* issues; that is, the debates they represent are active and ongoing.

Changes to this edition This third edition represents a considerable revision. There are 6 completely new issues: *Is Science a Faith?* (Issue 2); *Should the Theory of Evolution Be Replaced by Creationism?* (Issue 3); *Will Future Generations Have Enough to Eat?* (Issue 4); *Are Environmental Regulations Too Restrictive?* (Issue 8); *Are Computers Hazardous to Literacy?* (Issue 13); and *Is It Ethically Permissible to Clone Human Beings* (Issue 17).

In addition, for 5 of the issues retained from the previous edition, either the YES or the NO selection has been replaced to bring the debates up to date: *Are Electromagnetic Fields Dangerous to Your Health?* (Issue 7); *Should the Goals of the U.S. Space Program Include Manned Exploration of Space?* (Issue 9); *Is It Worthwhile to Continue the Search for Extraterrestrial Life?* (Issue 10); *Will the Information Revolution Benefit Society?* (Issue 12); and *Is It Ethical to Use Humans as "Experimental Animals"?* (Issue 16). In all, there are 15 new selections. The issue introductions and postscripts for the retained issues have been revised and updated where necessary.

A word to the instructor An *Instructor's Manual With Test Questions* (multiple-choice and essay) is available through the publisher for the instructor using *Taking Sides* in the classroom. It includes suggestions for stimulating in-class discussion for each issue. A general guidebook, *Using Taking Sides in the Classroom,* which discusses methods and techniques for integrating the pro-con approach into any classroom setting, is also available. An online version of *Using Taking Sides in the Classroom* and a correspondence service for *Taking Sides* adopters can be found at `www.cybsol.com/usingtakingsides/`. For students, we offer a field guide to analyzing argumentative essays, *Analyzing Controversy: An Introductory Guide,* with exercises and techniques to help them to decipher genuine controversies.

Taking Sides: Clashing Views on Issues in Science, Technology, and Society is only one title in the Taking Sides series. If you are interested in seeing the table of contents for any of the other titles, please visit the Taking Sides Web site at `http://www.dushkin.com/takingsides/`.

Acknowledgments A special thanks goes to those professors who responded to the questionnaire with specific suggestions for the third edition: John Bumpus, University of Northern Iowa; Thomas Liao, SUNY at Stony Brook; and Diane Long, California Polytechnic State University.

Special thanks are due to John Quigg of Thomas College.

Thomas A. Easton
Thomas College

CONTENTS IN BRIEF

PART 1 THE PLACE OF SCIENCE AND TECHNOLOGY IN SOCIETY 1

Issue 1. Should the Federal Government Point the Way for Science? **2**

Issue 2. Is Science a Faith? **24**

Issue 3. Should the Theory of Evolution Be Replaced by Creationism? **42**

PART 2 THE ENVIRONMENT 65

Issue 4. Will Future Generations Have Enough to Eat? **66**

Issue 5. Should Society Be Concerned About Global Warming? **88**

Issue 6. Is Ozone Depletion a Genuine Threat? **110**

Issue 7. Are Electromagnetic Fields Dangerous to Your Health? **130**

Issue 8. Are Environmental Regulations Too Restrictive? **146**

PART 3 THE CUTTING EDGE OF TECHNOLOGY 167

Issue 9. Should the Goals of the U.S. Space Program Include Manned Exploration of Space? **168**

Issue 10. Is It Worthwhile to Continue the Search for Extraterrestrial Life? **190**

Issue 11. Should Genetic Engineering Be Banned? **208**

PART 4 THE COMPUTER REVOLUTION 233

Issue 12. Will the Information Revolution Benefit Society? **234**

Issue 13. Are Computers Hazardous to Literacy? **252**

Issue 14. Will It Be Possible to Build a Computer That Can Think? **272**

PART 5 ETHICS 297

Issue 15. Is the Use of Animals in Research Justified? **298**

Issue 16. Is It Ethical to Use Humans as "Experimental Animals"? **318**

Issue 17. Is It Ethically Permissible to Clone Human Beings? **334**

CONTENTS

Preface **i**

Introduction: Analyzing Issues in Science and Technology **x**

PART 1 THE PLACE OF SCIENCE AND TECHNOLOGY IN SOCIETY **1**

ISSUE 1. Should the Federal Government Point the Way for Science? **2**

YES: David H. Guston and Kenneth Keniston, from "Updating the Social Contract for Science," *Technology Review* **4**

NO: National Academy of Sciences, from *Allocating Federal Funds for Science and Technology* **14**

Assistant professor of public policy David H. Guston and professor of human development Kenneth Keniston argue that public participation must be increased at all levels of decision making about science.The National Academy of Sciences, a private, nonprofit society of scholars engaged in scientific and engineering research, asserts that only scientists should decide what research is to be done.

ISSUE 2. Is Science a Faith? **24**

YES: Daniel Callahan, from "Calling Scientific Ideology to Account," *Society* **26**

NO: Richard Dawkins, from "Is Science a Religion?" *The Humanist* **35**

Bioethicist Daniel Callahan argues that science's domination of the cultural landscape unreasonably excludes other ways of understanding nature and the world. Biologist Richard Dawkins holds that science "is free of the main vice of religion, which is faith" because it relies on evidence and logic instead of tradition, authority, and revelation.

ISSUE 3. Should the Theory of Evolution Be Replaced by Creationism? **42**

YES: Jack Hitt, from "On Earth as It Is in Heaven: Field Trips With the Apostles of Creation Science," *Harper's Magazine* **44**

NO: Daniel C. Dennett, from *Darwin's Dangerous Idea: Evolution and the Meanings of Life* **52**

Jack Hitt, a contributing editor to *Harper's Magazine*, reports on a visit with a creationist who dreams of replacing much of human knowledge—particularly the theory of evolution—with something more consistent with the biblical Genesis. Philosopher Daniel C. Dennett argues that Charles Darwin's theory of evolution by means of natural selection has made its religious predecessors quite obsolete.

PART 2 *THE ENVIRONMENT* **65**

ISSUE 4. Will Future Generations Have Enough to Eat? **66**

YES: World Bank, from "Food Security for the World," Paper Prepared for the World Food Summit in Rome, Italy **68**

NO: Lester R. Brown, from "Can We Raise Grain Yields Fast Enough?" *World Watch* **77**

The World Bank argues that the only barriers to meeting the need for food of future generations are political and economic, not physical and biological. If we make the right decisions, agricultural productivity can keep up with population growth. Lester R. Brown, president of the Worldwatch Institute, argues that the physical and biological barriers are so significant that there is a strong chance that agricultural productivity will fall behind population growth.

ISSUE 5. Should Society Be Concerned About Global Warming? **88**

YES: Ross Gelbspan, from "The Heat Is On," *Harper's Magazine* **90**

NO: Wilfred Beckerman and Jesse Malkin, from "How Much Does Global Warming Matter?" *The Public Interest* **98**

Journalist Ross Gelbspan argues that the evidence for global warming is incontrovertible, despite the disinformation campaign being waged by the fossil fuels industry. Economists Wilfred Beckerman and Jesse Malkin argue that global warming, if it even occurs, will not be catastrophic and warrants no immediate action.

ISSUE 6. Is Ozone Depletion a Genuine Threat? **110**

YES: Mary H. Cooper, from "Ozone Depletion," *CQ Researcher* **112**

NO: James P. Hogan, from "Ozone Politics: They Call This Science?" *Omni* **121**

Mary H. Cooper, a staff writer for *CQ Researcher,* asserts that scientific findings in recent years indicate that the ozone layer is being depleted, exposing

Earth's living organisms to increasing levels of harmful ultraviolet radiation from the sun. James P. Hogan, a science fiction writer, maintains that reports of the ozone being destroyed by chlorofluorocarbons are unsupported by any valid scientific evidence.

ISSUE 7. Are Electromagnetic Fields Dangerous to Your Health? **130**

YES: Paul Brodeur, from *The Great Powerline Coverup: How the Utilities and the Government Are Trying to Hide the Cancer Hazard Posed by Electromagnetic Fields* **132**

NO: Edward W. Campion, from "Power Lines, Cancer, and Fear," *The New England Journal of Medicine* **140**

Writer Paul Brodeur argues that there is an increased risk of developing cancer from being exposed to electromagnetic fields (EMFs) given off by electric power lines and that the risk is significant enough to warrant immediate measures to reduce exposures to the fields. Physician Edward W. Campion argues that there is no credible evidence that there is any risk of developing cancer from EMF exposure.

ISSUE 8. Are Environmental Regulations Too Restrictive? **146**

YES: John Shanahan, from "Environment," in Stuart M. Butler and Kim R. Holmes, eds., *Issues '96: The Candidate's Briefing Book* **148**

NO: Paul R. Ehrlich and Anne H. Ehrlich, from "Brownlash: The New Environmental Anti-Science," *The Humanist* **157**

John Shanahan, vice president of the Alexis de Tocqueville Institution in Arlington, Virginia, argues that many government environmental policies are unreasonable and infringe on basic economic freedoms. Environmental scientists Paul R. Ehrlich and Anne H. Ehrlich argue that many objections to environmental protections are self-serving and based in bad or misused science.

PART 3 THE CUTTING EDGE OF TECHNOLOGY **167**

ISSUE 9. Should the Goals of the U.S. Space Program Include Manned Exploration of Space? **168**

YES: Robert Zubrin, from "Mars on a Shoestring," in Robert Zubrin and Richard Wagner, *The Case for Mars: The Plan to Settle the Red Planet and Why We Must* **170**

NO: John Merchant, from "A New Direction in Space," *IEEE Technology and Society Magazine* 179

Engineer Robert Zubrin argues that it is within our capabilities to establish a human presence on Mars and that doing so would benefit economies on Earth. John Merchant, a retired staff engineer at Loral Infrared and Imaging Systems, argues that it will be much cheaper to develop electronic senses and remotely operated machines that humans can use to explore other worlds.

ISSUE 10. Is It Worthwhile to Continue the Search for Extraterrestrial Life? 190

YES: Frank Drake and Dava Sobel, from *Is Anyone Out There? The Scientific Search for Extraterrestrial Intelligence* 192

NO: A. K. Dewdney, from *Yes, We Have No Neutrons: An Eye-Opening Tour Through the Twists and Turns of Bad Science* 200

Professor of astronomy Frank Drake and science writer Dava Sobel argue that scientists must continue to search for extraterrestrial civilizations because contact will eventually occur. Computer scientist A. K. Dewdney argues that although there may indeed be intelligent beings elsewhere in the universe, there are so many reasons why contact and communication are unlikely that searching for them is not worth the time or the money.

ISSUE 11. Should Genetic Engineering Be Banned? 208

YES: Andrew Kimbrell, from *The Human Body Shop: The Engineering and Marketing of Life* 210

NO: James Hughes, from "Embracing Change With All Four Arms: A Post-Humanist Defense of Genetic Engineering," *Paper Presented at the University of Chicago Health and Society Workshop* 219

Andrew Kimbrell, policy director of the Foundation on Economic Trends in Washington, D.C., argues that the development of genetic engineering is so marked by scandal, ambition, and moral blindness that society should be deeply suspicious of its purported benefits. James Hughes, assistant director of research at the MacLean Center for Clinical Medical Ethics in the Department of Medicine at the University of Chicago, argues that the potential benefits of genetic engineering greatly outweigh the potential risks.

PART 4 THE COMPUTER REVOLUTION 233

ISSUE 12. Will the Information Revolution Benefit Society? 234

YES: John S. Mayo, from "Information Technology for Development: The National and Global Information Superhighway," *Vital Speeches of the Day* 236

NO: Andrew L. Shapiro, from "Privacy for Sale: Peddling Data on the Internet," *The Nation* 244

John S. Mayo, president emeritus of Lucent Technologies Bell Laboratories, formerly AT&T Bell Laboratories, argues that the information revolution will aid economic development and improve access to education, health care, and other social services. Andrew L. Shapiro, a fellow of the Twentieth Century Fund, argues that information technology turns personal information into a commodity and threatens the basic human right to privacy.

ISSUE 13. Are Computers Hazardous to Literacy? 252

YES: Sven Birkerts, from *The Gutenberg Elegies: The Fate of Reading in an Electronic Age* 254

NO: Wen Stephenson, from "The Message Is the Medium: A Reply to Sven Birkerts and *The Gutenberg Elegies*," *Chicago Review* 262

Author Sven Birkerts argues that electronically presented information (i.e., via computer screens) threatens traditional conceptions of literacy, the literary culture, and the sense of ourselves as individuals. Wen Stephenson, editor of the *Atlantic Monthly*'s online edition, argues that the essence of literature, literacy, and the literary culture will survive the impact of computers.

ISSUE 14. Will It Be Possible to Build a Computer That Can Think? 272

YES: Hans Moravec, from "The Universal Robot," in *Vision-21: Interdisciplinary Science and Engineering in the Era of Cyberspace* 274

NO: John R. Searle, from "Is the Brain's Mind a Computer Program?" *Scientific American* 283

Research scientist Hans Moravec asserts that computers that match and even exceed human intelligence will eventually be developed. Professor of philosophy John R. Searle argues that artificial intelligence and human intelligence are so different that it is impossible to create a computer that can think.

PART 5 ETHICS **297**

ISSUE 15. Is the Use of Animals in Research Justified? **298**

YES: Elizabeth Baldwin, from "The Case for Animal Research in Psychology," *Journal of Social Issues* **300**

NO: Steven Zak, from "Ethics and Animals," *The Atlantic Monthly* **308**

Elizabeth Baldwin, research ethics officer of the American Psychological Association's Science Directorate, argues that the use of animals in scientific research is justified by the resulting benefits to both humans and animals. Research attorney Steven Zak argues that for society to be virtuous, it must recognize the rights of animals not to be sacrificed for human needs.

ISSUE 16. Is It Ethical to Use Humans as "Experimental Animals"? **318**

YES: Charles Petit, from "Sunday Interview: A Soldier in the War on AIDS," *San Francisco Chronicle* **320**

NO: Jonathan D. Moreno, from "The Dilemmas of Experimenting on People," *Technology Review* **325**

Science writer Charles Petit interviews an AIDS patient who underwent a highly experimental treatment and considers those who resist human experimentation to be far too cautious. Bioethicist Jonathan D. Moreno argues that although the requirements of informed consent may pose difficulties for research on human beings, "the simplicity and intuitive force of the ideas articulated in the Nuremberg Code ensure their lasting moral relevance."

ISSUE 17. Is It Ethically Permissible to Clone Human Beings? **334**

YES: John A. Robertson, from "A Ban on Cloning and Cloning Research Is Unjustified," Statement at the National Bioethics Advisory Commission Meeting, Washington, D.C. **336**

NO: Leon R. Kass, from "The Wisdom of Repugnance," *The New Republic* **342**

John A. Robertson, a medical ethics expert, argues that despite the various objections to cloning, "a ban on all human cloning is both imprudent and unjustified." Biochemist Leon R. Kass argues that human cloning is "so repulsive to contemplate" that it should be prohibited entirely.

Contributors **352**

Index **358**

INTRODUCTION

Analyzing Issues in Science and Technology

Thomas A. Easton

INTRODUCTION

As civilization approaches the dawn of the twenty-first century, it cannot escape science and technology. Their fruits—the clothes we wear, the foods we eat, the tools we use—surround us. Science and technology evoke in people both hope and dread for the future, for although new discoveries can lead to cures for diseases and other problems, new insights into the wonders of nature, and new toys (among other things), the past has shown that technological developments can also have unforeseen and terrible consequences.

Those consequences do *not* belong to science, for science is nothing more than a systematic approach to gaining knowledge about the world. Technology is the application of knowledge to accomplish things that otherwise could not be accomplished. Technological developments do not just lead to devices such as hammers, computers, and jet aircraft, but also to management systems, institutions, and even political philosophies. And it is, of course, such *uses* of knowledge that affect people's lives for good and ill.

It cannot be said that the use of technology affects people "for good *or* ill." As Emmanuel Mesthene said in 1969, technology is neither an unalloyed blessing nor an unmitigated curse.[1] Every new technology offers both new benefits and new problems, and the two sorts of consequences cannot be separated from each other. Automobiles, for example, provide rapid, convenient personal transportation, but precisely because of that benefit, they also cause suburban development, urban sprawl, crowded highways, and air pollution.

OPTIMISTS VS. PESSIMISTS

The inescapable pairing of good and bad consequences helps to account for why so many issues of science and technology stir debate in our society. Optimists tend to focus on the benefits of technology and to be confident that society will be able to cope with any problems that arise. Pessimists tend to fear the problems and to believe that the costs of technology will outweigh any possible benefits.

Sometimes the costs of new technologies are immediate and tangible. When new devices fail or new drugs prove to have unforeseen side effects, people can die. Sometimes the costs are less obvious. John McDermott, one of Mesthene's opponents, expressed confidence that technology led to the centralization of power in the hands of an educated elite; to his mind, technology was therefore antidemocratic.[2]

The proponents of technology answer that a machine's failure is a sign that it needs to be fixed, not banned. If a drug has side effects, it may need to be refined, or its list of permitted recipients may have to be better defined (the banned tranquilizer thalidomide, for example, is notorious for causing birth defects when taken early in pregnancy; it is apparently quite safe for men and nonpregnant women). And although several technologies that were developed in the 1960s seemed quite undemocratic at the time, one of them —computers—developed in a very different direction. Early on, computers were huge, expensive machines operated by an elite, but it was not long before they became so small, relatively inexpensive, and "user-friendly" that the general public gained access to them. Proponents lauded this as a true case of technological "power to the people."

CERTAINTY VS. UNCERTAINTY

Another source of debate over science and technology is uncertainty. Science is, by its very nature, uncertain. Its truths are provisional, open to revision.

Unfortunately, people are often told by politicians, religious leaders, and newspaper columnists that truth is certain. By this view, if someone admits uncertainty, then their position can be considered weak and they need not be heeded. This is, of course, an open invitation for demagogues to prey upon people's fears of disaster or side effects (which are always a possibility with new technology) or upon the wish to be told that greenhouse warming and ozone depletion are mere figments of the scientific imagination (they have yet to be proven beyond a doubt).

NATURAL VS. UNNATURAL

Still another source of controversy is rooted in the tendency of new ideas —in science and technology as well as in politics, history, literary criticism, and so on—to clash with preexisting beliefs or values. These clashes become most public when they pit science against religion and "family values." The battle between evolution and creationism, for example, still stirs passions a century and a half after naturalist Charles Darwin first said that human beings had nonhuman predecessors. It is nearly as provocative to some to suggest that homosexuality is a natural variant of human behavior (rather than a conscious choice); or that there might be a genetic component to intelligence or aggressiveness; or that the traditional mode of human reproduction might be supplemented with in vitro fertilization, embryo cloning, surrogate mother arrangements, and even genetic engineering.

Many new developments are rejected as "unnatural." For many people, "natural" means any device or procedure to which they have become accustomed. Very few realize how "unnatural" such seemingly ordinary things as circumcision, horseshoes, and baseball are.

However, humans do embrace change and are forever creating variations on religions, languages, politics, and tools. Innovation is as natural to a person as building dams is to a beaver.

PUBLIC VS. PRIVATE: WHO PAYS, AND WHY?

Finally, conflict frequently arises over the function of science in society. Traditionally, scientists have seen themselves as engaged solely in the pursuit of knowledge, solving the puzzles set before them by nature with little concern for whether or not the solutions to those puzzles might prove helpful to human enterprises such as war, health care, and commerce. Yet again and again the solutions discovered by scientists have proved useful—they have even founded entire industries.

Not surprisingly, society has come to expect science to be useful. When asked to fund research, society feels that it has the right to target research on issues of social concern, to demand results of immediate value, and to forbid research it deems dangerous or disruptive. Private interests such as corporations often feel that they have similar rights with regard to research that they have funded. For instance, tobacco companies have displayed a strong tendency to fund research that shows tobacco to be safe and to cancel funding for studies that come up with other results, which might interfere with profits. Another example is Boots Pharmaceuticals, which funded a university researcher to compare its version of synthetic thyroid hormone (Synthroid) with competing versions. When the researcher found that the versions were of comparable effectiveness, Boots tried to forbid publication of the report. See Dorothy S. Zinberg, "Editorial: A Cautionary Tale," *Science* (July 26, 1996).

One argument for public funding is that it avoids such conflict-of-interest issues. Yet politicians have their own interests, and their control of the purse strings—just like a corporation's—can give their demands a certain undeniable persuasiveness.

PUBLIC POLICY

The question of how to target research is only one way in which science and technology intersect the realm of public policy. Here the question becomes, How should society allocate its resources in general? Toward education or prisons? Health care or welfare? Research or trade? Encouraging new technologies or cleaning up after old ones? The problem is that money is limited—there is not enough to finance every researcher who proposes to solve some social problem. Faced with competing worthy goals, society must make choices. Society must also run the risk that the choices made will turn out to be foolish.

THE PURPOSE OF THIS BOOK

Is there any prospect that the debates over the proper function of science, the acceptability of new technologies, or the truth of forecasts of disaster will soon fall quiet? Surely not, for some issues will likely never die, and there will always be new issues to debate afresh. (For example, think of the population debate, which has been argued ever since Thomas Malthus's 1798 "Essay on the Principle of Population," and then consider the debate over the manned exploration of space and whether or not it is worthwhile for society to spend resources in this way.)

Since almost all technological controversies will affect the conditions of our daily lives, learning about some of the current controversies and beginning to think critically about them is of great importance if we are to be informed and involved citizens.

Individuals may be able to affect the terms of the inevitable debates by first examining the nature of science and a few of the current controversies over issues of science and technology. After all, if one does not know what science, the scientific mode of thought, and their strengths and limitations are, one cannot think critically and constructively about any issue with a scientific or technological component. Nor can one hope to make informed choices among competing scientific or technological priorities.

WOMEN AND MINORITIES IN SCIENCE

There are some issues in the area of science, technology, and society that, even though they are of vital importance, you will not find directly debated in this volume. An example of such an issue might be, "Should there be more women and minority members in science?" However, this is not a debate because no one seriously responds to this question in the negative. Although minorities compose 12 percent of college freshmen, they receive only 6 percent of the bachelor's degrees, 4 percent of the master's degrees, and 3.5 percent of the doctorates. The numbers have improved over recent decades, with one group—Asian Americans—gaining much more rapidly than African Americans, Hispanics, and American Indians. But whites still dominate science and engineering, holding 90 percent of the jobs, even more than they do the general workforce, where they hold 80 percent of the jobs.

Women earn more than half the bachelor's and master's degrees, but only a third of the doctorates and 16 percent of the science and engineering jobs, mostly in biology, psychology, and health. Although income disparities between men and women have diminished, they still remain. Women computer scientists and analysts, registered nurses, laboratory technicians and technologists, and health aides all make about 90 percent of what men in these occupations earn.

You should keep such considerations in mind as you read the issues in this book. And you should consider how the problems of discrimination and prejudice (based on race or class or gender) are played out in some of these debates; the debate over the use of humans as "experimental animals" is a prominent example.

Each year the American Association for the Advancement of Science (AAAS) publishes several issues of its journal *Science* that deal with careers in science and that pay frequent attention to women, minority, and foreign-born scientists. These issues contain a wealth of statistical information, interviews, and analyses invaluable to anyone who is considering a career in science. There are also, of course, vast amounts of other material available, such as *Women's Work: Choice, Chance or Socialization? Insights from Psychologists and Other Researchers* by Nancy Johnson Smith and Sylva K. Leduc (Detselig Enterprises, 1992).

People with Internet access can visit the AAAS and *Science* at `http://sci.aaas.org/`. In particular, watch for this Web page's "Next Wave" section at `http://sci.aaas.org/nextwave/`.

THE SOUL OF SCIENCE

The standard picture of science—a world of observations, hypotheses, experiments, theories, sterile white coats, laboratories, and cold, unfeeling logic—is a myth. This image has more to do with the way science is presented by both scientists and the media than with the way scientists actually perform their work. In practice, scientists are often less orderly, less logical, and more prone to very human conflicts of personality than most people suspect.

The myth remains because it helps to organize science. It provides labels and a framework for what a scientist does; it may thus be especially valuable to student scientists who are still learning the ropes. In addition, the image embodies certain important ideals of scientific thought. These ideals make the scientific approach the most powerful and reliable guide to truth about the world that human beings have yet devised.

THE IDEALS OF SCIENCE: SKEPTICISM, COMMUNICATION, AND REPRODUCIBILITY

The soul of science is a very simple idea: *Check it out*. Years ago, scholars believed that speaking the truth simply required prefacing a statement with "According to" and some ancient authority, such as Aristotle, or a holy text, such as the Bible. If someone with a suitably illustrious reputation had once said something was so, it was so.

This attitude is the opposite of everything that modern science stands for. As Carl Sagan says in *The Demon-Haunted World: Science as a Candle in the Dark* (Random House, 1995), "One of the great commandments of science is, 'Mistrust arguments from authority.'" Scientific knowledge is based not on authority but on reality. Scientists take nothing on faith; they are *skeptical*. When a scientist wants to know something, he or she does not look it up

in the library or take another's word for it. Scientists go in
or the forest, or the desert—wherever they can find the p
wish to know about—and they "ask" those phenomena dire
for answers in nature. And if they think they know the answ
not of books that they ask, "Are we right?" but of nature. Thi
of scientific experiments—they are how scientists ask nature wh ... or not
their ideas check out.

The concept of "check it out" is, however, an ideal. No one can possibly check everything out for himself or herself. Even scientists, in practice, look up information in books and rely on authorities. But the authorities they rely on are other scientists who have studied nature and reported what they learned. And, in principle, everything those authorities report can be checked. Experiments performed in the lab or in the field can be repeated. New theoretical or computer models can be designed. Information that is in the books can be confirmed.

In fact, a good part of the "scientific method" is designed to make it possible for any scientist's findings or conclusions to be confirmed. For example, scientists do not say, "Vitamin D is essential for strong bones. Believe me. I know." They say, "I know that vitamin D is essential for proper bone formation because I raised rats without vitamin D in their diet, and their bones became soft and crooked. When I gave them vitamin D, their bones hardened and straightened. Here is the kind of rat I used, the kind of food I fed them, the amount of vitamin D I gave them. Go and do likewise, and you will see what I saw."

Communication is therefore an essential part of modern science. That is, in order to function as a scientist, you must not keep secrets. You must tell others not just what you have learned but how you learned it. You must spell out your methods in enough detail to let others repeat your work.

Scientific knowledge is thus *reproducible* knowledge. Strictly speaking, if a person says, "I can see it, but you cannot," that person is not a scientist. Scientific knowledge exists for everyone. Anyone who takes the time to learn the proper techniques can confirm any scientific finding.

THE STANDARD MODEL OF THE SCIENTIFIC METHOD

As it is usually presented, the scientific method has five major components: *observation*, *generalization* (identifying a pattern), stating a *hypothesis* (a tentative extension of the pattern or explanation for why the pattern exists), *experimentation* (testing that explanation), and *communication* of the test results to other members of the scientific community, usually by publishing the findings. How each of these components contributes to the scientific method is discussed below.

..oservation
The basic units of science—and the only real facts that the scientist knows—are the individual *observations*. Using them, scientists look for patterns, suggest explanations, and devise tests for their ideas. Observations can be casual or they may be more deliberate.

Generalization
After making observations, a scientist tries to discern a pattern among them. A statement of such a pattern is a *generalization*. Cautious experimenters do not jump to conclusions. When they think they see a pattern, they often make a few more observations just to be sure the pattern holds up. This practice of strengthening or confirming findings by replicating them is a very important part of the scientific process.

The Hypothesis
A tentative explanation suggesting why a particular pattern exists is called a *hypothesis*. The mark of a good hypothesis is that it is *testable*. But there is no way to test a guess about past events and patterns and to be sure of absolute truth in the results, so a simple, direct hypothesis is needed. The scientist says, in effect, "I have an idea that X is true. I cannot test X easily or reliably. But if X *is* true, then so is Y. And I can test Y." Unfortunately, tests can fail even when the hypothesis is perfectly correct.

Many philosophers of science insist on *falsification* as a crucial aspect of the scientific method. That is, when a test of a hypothesis shows the hypothesis to be false, the hypothesis must be rejected and replaced with another. This is not to be confused with the falsification, or misrepresentation, of research data and results, which is a form of scientific misconduct.

In terms of the X and Y hypotheses mentioned above, if it has been found that Y is not true, can we say that X is false too? Perhaps, but bear in mind that X was not tested. Y was tested, and Y is the hypothesis that the idea of falsification says must be replaced, perhaps with hypothesis Z.

The Experiment
The *experiment* is the most formal part of the scientific process. The concept, however, is very simple: an experiment is a test of a hypothesis. It is what a scientist does to check an idea out. It may involve giving a new drug to a sick patient or testing a new process to preserve apples, tomatoes, and lettuce.

If the experiment does not falsify the hypothesis, that does not mean that the hypothesis is true. It simply means that the scientist has not yet come up with a test that falsifies the hypothesis. As the number of times and the number of different tests that fail to falsify a hypothesis increase, the likelihood that the hypothesis is true also increases. However, because it is impossible to conceive of and perform all the possible tests of a hypothesis, the scientist can never *prove* that it is true.

Consider the hypothesis that all cats are black. If you see a black cat, you do not really know anything at all about the color of all cats. But if you see a white cat, you certainly know that not all cats are black. You would have to look at every cat on Earth to prove the hypothesis, but only one (of a color other than black) to disprove it. This is why philosophers of science often say that *science is the art of disproving,* not proving. If a hypothesis withstands many attempts to disprove it, then it may be a good explanation of the phenomenon in question. If it fails just one test, though, it is clearly wrong and must be replaced with a new hypothesis.

Researchers who study what scientists actually do point out that most scientists do not act in accord with this reasoning. Almost all scientists, when they come up with what strikes them as a good explanation of a phenomenon or pattern, do *not* try to disprove the hypothesis. Instead, they design experiments to *confirm* it. If an experiment fails to confirm the hypothesis, then the researchers try another experiment, not another hypothesis.

The logical weakness in this approach is obvious, but it does not keep researchers from holding onto their ideas for as long as possible. Sometimes they hold on so long, even without confirming the hypothesis, that they wind up looking ridiculous. Other times the confirmations add up over the years, and any attempts to disprove the hypothesis fail to do so. The hypothesis may then be elevated to the rank of a theory, principle, or law. *Theories* are explanations of how things work (the theory of evolution *by means of* natural selection, for example). *Principles* and *laws* tend to be statements of things that invariably happen, such as the law of gravity (masses attract each other, or what goes up must come down) or the gas law (if you increase the pressure on an enclosed gas, the volume will decrease and the temperature will increase).

Communication

Each scientist is obligated to share her or his hypotheses, methods, and findings with the rest of the scientific community. This sharing serves two purposes. First, it supports the basic ideal of skepticism by making it possible for others to say, "Oh, yeah? Let me check that." It tells the skeptics where to look to see what the scientist saw and what techniques and tools to use.

Second, communication allows others to use in their work what has already been discovered. This is essential because science is a cooperative endeavor. People who work thousands of miles apart build with and upon each other's discoveries—some of the most exciting discoveries have involved bringing together information from very different fields.

Scientific cooperation stretches across time as well. Every generation of scientists both uses and adds to what previous generations have discovered. As Sir Isaac Newton said in 1675, in a letter to fellow scientist Robert Hooke, "If I have seen further than [other men], it is by standing upon the shoulders of Giants."

The communication of science begins with a process called "peer review," which typically has three stages. The first stage occurs when a scientist seeks funding—from government agencies, foundations, or other sources—to carry out a research program. He or she must prepare a report describing the intended work, laying out the background, hypotheses, planned experiments, expected results, and even the broader impacts on other fields. Committees of other scientists then go over the report to determine whether or not the applicant knows his or her area, has the necessary abilities, and is realistic in his or her plans.

Once the scientist has acquired funding, has done the work, and has written a report of the results, that report will be submitted to a scientific journal, which begins the second stage. Before publishing the report, the journal's editors will show it to other workers in the same or related fields and ask them whether or not the work was done adequately, the conclusions are justified, and the report should be published.

The third stage of peer review happens after publication, when the broader scientific community can judge the work.

It is certainly possible for these standard peer review mechanisms to fail. By their nature, these mechanisms are more likely to approve ideas that do not contradict what the reviewers think they already know. Yet unconventional ideas are not necessarily wrong, as German geophysicist Alfred Wegener proved when he tried to gain acceptance for his idea of continental drift in the early twentieth century. At the time, geologists believed that the crust of the Earth—which is solid rock, after all—did not behave like liquid. Yet Wegener was proposing that the continents floated about like icebergs in the sea, bumping into each other, tearing apart (to produce matching profiles like those of South America and Africa), and bumping again. It was not until the 1960s that most geologists accepted his ideas as genuine insights instead of harebrained delusions.

Currently, the ideal of communication is failing in another way as well. Modern science—especially in biotechnology—is producing a wealth of results with immediate applications in business. Many scientists are keeping their discoveries private until they can file for patents, strike lucrative deals with major corporations, or form their own companies. See W. Wayt Gibbs, "The Price of Silence," *Scientific American* (November 1996).

THE NEED FOR CONTROLS

Many years ago, I read a description of a "wish machine." It consisted of an ordinary stereo amplifier with two unusual attachments. The wires that would normally be connected to a microphone were connected instead to a pair of copper plates. The wires that would normally be connected to a speaker were connected instead to a whip antenna of the sort usually seen on cars.

To use this device, one put a picture of some desired item between the copper plates. It could be, for instance, a photo of a person with whom one wanted a date, a lottery ticket, or a college that one wished to attend. One test case used a photo of a pest-infested cornfield. The user then wished fervently for the date, the winning lottery ticket, a college acceptance, or whatever else one craved. In the test case, the testers wished that all the pests in the cornfield would drop dead.

Supposedly, the wish would be picked up by the copper plates, amplified by the stereo amplifier, and then sent via the whip antenna to wherever wish orders go. Whoever or whatever fills those orders would get the message and grant the wish. Well, in the test case, when the testers checked the cornfield after using the machine, there was no longer any sign of pests. What's more, the process seemed to work equally well whether the amplifier was plugged in or not.

You are probably now feeling very much like a scientist—skeptical. The true, dedicated scientist, however, does not stop with saying, "Oh, yeah? Tell me another one!" Instead, he or she says, "Let's check this out."[3]

Where must the scientist begin? The standard model of the scientific method says that the first step is observation. Here, our observations (as well as our necessary generalization) are simply the description of the wish machine and the claims for its effectiveness. Perhaps we even have the device itself.

What is our hypothesis? We have two choices, one consistent with the claims for the device and one denying those claims: the wish machine always works, or the wish machine never works. Both are equally testable and equally falsifiable.

How do we test the hypothesis? Set up the wish machine, and perform the experiment of making a wish. If the wish comes true, the device works. If the wish does not come true, the device does not work.

Can it really be that simple? In essence, yes. But in fact, no.

Even if you do not believe that wishing can make something happen, sometimes wishes do come true by sheer coincidence. Therefore, even if the wish machine is as nonsensical as most people think it is, sometimes it will *seem* to work. We therefore need a way to shield against the misleading effects of coincidence.

Coincidence is not, of course, the only source of error we need to watch out for. For instance, there is a very human tendency to interpret events in such a way as to agree with our preexisting beliefs, or our prejudices. If we believe in wishes, we therefore need a way to guard against our willingness to interpret near misses as not quite misses at all. There is also a human tendency not to look for mistakes when the results agree with our prejudices. The cornfield, for instance, might not have been as badly infested as the testers said it was, or a farmer might have sprayed it with pesticide between checks, or the testers may have accidentally checked the wrong field. The point is that correlation does not necessarily reflect cause. In other words, although an event seems to occur as the result of another, there may be other factors at work that negate the relationship.

We also need to check whether or not the wish machine does indeed work equally well when the amplifier is unplugged as when it is plugged in, and then we must guard against the tendency to wish harder when we know that it is plugged in. Furthermore, we would like to know whether or not placing a photo between the copper plates makes any difference, and then we must guard against the tendency to wish harder when we know that the wish matches the photo.

Coincidence is easy to protect against. All that is necessary is to repeat the experiment enough times to be sure that we are not seeing flukes. This is one major purpose of replication. Our willingness to shade the results in our favor can be defeated by having another scientist judge the results of our wishing experiments. And our eagerness to overlook errors that produce favorable results can be defeated by taking great care to avoid any errors at all; peer reviewers also help by pointing out such problems.

Other sources of error are harder to avoid, but scientists have developed a number of helpful *control* techniques. One technique is called "blinding." In essence, blinding requires setting up the experiment in such a way that the critical aspects are hidden from either the test subjects, the scientist who is physically performing the experiment, or both. This helps to prevent individuals' expectations from influencing the outcome of the experiment.

In the pharmaceutical industry, blinding is used whenever a new drug is tested. The basic process goes like this: A number of patients with the affliction that the drug is supposed to affect are selected. Half of them—chosen randomly to avoid any unconscious bias that might put sicker patients in one group[4]—are given the drug. The others are given a dummy pill, or a sugar pill, also known as a *placebo.* In all other respects, the two groups are treated exactly the same.

Although, placebos are not supposed to have any effect on patients, they can sometimes have real medical effects, apparently because people tend to believe their doctors when they say that a pill will cure them. That is, when we put faith in our doctors, our minds do their best to bring our bodies into line with whatever the doctors tell us. This mind-over-body effect is called the "placebo effect." To guard against the placebo effect, experimenters employ either single-blind or double-blind techniques.

Single-Blind With this approach, the researchers do not tell the patients what pill they are getting. The patients are therefore "blinded" to what is going on. Both placebo and drug then gain equal advantage from the placebo effect. If the drug seems to work better or worse than the placebo, then the researchers can be sure of a real difference between the two.

Double-Blind If the researchers know what pill they are handing out, they can give subtle, unconscious cues that let the patients know whether they are receiving the drug or the placebo. The researchers may also interpret any changes in the symptoms of the patients who receive the drug as being caused by the drug. It is therefore best to keep the researchers in the dark too; and when both researchers and patients are blind to the truth, the experiment is said to be "double-blind." Drug trials often use pills that differ only in color or in the number on the bottle, and the code is not broken until all the test results are in. This way nobody knows who gets what until the knowledge can no longer make a difference.

Obviously, the double-blind approach can work only when there are human beings on both sides of the experiment, as experimenter and as experimental subject. When the object of the experiment is an inanimate object (such as the wish machine), only the single-blind approach is possible.

With suitable precautions against coincidence, self-delusion, wishful thinking, bias, and other sources of error, the wish machine could be convincingly tested. Yet it cannot be perfectly tested, for perhaps it only works sometimes, such as when the aurora glows green over Copenhagen, in months without an *r*, or when certain people use it. It is impossible to rule out all the possibilities, although we can rule out enough to be pretty confident that the gadget is pure nonsense.

Similar precautions are essential in every scientific field, for the same sources of error lie in wait wherever experiments are done, and they serve very much the same function. However, no controls and no peer review system, no matter how elaborate, can completely protect a scientist—or science—from error. Here, as well as in the logical impossibility of proof (remember, experiments only fail to disprove) and science's dependence on the progressive growth of knowledge, lies the uncertainty that is the hallmark of science. Yet it is also a hallmark of science that its methods guarantee that uncertainty will be reduced (not eliminated). Frauds and errors will be detected and corrected. Limited understandings of truth will be extended.

Those who bear this in mind will be better equipped to deal with issues of certainty and risk.

NOTES

1. Mesthene's essay, "The Role of Technology in Society," *Technology and Culture* (vol. 10, no. 4, 1969), is reprinted in A. H. Teich, ed., *Technology and the Future*, 7th ed. (St. Martin's Press, 1997).

2. McDermott's essay, "Technology: The Opiate of the Intellectuals," *The New York Review of Books* (July 31, 1969), is reprinted in A. H. Teich, ed., *Technology and the Future*, 7th ed. (St. Martin's Press, 1997).

3. Must we, really? After all, we can be quite sure that the wish machine does not work because, if it did, it would likely be on the market. Casinos would then be unable to make a profit for their backers, deadly diseases would be eradicated, and so on.

4. Or patients that are taller, shorter, male, female, homosexual, heterosexual, black, white—there is no telling what differences might affect the test results. Drug (and other) researchers therefore take great pains to be sure groups of experimental subjects are alike in every way but the one way being tested.

On the Internet . . .

Science and Technology Policy
This page provides links to the United States of America's federal science and technology policy statements.
http://www.dtic.mil/lablink/areas_of_interest/policy.html

Reasons to Believe
According to the developers of this site, recent breakthrough discoveries in the sciences have virtually sealed the scientific case for the God of the Bible. This page exists to communicate this new evidence.
http://www.reasons.org/

The Creationism Connection
This page provides information and resources for biblical creationists and anyone else who is interested in creationism.
http://members.aol.com/dwr51055/Creation.html

The Institute for Creation Research
According to the developers of this site, the Institute for Creation Research (ICR) is a major center of scientific creationism. *http://www.icr.org/*

The Skeptical Inquirer
The Committee for the Scientific Investigation of Claims of the Paranormal encourages the critical investigation of paranormal and fringe-science claims from a responsible, scientific point of view and disseminates factual information about the results of such inquiries to the scientific community and the public. It also promotes science and scientific inquiry, critical thinking, science education, and the use of reason in examining important issues, and it publishes the *Skeptical Inquirer.*
http://www.csicop.org/

PART 1

The Place of Science and Technology in Society

The partnership between human society and science and technology is an uneasy one. Science and technology undoubtedly offer benefits, in both the short term and the long, but they also challenge received wisdom and present us with new worries. The issues in this section deal with the best way to ensure that society benefits from science and technology, the conflict between science and traditional elements of society, and the debate over creationism versus evolution.

- Should the Federal Government Point the Way for Science?
- Is Science a Faith?
- Should the Theory of Evolution Be Replaced by Creationism?

ISSUE 1

Should the Federal Government Point the Way for Science?

YES: David H. Guston and Kenneth Keniston, from "Updating the Social Contract for Science," *Technology Review* (November/December 1994)

NO: National Academy of Sciences, from *Allocating Federal Funds for Science and Technology* (National Academy Press, 1995)

ISSUE SUMMARY

YES: Assistant professor of public policy David H. Guston and professor of human development Kenneth Keniston argue that science can no longer set its own path and that public participation must be increased at all levels of decision making about science.

NO: The National Academy of Sciences, a private, nonprofit society of scholars engaged in scientific and engineering research, asserts that although the relationship between government or society and science needs changes, only scientists should decide what research is to be done.

What scientists do as they apply their methods is called *research.* Scientists who perform *basic* or *fundamental research* seek no specific result. Basic research is motivated essentially by curiosity. It is the study of some intriguing aspect of nature for its own sake. Basic researchers have revealed vast amounts of detail about the chemistry and function of genes, explored the behavior of electrons in semiconductors, revealed the structure of the atom, and discovered radioactivity. They have opened our minds to the immensity of both time and the space of the universe in which we live, and they have produced photos of the surface of Mars.

Applied or *strategic research* is more mission-oriented. Applied scientists turn basic discoveries into devices and processes, such as transistors, integrated circuits, and computers; genetic engineering; cures for diseases; and nuclear weapons and power plants. They have also used their basic understanding of the atmosphere to improve weather forecasting. And space research has yielded communications, weather, and earth-resource-survey satellites. There are thousands of such examples, all of which are answers to specific problems or needs, and many of which were quite surprising to the basic researchers who first gained the raw knowledge that led to these developments.

The unexpected fruits, or "spinoffs," of basic research have been used to justify steady growth in funding for this side of science ever since Vannevar

Bush's 1945 report *Science, the Endless Frontier* (reprint, National Science Foundation, 1990). However, the search for justification is an everlasting struggle. In 1953 Warren Weaver, in "Fundamental Questions in Science," *Scientific American* (September 1953), expressed concern that the successes of applied science meant the neglect of basic science. Today there is a movement to focus on "national goals" and "science in the national interest" and to emphasize applied research over basic research.

In 1997 Representative Vernon J. Ehlers (R-Michigan), a nuclear physicist and vice chair of the House Science Committee, began a study aimed at devising a new government science policy, justifying government support for research, linking basic and applied research more tightly, and speeding the movement of new discoveries into the marketplace to benefit society.

It is easy to see what drives the movement to put science to work. Society has a host of problems that cry out for immediate solutions. Yet there is also a need for research that is not tied to explicit need because such research undeniably supplies a great many of the ideas, facts, and techniques that problem-solving researchers use in solving society's problems. Basic researchers, of course, use the same ideas, facts, and techniques as they continue their probings into the way nature works.

There is also increasing pressure for public participation in deciding what scientists work on. Science, say some, must be removed from control by social, political, and intellectual elites, and scientists must be held accountable if they expect society to fund their work. Politicians expect a guarantee of results; the public demands guarantees of safety and of benefit to all society, not just the elite. David H. Guston and Kenneth Keniston recognize that such expectations are fed by the past successes of science and technology, the vast expense of the scientific enterprise, and a few spectacular technological failures. Yet, in the following selection, they say that in a democracy, the public should not be "excluded from decision making about science." That is, decisions should not be left to the experts alone. For many years, of course, the experts *have* made the decisions. And the National Academy of Sciences believes that they should continue doing so. "Panels of the nation's leading experts," says the academy in the second selection, should advise science policymakers, and working scientists—the peers in the peer review system—should decide what specific projects get funded.

YES

David H. Guston and
Kenneth Keniston

UPDATING THE SOCIAL CONTRACT FOR SCIENCE

In the years following World War II, the United States established a scientific enterprise that became the envy of the world. This enterprise rested on a vision of science as an "endless frontier" that would replace the American West as the font of economic growth, rising standards of living, and social change. The institutions that supported this frontier were a distinctively American blend of public and private enterprises, eventually including an array of national laboratories, mission agencies, and even a National Science Foundation. The practices that supported it entailed what Harvard political scientist Don K. Price called a new type of federalism: the provision of financial support to scientists at public and private research universities without co-opting their independence.

Research universities were the intellectual centerpiece of this enterprise, since it was there that most of the basic research was performed. At the heart of federal support for universities was the practice of competitive, peer-reviewed grants. The bargain that was struck between the federal government and university science—what is often called the "social contract for science" —can be stated concisely. On one hand, government promised to fund the basic science that peer reviewers found most worthy of support. Scientists, on the other hand, promised to ensure that the research was performed well and honestly, and to provide a steady stream of scientific discoveries that would be translated into new products, medicines, or weapons.

After five decades, the social contract for science shows signs of extreme duress. Scientists and politicians have serious complaints about each other. The issues are, by now, familiar: scientific fraud and dishonesty, the adequacy of science funding, indirect costs of research, administrative burdens in science, scientific priorities, big science, pork-barrel science, and so on. Reports by the congressional Office of Technology Assessment, the National Academy of Sciences, and the Carnegie Commission on Science, Technology, and Government have analyzed what some perceive as a "crisis" in science policy.

Despite this scrutiny, the underlying causes of today's conflicts in science policy remain obscure. We do not believe that the antagonism between science and politics signals either a new or a terminal crisis. But today's struggles do indicate that the old contract between science and government needs updating; they also point to enduring and irreducible tensions between the principles of science and those of democratic government.

CHANGED GOVERNMENT

Although scientists sometimes lament the passing of a golden age of government support for science, the history of postwar science policy fails to reveal a truly privileged past. Throughout the last 50 years, controversies between the political and the scientific communities have always been present—over the loyalty of scientists and the merits of military research, over financial accounting for grants, over applied versus basic research, over payment for the indirect costs of research, and above all, over how much money Washington should dedicate to scientific research.

The pattern of federal funding for research and development [R&D] also belies any image of a lost golden age. Those who pine for the good old days usually recall the mid-1060s, when federal R&D spending reached an all-time high, whether measured as a percentage of the gross national product (in which case 1964 was the maximum) or as a share of total federal spending (in which case the peak came in 1965). But measured in constant dollars, the situation is less clear. By the Office of Management and Budget's method of discounting for inflation, the peak of real federal spending was 1966 or 1967. By the National Science Foundation's method, R&D spending in 1990 was about 30 percent *higher* than the supposed 1966 peak.

In any event, the mid-60s spending levels are a problematic reference point, because federal spending for science and technology in those years was inflated by competition with the Soviets and by the Apollo program. From 1963 to 1972, defense R&D accounted for almost 54 percent of federal expenditures in science and technology. The Reagan defense buildup raised average defense R&D spending between 1983 and 1992 to about 56 percent of total federal R&D. But the average defense share has since fallen to less than 53 percent, and President Clinton has promised to reduce the defense share to 50 percent. Furthermore, space-related R&D, which accounted for 27 percent of federal expenditures between 1963 and 1972, accounted for only 7 percent between 1983 and 1992.

Another way to look at R&D spending is to compare it with the rest of the federal budget. Over the last decade the share of R&D in the domestic discretionary budget has risen, while almost all other items have fallen. That is, through the 1980s, R&D consumed a growing share of the shrinking pie of nondefense, nonentitlement spending. For this reason, calls for greatly increased science budgets are ill-starred from the beginning. The sufferings of scientists may be real, but in the words of Rep. George Brown (D-Calif.), one of the strongest patrons of science, they are not unique.

It nevertheless remains true that irreversible changes have occurred in the last five decades. Indeed, perhaps the simplest explanation for the heightening of tensions between government and science is that the original contract was made between a kind of government that no longer exists and a kind of scientific community that has long since disappeared.

In the postwar years, both the executive and legislative branches have changed in ways that affect the support of science. At the executive level, the "imperial presidency" has extended the chief executive's prerogatives far beyond their prewar limits, and the "management presidency," centered in the Office of Management and Budget, has emphasized control of the sprawling bureaucracy. The White House has added analytical capabilities: the special assistant to the president for science and technology, and the president's Science Advisory Committee. More recently, scientific advisory committees have proliferated in other departments and agencies. The executive branch increasingly tries to coordinate federal R&D in the various agencies, the most recent mechanism being the National Science and Technology Council, composed of Cabinet chiefs and the heads of independent agencies and chaired by the president.

In Congress, the power of committee chairs has declined through the postwar years and has been replaced by a radically decentralized organization, with participation from subcommittees as well as action outside of committees. There has been a resurgence of congressional oversight directed at maintaining accountability over burgeoning programs and agencies. In the early 1970s, Congress augmented its analytical capabilities by creating the Office of Technology Assessment and the Congressional Budget Office, expanding the Congressional Research Service, and increasing control over the General Accounting Office. Committee and personal staffs have increased in size and professional competence. Congress has also created an Office of Inspector General in each major department and agency to monitor the implementation of policy.

Such changes—even if not intentionally related to science—have given both the executive and the legislative branch greater motivation and competence to evaluate and oversee the scientific community.

CHANGED SCIENCE

If government has been transformed in the last five decades, so has science. The scientific enterprise has grown vastly in workforce, complexity, and size of projects, and it has therefore grown more expensive to fund. For example, the scientific workforce nearly doubled between 1965 and 1988, from 495,000 to about 950,000. And the proportion of the nation's workforce who are scientists and engineers engaged in R&D rose from its previous high of 67.9 per 10,000 in 1968 to 75.9 per 10,000 in 1987.

Federal funding of research has always sought to turn out more PhDs so as to provide the nation with a highly trained scientific workforce. But however commendable this goal, it has a bizarre consequence: the more successful the program is, the greater will be the future demand for research financing. It is rather as if a welfare program created a half-dozen new welfare applicants for every one who is given federal assistance. This steady increase in the number of scientists means that despite real growth in R&D funding, a smaller percentage of applications for grants can be funded each year. The scarcity of research funds felt by the scientific community is quite genuine on a per capita basis.

The size and complexity of scientific projects have also increased greatly. The Manhattan Project and other wartime endeavors inaugurated a trend toward "megascience." Research projects today involve more people and require more expensive equipment than ever before. Science has become a vastly more complex aggregate of new technologies and advanced education. As a result, the price of research has gone up much faster than inflation. For this reason, too, scarcity is felt even in the midst of generous funding.

Meanwhile, popular support for science has waned. The almost unqualified public enthusiasm that characterized the immediate postwar period has given way to a far more nuanced view of science and technology. Attitudes have been negatively influenced by conspicuous technological failures—Chernobyl, Bhopal, *Challenger*—which raise concerns about science by the reverse application of the logic that predicts technological benefits from scientific triumphs. It was President Eisenhower who appointed the first special assistant to the president for science and technology. But it was also Eisenhower who warned the American public in his farewell address that "public policy could itself become the captive of a scientific-technological elite." The apprehension of such an elite found expression through many voices: social critics like Theodore Roszak, environmental activists like Rachel Carson, and antimilitary movements that blossomed on the campuses of research universities.

Of all the changes since the postwar negotiation of the social contract for science, the end of the Cold War is probably the most consequential. Ever since 1945, the promise of military applications and the specter of Soviet competition has driven federal R&D expenditures in both military and civilian agencies. The expected usefulness of science and technology to the conduct of the Cold War—both in material terms of building effective weapons and in symbolic terms of conquering the new frontiers of space, the atom, and the cell—meant that governments and publics (in the former Soviet Union and the United States alike) viewed science in a favorable light. But today, without an implacable communist foe, the instrumental value of science and technology has lost some of its urgency.

The result, especially for the physical sciences, is that a new rationale for public support is needed. Previously, the goal upon which almost everyone agreed was countering the Soviet threat. Today, other goals for science are alleged—or, more precisely, revived. For the founders of the American system of science funding, the military rationale was only one among many, including human betterment through fuller employment, a rising standard of living, and better health. The health claim has never lost its persuasiveness, but the rationales of employment and living standards are now being resurrected and redefined.

This redefinition sometimes involves a claim that science-based innovation is the elixir that will stimulate the nation's economy and improve its international economic competitiveness. According to this argument, such innovation has produced entire new industries—consider the transistor and genetic engineering—and will give the United States a technological advantage in competing with other nations for markets and high-wage jobs. In its simplest form, the argument posits a direct causal link between the advances of science, success in the international marketplace, and a rising standard of living.

In this simple version, the argument is open to an obvious criticism: the United States is unquestionably the world's leading scientific power, but it lags by international standards in health, has fallen behind in productivity gains, and is being overtaken in standard of living and international trade. More sophisticated versions of the theory therefore argue that good science is a necessary but not sufficient condition for productivity. A primary point of this more subtle formulation maintains that the postwar research system, even though highly successful so far, has become less effective in today's environment because it was geared toward a different set of military, political, technological, and economic challenges.

Even this cursory analysis of changes in the last five decades suggests that the current strains in government-science relations were inevitable and necessary. Government has increased in size, complexity, competence, and capacity both to support and to oversee science. Science, too, has grown and now faces the consequences of its maturity. The old military rationale for public support has lost much of its cogency, and science faces a more critical public than it did 50 years ago. The old contract was written in simpler days. It has become more fragile today partly because the two parties that agreed to it have changed.

The contract clearly must be updated. But it must also confront the basic tensions between science and democracy.

SCIENCE VERSUS DEMOCRACY

Imagine members of Congress commissioning a National Academy of Sciences report on the organization of science-funding agencies, then gathering testimony from scientists on priorities in science funding, the role of different sectors and institutions in the scientific enterprise, the tension between centralization and pluralism in research, the merits of large-scale versus small-scale projects, and the financial accountability of researchers. Is this Rep. Brown's recent Task Force on the Health of Research? Rep. John Dingell's Subcommittee on Oversight and Investigations? Rep. Don Fuqua's Science Policy Task Force of the mid-1980s? The Fountain Committee, the Elliot Committee, or the Daddario Subcommittee of the 1960s?

Actually, it is the Allison Commission of the 1880s, a select congressional committee that examined all these questions with regard to the federal scientific establishment. Like some dysfunctional family, the science policy community in the United States seems to confront the same problems, never finally resolving them even over many years. Why do the same problems constantly arise? Why is it that no institutional arrangements seem capable of eliminating the tensions between government and science?

One can find only a partial answer in the complaints that scientists and politicians make about each other. Politicians are charged with a lack of knowledge and appreciation of the scientific enterprise; scientists, with arrogance, elitism, and political naïveté. But the dysfunction exists not simply because politicians can be ignorant or scientists arrogant. The deeper reason lies in fundamental and ineradicable differences between the organizing principles of a democratic polity and the organizing principles of a democratic polity and the organizing principles of the scientific community.

There are three fundamental tensions that make for an uneasy relation between government and science. The first is simply that popular tastes and preferences are different from, and sometimes antagonistic toward, those of the scientific community. One might call this the populist tension, and it can result in popular pressure for a more equitable geographic distribution of research funds, for more applied research, for a particular focus of programmatic research such as women's health, or for a greater emphasis on teaching and patenting than on research itself.

Scientists rightly ask whether public opinion should matter in science, because popular pressures could seriously reduce the long-term viability of the scientific enterprise, and at times can reflect "antiscientific" attitudes. But in a democratic society, citizens must be allowed to choose between the viability of science and the viability of other valued enterprises. Even though science is the pursuit of the truth, it is still only one pursuit among many that citizens value. What the populist tension really does is force the advocates of scientific research to articulate a publicly compelling rationale for their activities and then, like any beneficiary of public funds, to be accountable for the outcomes.

The second tension derives from the fact that the economic organization necessary for science to flourish may be at odds with the economic organization necessary for democracy to flourish. One might call it the plutocratic tension, because of the importance of wealth in determining the distribution of scientific resources. This tension is obvious in political concerns about the concentration of R&D funding at a small number of major research universities, as well as worries about the real growth of the R&D budget when most other domestic programs are contracting. It is also evident in concern over the growing fuzziness between public and private interests, as public employees and private firms benefit financially from the fruits of publicly funded research. Another expression of this tension is the fear that the benefits of science-based technology—from the profits yielded by new drugs to the conveniences of consumer technologies—more often accrue to the haves of society than to the have-nots. The basic question behind the plutocratic tension is whether science, because it is relatively rich and privileged, will become richer and more privileged still, and will mostly benefit the non-scientists who are already rich and privileged.

The third tension between democratic politics and scientific practice arises from the fact that democratic processes and goals are largely incompatible with scientific processes and goals. One might call this the exclusionary tension, because the requirements for membership in decision making within science are more exclusive—that is, being a scientist or an expert—than for membership in democratic decision making in general. Democratic decision making constantly seeks to encourage and expand participation; scientific decision making limits it. There is a risk that science may oppose democratic decisions that deviate from or deny some scientifically defined truth. But as political theorist Robert Dahl has written about the idea of allowing experts to guard democracy against incorrect decisions, scientific guardianship, if carried to an extreme, is simply a prettier name for dictatorship.

The tensions between democracy and science boil down to conflicting values: democratic politics cherishes participation and the pursuit of justice; science cherishes inquiry and the pursuit of truth. Because the gap between participation and truth can never be closed, the tensions will always exist.

Any two parties with different goals and structures require a carefully wrought contractual relation if they are to collaborate productively. It therefore follows that something like a social contract for science continues to be necessary. It follows, too, that this contract should give explicit attention to the details of the interaction between government and science, or, more precisely, between the public and scientists. An attempt to run science on democratic principles would destroy science; but that does not mean that the existing institutions and processes of science are democratic enough. An attempt to run government on scientific principles would destroy democracy; but similarly, that does not mean that our current politics is sufficiently informed by scientific knowledge. Only by deliberately designing institutions and processes that confront the inevitable tensions between democratic government and scientific practice can these tensions be minimized.

The old contract between government and science was fragile because it denied these tensions, attempting to keep politics and science as separate as possible. Such a contract has indeed outlived its usefulness. The new contract as it evolves must take into account the blurred boundaries between politics and science, all the while recognizing that the differences between them are intrinsic.

THE FUTURE OF THE CONTRACT

Scientists and politicians must be willing to concede to the other some role in each other's enterprise. The scientific community, in particular, must confront directly the fact that it is in competition for federal funding with other meritorious projects. Like it or not, if science expects public support, it moves into an arena where it must be political—in the best sense, and possibly the worst—in order to justify its claim to public support.

By being political, we do not simply mean joining the horde of lobbyists competing on behalf of clients for public boons—although in the United States, lobbying is a time-honored and appropriate activity. More than that, we mean recognizing and responding to the ways in which science and its support are embedded in public attitudes and public policy.

The scientific community and the research universities in which this community is rooted must undertake an educational role with a dual purpose: first, to make clear the nature and workings of science; and second, to bring to the greater community those scientific insights, findings, theories, outlooks, and facts that can indeed contribute to the public good.

In both regards, university science has only begun to explore its role. Academic scientists need to participate more actively in broadly educational activities such as training science and technology journalists, along with focused pedagogic activities like collaborating with educators in primary and secondary schools to improve scientific literacy.

Given the American science must compete with other good purposes and institutions for the favorable opinion and support of a democratic government, and given that the Cold War has ended, the future relationship between science and government depends heavily on the capacity of the scientific community to articulate a plausible rationale for public support and to demonstrate that rationale at every turn. As military preparedness yields to international economic competitiveness and domestic well-being on the list of national priorities, support for science will depend on the scientific community's willingness and capacity to help resolve economic and domestic problems.

What this requires is a program of vigorous outreach to the public, to public administrators, to leaders of the private sector, and to lawmakers. If academic science indeed has a contribution to make, it is no longer enough—if it ever was—for scientists to wait in their laboratories for the telephone to ring. More enterprising and collaborative projects are necessary. This change will be difficult for scientists whose talents lie in the laboratory rather than in public speaking. But there are others who are gifted teachers and interlocutors, and whose enthusiasm for science impels them to share its beauty and its relevance with others. The scientific community must treasure such individuals or risk undercutting public support for science.

At out own institution, we think of the Leaders in Manufacturing Program, an alliance of MIT [Massachusetts Institute of Technology] faculty with several major U.S. corporations, aimed at training a cohort of corporate leaders versed in the latest manufacturing technologies and management strategies. In the same vein is the creation of workshops for congressional staff members on science and technology. At a more general level, MIT's Knight Science Journalism Fellowship Program has expanded the knowledge of more than 100 leading science and technology journalists and media experts over the last 10 years.

The scientific community must initiate more activities like these: projects that move beyond lobbying to outreach and education, activities that constitute a series of "mini-contracts" between the needs of particular constituencies and the capacities of the scientific community to respond to those needs. It is not enough for the scientific community simply to claim that it is useful; the relevance of scientific knowledge and perspectives to the public interest must be demonstrated again and again in concrete projects.

Government, too, will require new strategies and perhaps new institutions if the contract with science is to be successfully renegotiated. One urgent and oft-noted need is for a more rational way to determine the level of overall federal spending for R&D and the priorities within those expenditures. Too often, public financing of science and technology is based on the political power of a particular disease lobby, the eagerness of members of Congress to earmark scientific and technological projects for their home districts, or intensive lobbying by a group of scientists for their own specialty. Needed instead is an orderly, open, and publicly accessible process. In this regard, the recently established White House National Science and Technology Council (NSTC) promises to be instrumental in drafting an overall R&D budget and in setting priorities within the budget. This body continues to rely on the tried-and-true process of peer review for evaluating individual projects.

What the NSTC needs is a reasonable and articulate strategy for choosing among projects and disciplines. Such a strategy might include giving priority to important disciplines in which the United States compares unfavorably with other nations (as a recent report of the National Academy of Sciences suggests) and inviting consumers of research in industry, education, health, and other fields to assess the output of federal research funding.

At the same time, however, the combination of political priority setting and scientific peer review must not shut out public input. Precisely because research is difficult and performing it can require many years of training, the temptation to confuse the performance of scientific research with the making of science policy is great. The making of science policy by the federal government, or for that matter by state and local governments, needs to be open and democratic. We have urged scientists to reach out to the public to explain what they do and to help ensure that their work is put to good use. This outreach goes for naught if the public is excluded from decision making about science. In this regard, public input, and not just expert advice, is essential at all levels of science policymaking. A "national forum" on science and technology priorities, such as that recently proposed by the Carnegie Commission on Science, Technology, and Government, could help provide such public input if properly constituted. Millions of Americans, not themselves scientists, have strong and legitimate opinions about the value to them and to the nation of space travel, local technology-development centers, and cancer research, among other scientific and technological projects. Their participation should be welcomed and respected.

A third major obligation of government is to preserve R&D as an example of the sturdy American principle of federalism—that decisions should be made and actions taken at the most local level possible. In science policy, this means resisting the temptation to micromanage scientific work, and the researchers and institutions that conduct it, from the distance of Washington. To be sure, government needs to establish standards: it may rightly impose exacting ethical and financial requirements upon researchers who receive public monies. But the only way to implement such requirements consistent with the federalism that inspired the social contract for science is to insist that universities and their researchers maintain primary responsibility. For example, an incentive system for dealing with indirect costs—in which the government sets the overall rate and universities can pocket the remainder if they come in under that rate—may be preferable on grounds of both principle and efficiency to either the preexisting system of making a separate agreement with each university, or any more invasive system in which government accountants would formulate budgets for overhead.

In science policy, as in other areas of governance, a primary responsibility of public officials is to preserve as many independent centers of initiative and locally governed activities as is consistent with the broad rules of accountability and fairness. In the long run, science and technology flourish when multiple independent centers of activity are encouraged; they fail to thrive under the heavy hand of centralized control and unified direction. This is just as it should be in a federal republic like the United States.

These amendments in the social contract for science will never resolve some of the tensions inherent between science and government. But in recognizing the tensions, the changes can make for a more robust and productive relationship. The American system of science and technology has been outstanding in the last half-century in good part because public policy was designed to foster a plurality of centers of scientific and technical excellence with the maximum possible autonomy and responsibility delegated to each local center. No better principle than federalism can be imagined for the new social contract for science.

NO National Academy of Sciences

ALLOCATING FEDERAL FUNDS FOR SCIENCE AND TECHNOLOGY

In a report accompanying funding for the National Institutes of Health [NIH] for Fiscal Year 1995, the Senate Appropriations Committee requested a study from the National Academy of Sciences, the National Academy of Engineering, and the Institute of Medicine. The study was to address "the criteria that should be used to judging the appropriate allocation of funds to research and development activities, the appropriate balance among different types of institutions that conduct such research, and the means of assuring continued objectivity in the allocation process." The study originated from the Appropriations Committee's concern "that at a time when there is much opportunity to understand and cure disease, funding for health research supported by NIH in the next fiscal year is held to below the inflation rate for medical research due to budget constraints. Similarly, other Federal research agencies are confronted with constrained resources resulting from the virtual freeze in discretionary outlays."

The charge was daunting when it was requested by the Appropriations Committee and is even more so now. With a year's passage, the concern with a "virtual freeze in discretionary outlays" seems an understatement. The efforts by both the Administration and the Congress to reduce the federal deficit have prompted proposals to cut programs, consolidate or abolish agencies, and even do away with whole departments. The federal research and development enterprise has not been exempt from examination, nor should it be. Since the end of World War II, this enterprise has become vast and complex, and it accounts for a significant part of the discretionary outlays of the federal government. It is thus important that the nature and structure of federal support for research and development, as well as the benefits it brings, be understood to assure that as budgets are reduced, the strengths of U.S. science and technology are maintained, while the anachronistic or weak aspects are pruned.... The theme of the committee's report is continuance in the face of change.

Continuance builds on the spectacularly successful results of postwar federal investments in research and development. By any measure, these investments have been recouped many times over in contributing to a strong and globally competitive U.S. economy, hastening the end of the Cold War, providing continuing national security against new enemies, advancing the fight against disease, improving our environment, and producing revelations about ourselves, our world, and our cosmos.

Change comes in acknowledging that the federal research and development enterprise must adapt to a new world. The Cold War is over. Global competition is both economic and military, involving many more nations than did the past bipolar confrontation of nuclear superpowers. These problems create opportunities. Indeed, science and technology will be even more important in the future than they are today. Change is also reflected in the very doing of science, as computers and high-speed communication networks expand access to databases and facilities throughout the world and enable daily collaboration among scientists and engineers separated by great distances.

Over time, institutions and programs have been created that no longer serve us well. Even good programs and institutions must give way to successors that are better and are more closely linked to new national needs. These are painful messages. Some of the committee's members have built their professional lives through programs and institutions that may not survive application of the principles the committee proposes for judging future expenditures. At the same time, the committee believes strongly that failure to make these choices will prove costly, serving neither the nation nor the scientific community. That said, the committee appreciates that its principles for judging programs and institutions are, by necessity, general and must be given more specificity when applied to particular programs and institutions. As a practical matter, the committee did not offer specific details for implementing the judgments that must be made. The committee believes that those who must make the decisions and execute them should be given the latitude to apply these principles sensibly....

Some will think it politically unwise that we recommend a process and guidelines for identifying activities that can be reduced or eliminated and for reallocating the savings to ones more essential to preserving U.S. leadership in science and technology. We have been told that our advice will be only partially followed—that the cuts will be made but that the savings will not be reallocated to federal science and technology. Perhaps. But we see no alternative. We can only hope that the case we have made is convincing, and trust that our recommendations to maintain U.S. strength in science and technology will be accepted. The committee believes that the political wisdom that created the remarkably successful U.S. research and development enterprise will endure, driven by the U.S. public's strong and abiding support for federal science and technology....

— Frank Press
Chair
Committee on Criteria for Federal Support of Research and Development

DETERMINING PRINCIPLES FOR ALLOCATING FEDERAL FUNDS

The federal government has played a pivotal role in developing the world's most successful system of research and development. Over the past 5 decades the U.S. scientific and technical enterprise has expanded dramatically, and the federal investments in it have produced enormous benefits for the nation's economy, national defense, health, and social well-being. Science and technology will be at least as important for our nation's future as they have been for our past, but further expansion of federal funding for research and development is unrealistic in the next several years. Both the current administration's 10-year budget plan and the 7-year plans passed by the House and Senate propose significant reductions in federal discretionary spending. Maintaining the vigor of research and development is important—indeed essential—to the nation's future and will require the ability to increase funding for new opportunities selectively, even while reducing the overall budget.

The Committee on Criteria for Federal Support of Research and Development believes that it will be possible to sustain this country's scientific and technological preeminence and the strong federal role within the current fiscal constraints if the recommendations in this report are adopted. Ensuring the nation's future health, however, may well require augmented investments later—after the current period of reorganization and consolidation has helped control costs and sharpen focus.

As we consider how to restructure federally funded research and development to meet today's budget realities, it is important to recognize the considerable strengths of the current system. Those strengths should not be lost. "Top-down" mission-oriented management and "bottom-up" investigator-initiated research projects have combined to create a powerful research and development engine that is the envy of the world. Computer science, surface science, molecular biology, and other fields have emerged in response to new opportunities, and widely disparate fields have been combined to create entirely new applications. Competitively funded research and development projects subject to national merit review and conducted in every state of our nation have proven particularly effective. Federally funded university science and engineering, in addition to yielding new discoveries, has produced new generations of scientists and engineers who serve in academia, industry, and government and also fill critical management positions there. Investments in science have dramatically expanded our knowledge of ourselves and our universe, and new technologies have improved our daily lives. The fruits of federally funded research and development have been applied effectively by U.S. industry. Drawing on the support provided by many sponsoring agencies and the results from a wide range of performing institutions, the American entrepreneurial spirit has tapped federally funded research and development to form entirely new industries in areas such as microelectronics, biotechnology, and communications and information technology, among others....

The extraordinary success of U.S. research and development can be continued within current budget constraints. However, ensuring continuing success will require rigorous discipline and a coherent and comprehensive approach for deciding how resources are used. This report proposes a new process for allocating and monitoring federal spending for science and technology across disciplines and government agencies. With an integrated view and a coherent federal science and technology budget, it will be possible to make selective reductions in some areas, so as to free badly needed resources for more productive investments and new opportunities that arise....

The United States Should Strive to Continue as the World Leader in Science and Technology

RECOMMENDATION. The President and Congress should ensure that the FS&T [Federal Science and Technology] budget is sufficient to allow the United States to achieve preeminence in a select number of fields and to perform at a world-class level in the other major fields.

The pool of approximately $35 billion to $40 billion in annual public support for FS&T is large and diverse. The committee believes that it is possible within that budget to reduce some programs, eliminate others, increase support of high-opportunity fields, and restrain federal spending—all while maintaining our nation's tradition of excellence in science and technology. To continue as a world leader, the United States should strive for clear leadership in the most promising areas of science and technology and those deemed most important to our national goals. In other major fields, the United States should perform on a par with other nations so that it is "poised to pounce" if future discoveries increase the importance of one of these fields. If the nation sets priorities in this way (see bulleted items below) and uses them in conjunction with the FS&T budget process, the result will be better decisions about reallocating and restructuring the U.S. research and development enterprise, preserving its core strengths, and positioning it well for strong future performance.

The international comparisons needed to assess U.S. achievement of its goals for leadership in research and development should be conducted by panels of the nation's leading experts under White House auspices. Reallocation decisions should be made with the advice and guidance of these expert panels, capable of determining the appropriate scope of the fields to assess and to judge the international stature of U.S. efforts in each field. These panels would recommend to the President, his advisors, and Congress:

- Which fields must attain or maintain preeminence, based on goals such as economic importance, national security, unusual opportunity for significant discoveries, global resource or environmental issues, control of disease, mitigation of natural disasters, food production, a presidential initiative (such as human space-flight), or an unanticipated crisis;
- Which fields require increases in funding, changes in direction, restructuring, or other actions to achieve these goals; and
- Which fields have excess capacity (e.g., are producing too many new investigators, have more laboratories or facilities than needed) relative to national needs and international benchmarks.

The committee believes that designing the budget process so as to secure an FS&T budget sufficient to ensure preeminence in select fields and world status in others will allow the United States to maintain continued world leadership. The FS&T budget process must be coupled to systematic review of investments by the nation's best scientific and technical experts, reporting to the highest reaches of government, to produce an appropriately balanced mix of activities. The committee emphasizes that wise federal investments will lead to the creation of new wealth in the future to an even greater extent than they have in the past. As a result, these investments will help reduce the federal deficit in the long run. After a period of budget constraints, reconfiguration, and adjustment, national needs may justify increased investments in FS&T....

Maintaining U.S. Leadership in Science and Technology Despite Budget Constraints Will Require Discipline in the Allocation of Resources for Federal Investments....

RECOMMENDATION. *The federal government should encourage, but not directly fund, private-sector commercial technology development, with two limited exceptions:*

- *Development in pursuit of government missions, such as weapons development and spaceflight; or*
- *Development of new enabling, or broadly applicable, technologies for which government is the only funder available.*

The federal government has long sponsored research and education as a means of developing technologies for its own use and has also encouraged the development of state-of-the-art technologies in its capacity as a customer. The histories of the development of airframes and aircraft engines, missiles and satellites, advanced materials, semiconductors, and computers are replete with examples of federal procurement and research support that have contributed to the creation of commercially important technology. Indeed, the government was the first purchaser of key pieces of equipment used to build the components of what has become the Internet. Both FS&T funding and federal procurement will continue to be important in these and other emerging growth sectors linked to federal missions such as health and environmental cleanup. In the future, however, funding for the nation's science and technology base may contribute more to stimulating new sectors of economic growth than will federal procurement and the "demand pull" on an emerging technology.

Even before the end of the Cold War, high-technology spin-offs from federally funded R&D [research and development] in defense and space had diminished. Efforts have been under way for some time to foster the development of dual-use technologies or to use off-the-shelf commercial technologies in federal programs that develop products for government use. In many cases, civilian applications have now surpassed military ones.

As the Academies' Committee on Science, Engineering, and Public Policy pointed out in its 1993 report, U.S. leadership in high-technology markets cannot be achieved or maintained primarily through federal actions. Commercial technology development will occur largely in the private sector. Firms motivated by market forces and judged by their performance in satisfying demand have a better record than governments of investing in new technologies with large commercial payoffs. As the presumptive owner of the results, the private sector should be the funder of such commercial technology development projects.

The federal government's main role in encouraging commercial technology development and ensuring economic success is to maintain an environment conducive to private-sector development and adoption of new technologies. Such an environment depends on a range of federal policies that influence taxation, macroeconomic stability, national savings, and the volume of international trade. Economic success also is determined by legislation concerned with unfair monopolies, patent protection, product liability, and environmental and consumer protection. Although examination of these critical issues is beyond the scope of this report, the committee believes that government policies, such as those related to taxation, regulation, intellectual property rights protection, social mandates, and others, are usually more important to commercial outcomes than is direct government funding to industry.

The government should not subsidize specific private firms for projects that they would undertake anyway. In a suitable economic context, a firm engaged in product or process innovation will capture or "appropriate" a large fraction of the benefits that it creates. If so, market incentives will guide firms to undertake the right kinds of innovations without any central planning or guidance.

In many cases, however, no one firm can capture the full benefits of its investment. This is generally the case for investment in basic research and can also apply in development related to emerging technologies. One approach to addressing this problem is represented by Sematech, an industry consortium created to improve semiconductor manufacturing, and for which the federal government provided some initial funding. Federal funding may help to establish such consortia in limited and highly specific areas and can be appropriate to support research in consortia formed by industry.

In addition, the government may still have a role in fostering new enabling technologies. Many people believe that nanotechnology (i.e., at scales of one-billionth of a meter) and micromanufacturing, for example, offer exciting commercial opportunities. Government should support training and research that will establish the general scientific and technical principles that firms will ultimately exploit to develop new commercial products and processes. Such investments are appropriate for the federal government because they can generate large benefits that accrue to the nation but would not be captured by any one firm. For example, federal support for research as a component in the education of individuals entering careers in electrical engineering and computer science has helped to produce the skilled people who have developed our modern information technology industries. Support for the work at universities has resulted in the development of the protocols used to exchange information over computer networks, a crucial piece of intellectual capital that all firms have been able to exploit as they enter this new field. Transfer to industry of state-of-the-art technical knowledge produced at science and engineering schools occurs most effectively when faculty, graduate students, and postdoctoral fellows move to the private sector.

Federal funding that improves graduate and undergraduate education is an example of another way to encourage commercial development indirectly, while also supporting R&D in the national interest. In addition to helping stimulate the development and transfer of new enabling technologies into the private sector, the engineering research centers funded by NSF [National Science Foundation], for instance, have helped change the nature of graduate engineering education. By working in close collaboration with their counterparts in industry, graduate students and faculty have become more aware of the specific technology needs and practices of industry. As a consequence, engineering research programs are more focused and students are better prepared to work in industrial research and development laboratories.

The government also sponsors research and development with potential commercial applications in its own laboratories, in FFRDCs, including the national laboratories, and in independent medical research institutes and other nonprofit organizations (almost half of FS&T funding goes to those organizations, the rest to universities and industrial laboratories). Education is not a central mission of those organizations —an important consideration given that movement of people is one of the most effective ways to transfer new ideas and technologies into the private sector. Several recent reports have noted other reasons that federal laboratories, whether operated by the government or contractors, generally have been less successful than they could be at transferring new enabling technologies to potential users in the private sector. New mechanisms such as cooperative research and development agreements (CRADAs) between firms and the government laboratories were introduced to address this problem. Many successful collaborations have been forged between federal laboratories and industry. Several recent reports argue, however, that CRADAs may be less effective than alternatives, that they are difficult to evaluate because of inadequate data, that ownership of intellectual property is often uncertain, and that they create few jobs. Under some CRADAs, the government may be performing research that the partner firm would have done on its own in the absence of a cooperative research agreement. The committee believes that in many cases the government resources that support CRADA research could be better spent on other, more productive items in the FS&T budget.

In addition to providing funds for research and graduate education at universities and government laboratories, the federal government also supports a variety of other programs that promote the development of commercial technologies in the private sector. They include the Advanced Technology Program [ATP], the Technology Reinvestment Program [TRP], the Manufacturing Extension Partnerships program [MEP], Small Business Innovation Research grants and other small business set-asides, and direct government subsidy to private firms. Those programs have different goals and structures but share in their intention to cultivate industrial innovation. The ATP and the TRP involve funding of private-sector projects; the MEP program is modeled after the agricultural extension service program and primarily helps small businesses to incorporate new technologies. Most of these programs are too new to be carefully evaluated, and, because of inherent features in program design and prospects of unstable funding, we may never be able to tell whether some of them achieved their goals.

At this time, the very concept of a government role in subsidizing the development of private-sector product and process development is controversial. Some difficult questions arise with subsidized partnership programs such as the ATP— will they succeed in fostering new, commercially relevant technologies that otherwise would not develop as quickly, and are they the most efficient uses of increasingly scarce federal R&D dollars? The committee is skeptical that the answer to these questions is yes. It therefore believes that these subsidized industrial partnership programs should be continued only if the case is convincingly made that the government is the funder of last resort for an important enabling technology, and they should be pursued only on an experimental basis, with careful attention to their goals, the distribution of proprietary rights, and how they will be evaluated. Where a new technology is needed to address a specific mission such as a military need, however, federal leadership is better justified....

Within the General Constraints Determined by National Priorities, the Selection of Individual Projects Must Reflect the Standards of the Scientific and Technical Community

RECOMMENDATION. *Because competition for funding is vital to maintain the high quality of FS&T programs, competitive merit review, especially that involving external reviewers, should be the preferred way to make awards.*

The highest-quality projects and people should be supported with FS&T funds. The best-known mechanism to accomplish that is some form of open competition involving evaluation of merit by peers. Competitive merit review involves the use of criteria that include technical quality, the qualifications of the proposer, relevance and educational impacts of the proposed project, and other factors pertaining to research goals rather than to political or other nonresearch considerations. Open competition means that, at some level within the framework of an agency's mission, researchers propose their best ideas and anyone may apply and be funded regardless of institution or geographic location. However, in the case of highly targeted missions, quality can also be maintained by knowledgeable program managers who have established external scientific and technical advisory groups to help assess quality and to help monitor whether agency needs are met.

The committee believes that the principle of merit review—which emphasizes competition among ideas, diversity of funders and performers of research and development, and organizational flexibility—has been largely responsible for the remarkable quality, productivity, and originality of U.S. science and technology in the past. Competitive merit review should be the method of choice for making future decisions about FS&T funding.

Many federal research and development agencies have developed some form of competitive merit review process to use in making extramural awards for research, training, and facilities. They have also worked to develop equivalent systems of review for allocating intramural funding, but merit review of in-house research is much more difficult because federal research scientists and engineers are in the civil service and still retain salary and benefits even if they are not productive or their area has lower priority or has become obsolete. That problem is a perennial one in the periodic reviews of federal laboratories. The FFRDCs, including the national laboratories, also have procedures for allocating research funding competitively based on performance. Some do it well, but overall the results have been uneven.

There are other approaches to promoting high quality in federally supported research and development. Some programs try to identify top researchers and give them long-term support rather than require them to submit specific proposals to compete every few years. Some funding for agricultural research is allocated to state agricultural experiment stations and land-grant colleges on a formula basis, and the supported institutions choose the researchers and their projects. Evaluations of that system of formula-grant allocation have not given high marks to its responsiveness or the quality of the resulting research. Other federal funding is awarded competitively to research centers, which in turn distribute the funding among individual researchers and groups.

There is benefit to having a variety of approaches to supporting FS&T, especially because mission agencies have specialized assignments to fulfill. However, the committee believes that fiscal constraint makes it important to level the playing field. Competitive merit review should therefore be increased relative to other mechanisms for awarding FS&T funds. Merit review is best exemplified by the processes used at the NSF and NIH, that is, the use of external peer review to identify and select the best proposals for individual research projects as part of a review process based on competition and expert evaluation of merit criteria. That approach enables those two agencies to choose the best performers. Accordingly, use of competitive merit review to allocate federal funding should be the default presumption, supplemented with other mechanisms for inherently governmental functions that cannot be accomplished through competitive merit review....

LOOKING TO THE FUTURE

A robust national system of innovation lies at the heart of our economy, our health, and our national security. That system of innovation depends on federal investments. The committee believes that its recommendations address a crucial need: maintaining the strength and vigor of U.S. research and development despite the prospect of declining federal discretionary spending over the next several years. Seeing the science and technology enterprise through the lens of a unified FS&T budget can help leaders in government and the American public to gauge its fiscal health. A carefully constructed comprehensive budget offers a unitary view, not artificially balkanized into agency budgets, but sensitive to the complexities and relationships among government programs vital to maintaining the United States at the forefront of world-class science and technology. The corollary proposals provide the basis for continuing excellence—emphasizing programs and people rather than institutions, subjecting all federal science and technology activities to competitive merit review, linking science and engineering research to education, and maintaining a pluralistic system of research and development tied to public missions. The committee's recommendations are designed to help root out obsolete or noncompetitive activities, allowing good programs to be replaced by even better ones.

Science and technology have utterly transformed our world over the past 50 years, touching almost every aspect of our daily lives—from communication to transportation to health. They will be at least as important over the next half century. Preeminence in science and technology has become a national asset, at once a point of pride and an immensely practical investment. Prudent stewardship of science and technology, as much as any other area of federal policy, will dictate how our children and our grandchildren live.

POSTSCRIPT

Should the Federal Government Point the Way for Science?

The debate about the extent to which the government should be involved in scientific research is not new. Before the publication of Vannevar Bush's report *Science, the Endless Frontier* in 1945, Senator Harley M. Kilgore (D-West Virginia) said that he wanted "federal research activities to be planned in accordance with liberal social purposes" (see Daniel J. Kevles, "The Changed Partnership," *The Wilson Quarterly*, Summer 1995). Bush's report was in large part an effort to head off any attempt to put science under the explicit control of society, saying that it would surely pay off more handsomely if left to itself, though with generous public funding. On the record, Bush was quite right: Science and technology had helped the United States to win World War II, antibiotics were the miracle drugs of the time, television was just around the corner, and computers were just being built. Science and technology wore a definite shine.

As Kevles notes, much of the shine wore off over the next few decades. Because so much of their funding came from the Defense Department and was aimed at winning wars, science and technology soon began to smell of death, not life. In the 1960s, when thousands of young people were rejecting established authority (particularly the government) for various reasons, government funding stripped science of legitimacy in many people's eyes. Technological disasters such as the Three Mile Island nuclear power plant failure did not help. And then science and technology proved helpless in the face of increasing poverty (psychiatric medications were even held responsible, in part, for the rise in numbers of the homeless) and new diseases, such as AIDS.

J. Michael Bishop, in "Enemies of Promise," *The Wilson Quarterly* (Summer 1995), states that a good part of the problem lies in the public's sense of betrayal when science fails to solve problems, due largely to the public's failure to understand just what science can and cannot do.

Does Bishop's assertion indicate that the National Academy of Sciences is correct that experts should make the decisions regarding scientific research? Such a "social contract," however, cannot be one-sided. Radford Byerly, Jr., and Roger A. Pielke, Jr., in "The Changing Ecology of United States Science," *Science* (September 15, 1995), argue that when the parties to such a contract change their interpretations, it must be revised. "To be sustainable," they say, "science must meet two related conditions: (i) democratic accountability, including accountability to societal goals, and (ii) sustained political support." If science fails to meet these conditions, it will lose public funding.

ISSUE 2

Is Science a Faith?

YES: Daniel Callahan, from "Calling Scientific Ideology to Account," *Society* (May/June 1996)

NO: Richard Dawkins, from "Is Science a Religion?" *The Humanist* (January/February 1997)

ISSUE SUMMARY

YES: Bioethicist Daniel Callahan argues that science's domination of the cultural landscape unreasonably excludes other ways of understanding nature and the world and sets it above any need to accept moral, social, and intellectual judgment from political, religious, and even traditional values.

NO: Biologist Richard Dawkins holds that science "is free of the main vice of religion, which is faith" because it relies on evidence and logic instead of tradition, authority, and revelation.

Science and technology have come to play a huge role in human culture, largely because they have led to vast improvements in nutrition, health care, comfort, communication, transportation, and mankind's ability to affect the world. However, science has also enhanced understanding of human behavior and of how the universe works, and in this it frequently contradicts what people have long thought they knew. Furthermore, it actively rejects any role of God in scientific explanation.

Many people therefore reject what science tells us. They see science as just another way of explaining how the world and humanity came to be; in this view, science is no truer than religious accounts. Indeed, some say science is just another religion, with less claim on followers' allegiance than other religions that have been divinely sanctioned and hallowed by longer traditions. Certainly, they see little significant difference between the scientist's faith in reason, evidence, and skepticism as the best way to achieve truth about the world and the religious believer's faith in revelation and scripture.

The antipathy between science and religion has a long history. In 1616 the Catholic Church attacked the Italian physicist Galileo Galilei (1564–1642) for teaching Copernican astronomy and, thus, contradicting the teachings of the Church; when invited to look through the telescope and see the moons of Jupiter for themselves, the Church's representatives reportedly refused (Pope John Paul II finally pardoned Galileo in 1983). On the other side of the conflict, the French Revolution featured the destruction of religion in the name of rationality and science, and the worship of God was officially abolished on November 10, 1793.

To many people, the conflict between science and religion is really a conflict between religions, or faiths, much like those between Muslims and Hindus or between conservative and liberal Christians. This view often becomes explicit in the debates between creationists and evolutionists.

The rejection of science is also evident among those who see science as denying both the existence of God and the importance of "human values" (meaning behaviors that are affirmed by traditional religion). This leads to a basic antipathy between science and religion, especially conservative religion, and especially in areas—such as human origins—where science and scripture seem to be talking about the same things but are contradicting each other. This has been true ever since evolutionary theorist Charles Darwin first published *On the Origin of Species by Means of Natural Selection* in 1859.

Religious people are not the only ones who see in science a threat to "human values." Science also contradicts people's preferences, which are often based less on religion than on tradition and prejudice. For instance, science insists that no race or gender is superior to another; that homosexuality is natural, not wicked; that different ways of living deserve respect; and that it is possible to have too many children and to cut down too many trees. It also argues that religious proscriptions that may have once made sense are no longer relevant (the Jewish practice of not eating pork, for example, is a good way to avoid trichinosis; however, says science, so are cooking the meat at higher temperatures and not feeding pigs potentially contaminated feed).

Many people feel that there is a baby in the bathwater that science pitches out the window. Science, they say, neglects a very important side of human existence embodied in that "human values" phrase. Daniel Callahan, a bioethicist and head of the Hastings Institute, sees this side as the source of moral, political, and intellectual judgment, which science by its dominance of society tends to evade. Science, he argues in the following selection, has become an ideology in its own right, as intolerant as any other, and it sorely needs judgment or criticism to keep it from steamrollering the more human side of life.

In the second selection, Richard Dawkins, the Charles Simonyi Professor of the Public Understanding of Science at Oxford University, maintains that science differs profoundly from religion in its reliance on evidence and logic —not on tradition, authority, and revelation—and is therefore to be trusted much more.

YES

Daniel Callahan

CALLING SCIENTIFIC IDEOLOGY TO ACCOUNT

I come to the subject of science and religion with some complex emotions and a personal history not irrelevant to my own efforts to think about this matter. It seems appropriate for me to lay this history out a bit to set the stage for the argument I want to make. For the first half of my life, from my teens through my mid-thirties, I was a serious religious believer, a church member (Roman Catholic), and someone whose identity as both a person and as an intellectual had a belief in God at its center. During that time I had little contact with the sciences; literature and philosophy caught my imagination. I was a fine example, for that matter, of the gap between the two cultures that C. P. Snow described, caught up as I was in the humanities and generally ignorant about science. I spent most of my time among humanists and religious believers (though believers of a generally liberal kind).

All of that changed in my late thirties. Two events happened simultaneously. The first was a loss of my religious faith, utterly and totally. I ceased to be a theist, became an atheist, and so I remain today. I did not, however, have any revolt against organized religion (as it is sometimes pejoratively called) or the churches; nor did I lose respect for religious believers. They just seem to me wrong in their faith and mistaken in their hope. The second event was my discovery of the field of biomedical ethics, seemingly a fertile area for my philosophical training and an important window into the power of the biomedical sciences to change the way we think about and live our lives. With this new interest I began spending much of my time with physicians and bench scientists and worked hard to understand the universe of science that I was now entering (through the side door of biomedical ethics).

Meanwhile, as I was undergoing my own personal changes, the relationship between science and religion was shifting in the country as well. When I was growing up, there was still considerable debate about religion and science, with some believers arguing that there was a fundamental incompatibility between them and others holding that they were perfectly congenial. Some scientists, for their part, wrote books about religion, saying that they had found God in their science. Others, of a more positivistic bent, thought

From Daniel Callahan, "Calling Scientific Ideology to Account," *Society*, vol. 33, no. 4 (May/June 1996).

that science had forever expunged the notion of a God and that science would eventually offer an explanation of everything.

This debate seemed to subside significantly in the 1970s and 1980s. Science came almost totally to win the minds and emotions of educated Americans, and technological innovation was endlessly promoted as the key to both human progress and economic prosperity, a most attractive combination of doing good and doing well. While public opinion polls and church attendance figures, not to mention the gestures of politicians, showed the continuing popularity of religion, it was science that had captured the academy, the corridors of economic power, and high-brow prestige in the media. There remained, to be sure, skirmishes here and there over such issues as the teaching of creationism in the schools, particularly in the Bible Belt, and mutterings about the "religious Right" and its opposition to abortion, embryo and fetal research, and the like. Although there had been some bursts of anti-technology sentiments as part of the fallout of the 1960s culture wars, they had little staying power. The "greening of America" soon ran into a drought.

Science, in short, finally gained the ascendancy, coming to dominate the cultural landscape as much as the economic marketplace. This was the world of science I entered and in which I still remain enmeshed. My reaction to the news in May 1995 that a religious group, with the help of Jeremy Rifkin, was entering a challenge to the patenting of life was one of rueful bemusement: what a quixotic gesture, almost certainly doomed to failure but not, perhaps, before a round of media attention. Such battles make good copy, but that's about it.

The specific issue of the patenting of life deserves discussion, and someone or other would have raised it. Yet it hardly signals a new struggle between science and religion. It is neither that central an issue, nor did it appear even to galvanize a serious follow-up response among most religious groups. Congress, moreover, has given no indication that it will take up the issue in any serious way. In other words, it appears to have sunk as an issue as quickly as it arose.

Yet I confess to a considerable degree of uneasiness here. Science should not have such easy victories. It needs to have a David against its Goliath. This is only to say that scientific modernism—that is, the cultural dominance of science—desperately needs to have a serious and ongoing challenger. By that I mean the challenge of a different way of looking at nature and the world, one capable of shaking scientific self-satisfaction and complacency and resisting its at-present overpowering social force. Science needs, so to speak, a kind of loyal opposition.

This kind of opposition need not and should not entail hostility to the scientific method, to the investment of money in scientific research, or to the hope that scientific knowledge can make life better for us. Not at all. What it does entail is a relentless skepticism toward the view that science is the single and greatest key to human progress, that scientific knowledge is the only valid form of knowledge, and that some combination of science and the market is the way to increased prosperity and well-being for all. When religion can only fight science with the pea-shooters of creationism and antipatenting threats, it has little going for it. That response surely does not represent a thoughtful, developed, and articulate counterbalance to the hold of science on modern societies.

I say all of this because what I discovered upon entering the culture of science—that is, scientism—was something more than a simple commitment to the value and pursuit of scientific knowledge. That is surely present, but it is also accompanied socially by two other ingredients, science as ideology and science as faith.

SCIENCE AS IDEOLOGY

By science as ideology I mean that constellation of values that, for many, constitutes a more or less integrated way of interpreting life and nature, not only providing a sense of meaning but also laying out a path to follow in the living of a life. At the core of that ideology is a commitment to science as the most reliable source of knowledge about the nature of things and to technological innovation as the most promising way to improve human life. Closely related features of that ideology are an openness to untrammeled inquiry, limited by neither church nor state, skepticism toward all but scientifically verifiable claims, and a steady revision of all knowledge. While religion should be tolerated in the name of toleration rather than on grounds of credibility, it should be kept in the private sphere, out of the public space, public institutions, and public education. The ideology of scientism is all-encompassing, a way of knowing, and, culturally embodied, a way of living.

By science as faith I mean the ideology of science when it includes also a kind of non-falsifiable faith in the capacity of science not simply to provide reliable knowledge but also to solve all or most human problems, social, political, and economic. It is non-falsifiable in the sense that it holds that any failure to date of science to find solutions to human problems says nothing at all about its future capacity to do so; such solutions are only a matter of time and more refined knowledge. As for the fact that some of the changes science and technology have wrought are not all good, or have both good and bad features, science as faith holds that there is no reason in principle that better science and new knowledge cannot undo earlier harm and avoid future damage. In a word, no matter what science does, better science can do even better. No religious believer, trying to reconcile the evil in the world with the idea of a good and loving God, can be any more full of hope that greater knowledge will explain all than the scientific believer. And there is no evidence that is allowed to count against such a belief, and surely not religious arguments.

It is at just this point that I, the former religious believer, find it hard to confidently swallow the ideology of science, much less the serene faith of many of its worshippers. I left one church but I was not looking to join another. Nonetheless, when I stepped into the territory of science that appeared to be exactly the demand: If you want to be one of us, have faith. Yet a perspective that aims to supply the kind of certain metaphysical and ethical knowledge once thought limited to religion and to provide the foundations for ways of life seems to me worthy of the same kind of wariness that, ironically, science first taught me to have about religion. If science warns us to be skeptical of traditionalism, of settled but unexamined views, of knowledge claims poorly based on hard evidence, on acts of faith that admit of no falsifiability, why should I not bring that same set of attitudes to science itself? That interesting magazine, *The Skeptical Inquirer*, dedicated to getting the hard facts to debunk superstition, quackery, and weird claims by strange groups, does not run many articles devoted to debunking science or claims made in behalf of the enlightenment it can bring us. (I believe it has yet to publish even one such article, but I may be wrong about that.)

Maybe that is not so surprising. Such rebelliousness seems utterly unacceptable to scientism, utterly at odds with its solemn pieties and liturgical practices. To question the idea of scientific progress, to suggest that there are valid forms of nonscientific knowledge, to think that societies need something more than good science and high technology to flourish is to risk charges of heresy in enlightened educated circles every bit as intimidating as anything that can be encountered in even the most conservative religious groups. The condescension exhibited toward the "religious Right" surely matches that once displayed by Christianity toward "pagans." Even a Republican-dominated, conservative Congress knows it can far better afford politically to drastically cut or eliminate funding for the National Endowments for the Humanities and the Arts than for the National Science Foundation or the National Institutes of Health.

Now I come to the heart of my problem with the ideology and faith of scientism. Like any other human institution and set of practices, science needs to be subject to moral, social, and intellectual judgment; it needs to be called to task from time to time. Ideally that ought to be done by institutions that have the cultural clout to be taken seriously and by means of criteria for judgment that cannot themselves easily be called into question. Religion itself has always had this notion as part of its own self-understanding: It believes that it—churches, theologies, creeds—stands under the higher judgment of God and recognizes that it can itself fall into idolatry, the worship of false gods. One might well complain that the churches have seemed, in fact, exceedingly slow in rendering negative judgment upon themselves. Even so, they have the idea of such judgment and on occasion it has indeed been exercised.

Unfortunately—and a profound misfortune it is—science no longer has seriously competitive ways of thinking or institutions that have a comparable prestige and power. Science no longer has a counterweight with which it must contend, no institution or generally persuasive perspective that can credibly pass judgment on scientific practices and pretensions. No secular force or outlook or ideology exists to provide it. Religion once played that role: Popes, prelates, and preachers could once rain some effective fire and brimstone down on science, often enough mistakenly yet sometimes helpfully. But religion, too concerned to protect its own turf, too unwilling to open its eyes to new possibilities and forms of knowledge, offered mainly condemnation along with, now and then, some lukewarm support. Moreover, the gradual secularizing of the cultures of the developed countries of the world, relegating religion to the domestic sphere, took away religion's platform to speak authoritatively to public life. Scientific modernism was there to fill the gap, and it has been happy to do so. It is not possible to utter prayers in public schools, but there are no limits to the homage that can be lavished upon science and its good works.

The absence of a counterweight to the ideology of science has a number of doleful effects. It helps to substantiate the impression that there is no alternative, much less higher, perspective from which to judge science and its works. If you are the king of the hill, all things go your way and those below you are fearful or hesitant to speak out. It helps as well to legitimate the mistaken belief that all other forms of knowledge are not only inferior but that they are themselves always subject to the superior judgment of science. Accordingly, claims of religious knowledge of a credible kind were long ago dismissed by science. At its best, science is benignly tolerant of religion, patting it on the head like a kindly but wiser grandparent. At its worst, it can be mocking and dismissive. The kinds of knowledge generated by the humanities fare a little better, but not all that much.

From the perspective of my own field, bioethics, it is distressing to see the way that claims for the value or necessity of scientific research are treated with an extraordinary deference, usually going unquestioned. A recent federal panel on embryo research, for instance, set the issue up as a struggle between the moral status of the embryo, on the one hand, and that of the "need" (not just desire on the part of researchers) for embryo research, on the other. In a fine display of nuanced, critical thinking, the panel took apart excessive claims for the rights of embryos, urging "respect" but allowing research. As for the claims of research, they were accepted without any doubts or hesitations at all; they seemed self-evident to the panel, not in need of justification. Even Henry VIII, the king of his hill, hardly got that kind of deference, even from those luckless wives he had beheaded. In a culture saturated with the ideology of science, there seems hardly any forceful voice to call it to account.

If there was a loyal opposition, it would not let the claims and triumphalism of the scientific establishment go unchallenged. It would treat that establishment with respect, but it would fully understand that it is an *establishment*, intent on promoting its own cause and blowing its own horn, critical of its opponents and naysayers, and of course never satisfied with the funds available to it (funds that, if forthcoming in greater quantity, will someday find a cure for cancer, discover the molecular basis for disease, give us cheap energy generated by cold fusion, etc., etc.). A loyal opposition would bring to science exactly the same cool and self-critical eye that science itself urges in the testing of scientific ideas and hypotheses. One of the great intellectual contributions of science has been its methodological commitment to self-criticism and self-revision; and that is one reason it came to triumph over religion, which has not always shown much enthusiasm for skepticism about its key doctrines.

But if self-criticism and self-revision are at the heart of the scientific method, then a good place to begin employing them is at home, on the scientific ideology that culturally sustains the whole apparatus. A loyal opposition would do this not only to temper exaggerated self-congratulations on the part of science but also to keep science itself scientific.

The insuperable limitation of the scientific method is that it cannot be used to criticize the ideology of science or its methods. To try to do so only begs the question of its validity. In the end, we judge that method more by its fruits and consequences than by its a priori validity. The problem here is that science cannot tell us what consequences we ought to want, what kind of knowledge we need, or what uses are best for the knowledge that science demonstrates. Science, that is, is far more helpful with our means than our ends. Good science cannot tell us how to organize good societies or develop good people (or even tell us how to define "good") or tell us what is worth knowing. There is no scientific calculus to tell us how much a society should invest in scientific research; that is a matter of prudence.

It is here that the other forms of knowledge ought and must come into play: the knowledge developed by the humanities or the "soft" social sciences; the political values and structures created by democratic societies, built upon argument, some consensus, and some compromise. My own domain, that of the humanities, was long ago intimidated by science. It does not complain about the grievous disparity between research resources lavished upon it in comparison with science. Those humanists who dare enter the church of science and mutter to its high priests are given the back of the scientific hand, quickly labeled as cranks or, black mark of black marks, Luddites. The scientific establishment should help to encourage and support other forms of knowing and should be willing to learn from them; that would be to display the openness and creativity it touts as its strength. It does not, however, take the fingers of even one hand to count the number of Nobel laureates in science who have petitioned Congress for stronger support for the humanities.

What is a proper role for religion in a society captured by the ideology of science? Its most important role, the one it has played from time to time with other principalities and powers, would be simply to urge some humility on science and to call it to task for pretentiousness and power grabbing. Science ought to stand under constant moral judgment, and there is an important role for religion to play in formulating some of the criteria for such judgment. It is thus proper for religion to remind science of something religion should always be reminding itself of as well: Neither science nor religion are whole and entire unto themselves. Religion stands under the judgment of God (it tells us), and science stands under the judgment of the collective conscience of humankind (which religion does *not* tell us). Religion can remind the world, and those in science, that the world can be viewed from different perspectives. And it can remind that world, including science, what it means to attempt, as does religion, to make sense of everything in some overall coherent way. There is no need to agree with the way in which religion comprehends reality in order to be reminded of the human thirst for some sense of coherence and meaning in the world.

There has always been an aspect of science that overlaps with supernatural religion. That is the kind of natural piety and awe that many scientists feel in the face of the mysteries and beauty of the natural world. This can be called a kind of natural religion, and some scientists easily make the move from the natural to the supernatural, even if many of their more skeptical colleagues—who also share the sense of natural awe—do not follow them in taking that step.

This natural awe frequently expresses itself in a hesitation to manipulate nature for purely self-interested ends, whether economic or medical. The concern of ecologists for the preservation of biodiversity, the hesitations of population geneticists about germ-line therapy, the worry of environmentalists about the protection of tropical forests or of biologists for the preservation of even rare species, all testify to that kind of natural piety. It is here that there is room for an alliance between science and religion, between that science that sees the mystery and unprobed depths of the natural world and that religion that sees nature as the creation and manifestation of a beneficent god.

It is important, for that matter, that science find allies in its desire to keep its natural piety alive and well. The primary enemies of that piety are the casual indifference of many human beings to nature and the more systematic despoiling of nature carried out in the name of the market, human betterment, or the satisfying of private fantasies and desires. Environmentalism has long been torn by a struggle that pits conservationists against preservationists. Conservationists believe that the natural world can be cultivated for human use and its natural resources protected if care is taken. Preservationists, and particularly the "deep ecologists," are hostile to that kind of optimism, holding that nature as it is needs to be protected, not manipulated or exploited. Conservationism has a serious and sober history and has been by no means oriented toward a crude exploitation of nature. But it is a movement that has often been allowed to shade off into that kind of technological optimism that argues that whatever harm scientific progress and technological innovation cause, it can just as readily be undone and corrected by science.

This is the ideology of science taken to extremes, but a common enough viewpoint among those who see too much awe of nature, too much protectionism, as a threat to economic progress. Religion could well throw its weight behind responsible conservation, and it would not hurt a bit if some theologians and church groups took up the cause of deep ecology. That is an unlikely cause to gain great support in an overcrowded world, and particularly in the poverty-stricken parts of that world. But it is a strong countercurrent worth introducing into the larger stream of efforts to preserve and respect nature. A little roughage in the bowels helps keep things moving.

Perhaps the cultural dominance of science is nowhere so evident as in a feature of our society frequently overlooked: the powerful proclivity to look to numbers and data as the key to good public policy. Charts, tables, and graphs are the standard props of the policy analyst and the legislator. This is partly understandable and justifiable. With issues of debate and contention, hard data is valuable. It can help to determine if there is a real problem, the dimensions of that problem, and the possible consequences of different solutions. But the soft underside of the deification of data is the too frequent failure to recognize that data never tells its own story, that it is always subject to, and requires, interpretation.

There is no data that can carry out that work. On the contrary, at that point we are thrown back upon our values, our way of looking at the world and society, and our different social hopes and commitments. The illusion of the inherent persuasiveness of data is fostered by scientism, which likes to think that there can be a neutral standpoint from which to assess those matters that concern us, that scientific information plays that role, and that the answer to any moral and social battles is simply more and better information.

The dominance of the field of economics in social policy itself tells an interesting story: the need to find a policy discipline that has all the trapping of science in its methods and that can capture its prestige. It is a field that aspires to be a science and that speaks the culturally correct language of modeling, hypothesis testing, and information worship. And it has been amply rewarded for its troubles, recently gaining the blessing of a Nobel prize for its practitioners to signal its status as a science, and for many years capturing the reins of public power and office in a way unmatched by any other academic discipline.

There is a prestigious government Council of Economic Advisors. There is not now, and probably never will be, a Council of Philosophical Advisors, or Historical Advisors, or Humanistic Advisors. But then, that is likely to be the fate of any field that cannot attach itself to the prestige of science. It will lack social standing, just as religion now lacks serious intellectual standing. Note that I say "intellectual standing." There is no doubt that religion can still have a potent political status or that religion can from time to time make trouble for science (or, more accurately, make trouble for the agendas of some scientists, for example, for those who would like to do embryo research). But in the larger and more enduring world of dominant ideas and ideologies, science sits with some serenity, and much public adulation, in an enviable position. It is interesting to note what no one seems to have noticed. In the demise of communism as a political philosophy and a set of political regimes, one of its features has endured nicely: its faith in science. That is the one feature it shared with the Western capitalist democracies that triumphed over it. It is also, let it be noted, a key feature of a market ideology, the engine of innovation, a major source of new products, and—in its purported value neutrality—a congenial companion for a market ideology that just wants to give people the morally neutral gift of freedom of economic choice, not moralisms about human nature and the good society of a kind to be found in the now-dead command economies of the world.

Allow me to end as I began. There was a time when I hoped my own field, bioethics, might serve as the loyal opposition to scientific ideology, at least its biomedical division. In its early days, in the 1960s and 1970s, many of those first drawn to it were alarmed by the apparently unthinking way in which biomedical knowledge and technologies were being taken up and disseminated. It seemed important to examine not only the ethical dilemmas generated by a considerable portion of the scientific advances but also to ask some basic questions about the moral premises of the entire enterprise of unrelenting biomedical progress. That latter aspiration has yet to be fulfilled. Most of those who have come into the field have accepted scientific ideology as much as most scientists, and they have no less been the cultural children of their times, prone to look to medical progress and its expansion of choice as a perfect complement to a set of moral values that puts autonomy at the very top of the moral hierarchy. Nothing seems to so well serve the value of autonomy as the expanded range of human options that science promises to deliver, whether for the control of procreation or the improvement of health or the use of medical means to improve our lives. Not many people in bioethics, moreover, care to be thought of as cranks, and there is no faster way to gain that label than to raise questions about the scientific enterprise as a whole. Bioethicists have, on the whole, become good team players, useful to help out with moral puzzles now and then and trustworthy not to probe basic premises too deeply. Unless one is willing to persistently carry out such probes, the idea of a loyal opposition carries no weight.

Can religion, or bioethics, or some other social group or force in our society call science to account when necessary? Can it do so with credibility and serious credentials? Can it do so in a way that helps science to do its own work better, and not simply to throw sand in the eyes of scientists? I am not sure, but I surely hope so. I can only say, for my part, that I left one church and ended in the pews of another one, this one the Church of Science. In more ways than one—in its self-confidence, its serene faith in its own value, and its ability to intimidate dissenters—it seems uncomfortably like the one I left. How can it be made to see that about itself?

NO

Richard Dawkins

IS SCIENCE A RELIGION?

It is fashionable to wax apocalyptic about the threat to humanity posed by the AIDS virus, "mad cow" disease, and many others, but I think a case can be made that *faith* is one of the world's great evils, comparable to the smallpox virus but harder to eradicate.

Faith, being belief that isn't based on evidence, is the principal vice of any religion. And who, looking at Northern Ireland or the Middle East, can be confident that the brain virus of faith is not exceedingly dangerous? One of the stories told to young Muslim suicide bombers is that martyrdom is the quickest way to heaven—and not just heaven but a special part of heaven where they will receive their special reward of 72 virgin brides. It occurs to me that our best hope may be to provide a kind of "spiritual arms control": send in specially trained theologians to deescalate the going rate in virgins.

Given the dangers of faith—and considering the accomplishments of reason and observation in the activity called science—I find it ironic that, whenever I lecture publicly, there always seems to be someone who comes forward and says, "Of course, your science is just a religion like ours. Fundamentally, science just comes down to faith, doesn't it?"

Well, science is not religion and it doesn't just come down to faith. Although it has many of religion's virtues, it has none of its vices. Science is based upon verifiable evidence. Religious faith not only lacks evidence, its independence from evidence is its pride and joy, shouted from the rooftops. Why else would Christians wax critical of doubting Thomas? The other apostles are held up to us as exemplars of virtue because faith was enough for them. Doubting Thomas, on the other hand, required evidence. Perhaps he should be the patron saint of scientists.

One reason I receive the comment about science being a religion is because I believe in the fact of evolution. I even believe in it with passionate conviction. To some, this may superficially look like faith. But the evidence that makes me believe in evolution is not only overwhelmingly strong; it is freely available to anyone who takes the trouble to read up on it. Anyone can study the same evidence that I have and presumably come to the same conclusion. But if you

From Richard Dawkins, "Is Science a Religion?" *The Humanist* (January/February 1997).

have a belief that is based solely on faith, I can't examine your reasons. You can retreat behind the private wall of faith where I can't reach you.

Now in practice, of course, individual scientists do sometimes slip back into the vice of faith, and a few may believe so single-mindedly in a favorite theory that they occasionally falsify evidence. However, the fact that this sometimes happens doesn't alter the principle that, when they do so, they do it with shame and not with pride. The method of science is so designed that it usually finds them out in the end.

Science is actually one of the most moral, one of the most honest disciplines around—because science would completely collapse if it weren't for a scrupulous adherence to honesty in the reporting of evidence. (As [famous magician] James Randi has pointed out, this is one reason why scientists are so often fooled by paranormal tricksters and why the debunking role is better played by professional conjurors; scientists just don't anticipate deliberate dishonesty as well.) There are other professions (no need to mention lawyers specifically) in which falsifying evidence or at least twisting it is precisely what people are paid for and get brownie points for doing.

Science, then, is free of the main vice of religion, which is faith. But, as I pointed out, science does have some of religion's virtues. Religion may aspire to provide its followers with various benefits—among them explanation, consolation, and uplift. Science, too, has something to offer in these areas.

Humans have a great hunger for explanation. It may be one of the main reasons why humanity so universally has religion, since religions do aspire to provide explanations. We come to our individual consciousness in a mysterious universe and long to understand it. Most religions offer a cosmology and a biology, a theory of life, a theory of origins, and reasons for existence. In doing so, they demonstrate that religion is, in a sense, science; it's just bad science. Don't fall for the argument that religion and science operate on separate dimensions and are concerned with quite separate sorts of questions. Religions have historically always attempted to answer the questions that properly belong to science. Thus religions should not be allowed now to retreat from the ground upon which they have traditionally attempted to fight. They do offer both a cosmology and a biology; however, in both cases it is false.

Consolation is harder for science to provide. Unlike Religion, science cannot offer the bereaved a glorious reunion with their loved ones in the hereafter. Those wronged on this earth cannot, on a scientific view, anticipate a sweet comeuppance for their tormentors in a life to come. It could be argued that, if the idea of an afterlife is an illusion (as I believe it is), the consolation it offers is hollow. But that's not necessarily so; a false belief can be just as comforting as a true one, provided the believer never discovers its falsity. But if consolation comes that cheap, science can weigh in with other cheap palliatives, such as pain-killing drugs, whose comfort may or may not be illusory, but they do work.

Uplift, however, is where science really comes into its own. All the great religions have a place for awe, for ecstatic transport at the wonder and beauty of creation. And it's exactly this feeling of spine-shivering, breath-catching awe—almost worship—this flooding of the chest with ecstatic wonder, that modern science can provide. And it does so beyond the wildest dreams of saints and mystics. The fact that the supernatural has no place in our explanations, in our understanding of so much about the universe and life, doesn't diminish the awe. Quite the contrary. The merest glance through a microscope at the brain of an ant or through a telescope at a long-ago galaxy of a billion worlds is enough to render poky and parochial the very psalms of praise.

* * *

Now, as I say, when it is put to me that science or some particular part of science, like evolutionary theory, is just a religion like any other, I usually deny it with indignation. But I've begun to wonder whether perhaps that's the wrong tactic. Perhaps the right tactic is to accept the charge gratefully and demand equal time for science in religious education classes. And the more I think about it, the more I realize that an excellent case could be made for this. So I want to talk a little bit about religious education and the place that science might play in it.

I do feel very strongly about the way children are brought up. I'm not entirely familiar with the way things are in the United States, and what I say may have more relevance to the United Kingdom, where there is state-obliged, legally enforced religious instruction for all children. That's unconstitutional in the United States, but I presume that children are nevertheless given religious instruction in whatever particular religion their parents deem suitable.

Which brings me to my point about mental child abuse. In a 1995 issue of the *Independent*, one of London's leading newspapers, there was a photograph of a rather sweet and touching scene. It was Christmas time, and the picture showed three children dressed up as the three wise men for a nativity play. The accompanying story described one child as a Muslim, one as a Hindu, and one as a Christian. The supposedly sweet and touching point of the story was that they were all taking part in this nativity play.

What is not sweet and touching is that these children were all four years old. How can you possibly describe a child of four as a Muslim or a Christian or a Hindu or a Jew? Would you talk about a four-year-old economic monetarist? Would you talk about a four-year-old neo-isolationist or a four-year-old liberal Republican? There are opinions about the cosmos and the world that children, once grown, will presumably be in a position to evaluate for themselves. Religion is the one field in our culture about which it is absolutely accepted, without question—without even noticing how bizarre it is—that parents have a total and absolute say in what their children are going to be, how their children are going to be raised, what opinions their children are going to have about the cosmos, about life, about existence. Do you see what I mean about mental child abuse?

Looking now at the various things that religious education might be expected to accomplish, one of its aims could be to encourage children to reflect upon the deep questions of existence, to invite them to rise above the humdrum preoccupations of ordinary life and think *sub specie alternitatis.*

Science can offer a vision of life and the universe which, as I've already remarked, for humbling poetic inspiration far outclasses any of the mutually contradictory faiths and disappointingly recent traditions of the world's religions.

For example, how could any child in a religious education class fail to be inspired if we could get across to them some inkling of the age of the universe? Suppose that, at the moment of Christ's death, the news of it had started traveling at the maximum possible speed around the universe outwards from the earth? How far would the terrible tidings have traveled by now? Following the theory of special relativity, the answer is that the news could not, under any circumstances whatever, have reached more than one-fiftieth of the way across one galaxy —not one-thousandth of the way to our nearest neighboring galaxy in the 100-million-galaxy-strong universe. The universe at large couldn't possibly by anything other than indifferent to Christ, his birth, his passion, and his death. Even such momentous news as the origin of life on Earth could have traveled only across our little local cluster of galaxies. Yet so ancient was that event on our earthy time-scale that, if you span its age with your open arms, the whole of human history, the whole of human culture, would fall in the dust from your fingertip at a single stroke of a nail file.

The argument from design, an important part of the history of religion, wouldn't be ignored in my religious education classes, needless to say. The children would look at the spell-binding wonders of the living kingdoms and would consider Darwinism alongside the creationist alternatives and make up their own minds. I think the children would have no difficulty in making up their minds the right way if presented with the evidence. What worries me is not the question of equal time but that, as far as I can see, children in the United Kingdom and the United States are essentially given *no* time with evolution yet are taught creationism (whether at school, in church, or at home).

It would also be interesting to teach more than one theory of creation. The dominant one in this culture happens to be the Jewish creation myth, which is taken over from the Babylonian creation myth. There are, of course, lots and lots of others, and perhaps they should all be given equal time (except that wouldn't leave much time for studying anything else). I understand that there are Hindus who believe that the world was created in a cosmic butter churn and Nigerian peoples who believe that the world was created by God from the excrement of ants. Surely these stories have as much right to equal time as the Judeo-Christian myth of Adam and Eve....

When the religious education class turns to ethics, I don't think science actually has a lot to say, and I would replace it with rational moral philosophy. Do the children think there are absolute standards of right and wrong? And if so, where do they come from? Can you make up good working principles of right and wrong, like "do as you would be done by" and "the greatest good for the greatest number" (whatever that is supposed to mean)? It's a rewarding question, whatever your personal morality, to ask as an evolutionist where morals come from; by what route has the human brain gained its tendency to have ethics and morals, a feeling of right and wrong?

Should we value human life above all other life? Is there a rigid wall to be built around the species *Homo sapiens*, or should we talk about whether there are other species which are entitled to our humanistic sympathies? Should we, for example, follow the right-to-life lobby, which is wholly preoccupied with *human* life, and value the life of a human fetus with the faculties of a worm over the life of a thinking and feeling chimpanzee?

What is the basis of this fence we erect around *Homo sapiens*—even around a small piece of fetal tissue? (Not a very sound evolutionary idea when you think about it.) When, in our evolutionary descent from our common ancestor with chimpanzees, did the fence suddenly rear itself up?

...[S]cience could give a good account of itself in religious education. But it wouldn't be enough. I believe that some familiarity with the King James version of the Bible is important for anyone wanting to understand the allusions that appear in English literature. Together with Book of Common Prayer, the Bible gets 58 pages in the *Oxford Dictionary of Quotations.* Only Shakespeare has more. I do think that not having any kind of biblical education is unfortunate if children want to read English literature and understand the provenance of phrases like "through a glass darkly," "all flesh is as grass," "the race is not to the swift," "crying in the wilderness," "reaping the whirlwind," "amid the alien corn," "Eyeless in Gaza," "Job's comforters," and "the widow's mite."

I want to return now to the charge that science is just a faith. The more extreme version of this charge—and one that I often encounter as both a scientist and a rationalist—is an accusation of zealotry and bigotry in scientists themselves as great as that found in religious people. Sometimes there may be a little bit of justice in this accusation; but as zealous bigots, we scientists are mere amateurs at the game. We're content to *argue* with those who disagree with us. We don't kill them.

But I would want to deny even the lesser charge of purely verbal zealotry. There is a very, very important difference between feeling strongly, even passionately, about something because we have thought about and examined the evidence for it on the one hand, and feeling strongly about something because it has been internally revealed to us, or internally revealed to somebody else in history and subsequently hallowed by tradition. There's all the difference in the world between a belief that one is prepared to defend by quoting evidence and logic and a belief that is supported by nothing more than tradition, authority, or revelation.

POSTSCRIPT

Is Science a Faith?

The conflict between science and religion is deep and broad. The root reason may be simply that science says, "Check it out—don't take anyone's word for the truth," while religion says, "Take the word of your preacher or your scripture. Believe—but don't even *think* about checking." Scientific skepticism is always a threat to established authority. It challenges old truths. It revises and replaces beliefs, traditions, and power structures.

Does this mean that science is a threat to society? Those who share the beliefs under attack often think so. They may believe that the Bible or the Koran is a much better guide to the nature of the world than science is. They may believe in crystal power and magic spells. They may tie knots in their electric cords to trim the size of their electric bills. They may even be postmodernist university professors who say that science is just a "useful myth," no different from any other fiction. Or they may, like Callahan, wish that there were some segment of society with sufficient stature to sit in judgment over science, to criticize it, and perhaps to rein it in, certainly to keep it from arrogantly quashing other views, such as those of religion. And although most Americans welcome the benefits of science and technology, they are often very leery of the unrestricted inquiry that characterizes science and challenges tradition. See, for example, Janet Raloff, "When Science and Beliefs Collide," *Science News* (June 8, 1996); Gerald Holton, *Einstein, History, and Other Passions: The Rebellion Against Science at the End of the Twentieth Century* (Addison- Wesley, 1996); and "Science Versus Antiscience?" *Scientific American* (January 1997). Even some scientists feel threatened by the conflict between their professional and private beliefs. Some have therefore spent a great deal of effort searching for ways to reconcile science and religion. For instance, Leon Lederman and Dick Teresi write about the quest for the most fundamental fragment of the atom in *The God Particle* (Dell, 1994). Stephen Hawking, in *A Brief History of Time* (Bantam Books, 1988), expresses the thought that science might lead humanity to "know the mind of God."

Can these scientists be speaking in more than metaphorical terms? Perhaps not, for science deals in observable reality, which can provide at best only hints of a designer, creator, or God. Science cannot provide *direct* access to God, at least as people currently understand the nature of God. Still, it is not only creationists who see signs of design. Some scientists find the impression of design quite overwhelming, and many feel that science and religion actually have a great deal in common. Harvard University astronomer and evangelical Christian Owen Gingerich says that both are driven by human beings' "basic wonder and desire to know where we stand in the universe." It is therefore not terribly surprising to find the two realms of human thought intersecting very frequently or to find many people in both realms concerned with reconciling differences. See Gregg Easterbrook, "Science and God: A Warming Trend?" *Science* (August 15, 1997).

On the other hand, some scientists find attempts to reconcile science and religion strange at best. Eugenie Scott, of the National Center for Science Education, insists that "science is just a method" and that people who see God in the complexity of biology or astronomy are "going beyond their data" and misusing science "to validate their positions." Paul Gross, former director of the Woods Hole Marine Biological Laboratory and coauthor of *Higher Superstition: The Academic Left and Its Quarrels With Science* (Johns Hopkins University Press, 1994), even finds those who see God in science frightening. More recently, Gross, Norman Levitt, and Martin W. Lewis coedited *The Flight from Science and Reason* (New York Academy of Sciences, 1997) to consider the opposition to the scientific, rational approach to the world that now finds wide expression in many nonscientific academic areas.

Are such views no more than an illustration of Callahan's claim that science —or "scientism"—has become an ideology and a faith as intolerant of others as any religion? Check out the journal *Zygon*, whose purview is the continuing debate between science and religion.

ISSUE 3

Should the Theory of Evolution Be Replaced by Creationism?

YES: Jack Hitt, from "On Earth as It Is in Heaven: Field Trips With the Apostles of Creation Science," *Harper's Magazine* (November 1996)

NO: Daniel C. Dennett, from *Darwin's Dangerous Idea: Evolution and the Meanings of Life* (Simon & Schuster, 1995)

ISSUE SUMMARY

YES: Jack Hitt, a contributing editor to *Harper's Magazine,* reports on a visit with a creationist who dreams of replacing much of human knowledge—particularly the theory of evolution—with something more consistent with the biblical Genesis.

NO: Philosopher Daniel C. Dennett argues that Charles Darwin had in his theory of evolution by means of natural selection the single best idea of all time. Far from being replaceable, he asserts, the theory has made its religious predecessors quite obsolete.

Before science came along, the usual answer to questions such as "Why do elephants have trunks?" or "Why is the sky blue?" was "Because God made it that way." No one could add any more. Today children still hear such answers in Sunday School, but the rest of us generally believe that more complete and satisfying answers have come from scientists who were not satisfied with "God's will" as an answer. It has long been a dogma of scientific faith that "why" questions are unreasonable to ask. They are teleological; that is, they presume that there is an intent or design behind the phenomena we wish to explain. As an answer, "God's will" is out of bounds largely because accepting it means accepting that it is a waste of time to look for other answers. Outside science, on the other hand, "God's will" is very much *in* bounds. This leads to a continuing struggle between the forces of faith and the forces of reason. Conservative Christians in the southern United States, Texas, and California have mounted vigorous campaigns to require public school biology classes to give equal time to both biblical creationism and Darwinian evolution. For many years, this meant that evolution was hardly mentioned in high school biology textbooks. On this, see Dorothy Nelkin, "The Science-Textbook Controversies," *Scientific American* (April 1976), pp. 33–39.

For a time, it looked like evolution had scored a decisive victory. In 1982 federal judge William K. Overton struck down an Arkansas law that would have required the teaching of straight biblical creationism, with its explicit talk of God the Creator, as an unconstitutional intrusion of religion into a government activity: education. But the creationists have not given up. They have returned to the fray with something they call "scientific creationism," and they have shifted their campaigns from state legislatures and school boards to local school boards, where it is harder for lawyers and biologists to mount effective counterattacks. See Gary Stix, "Postdiluvian Science," in "Science Versus Antiscience?" *Scientific American* (January 1997). "Scientific creationism" tries to show that the evolutionary approach is incapable of providing satisfactory explanations. For one thing, it says that natural selection relies on random chance to produce structures whose delicate intricacy really could only be the product of deliberate design. Therefore, there must have been a designer. There is no mention of God—but, of course, that is the only possible meaning of "designer" (unless one believes in ancient extraterrestrial visitors). Scientific creationists reinforce their claim that evolution is inadequate by seeking weaknesses in the evidence—fossils, anatomy, embryology, DNA, and more—that more conventional biologists cite in their own discussions of evolution and natural selection. They hope to thereby weaken the credibility of evolutionists. At the same time, scientific creationists can present the quest for weaknesses in the evolutionists' argument as perfectly appropriate scientific skepticism—after all, they are *scientific* creationists.

William Johnson, associate dean of academic affairs at Ambassador University in Big Sandy, Texas, offered another argument for replacing the theory of evolution in a 1994 speech reprinted in "Evolution: The Past, Present, and Future Implications," *Vital Speeches of the Day* (February 15, 1995). He argued that the triumph of Darwin's theory "meant the end of the traditional belief in the world as a purposeful created order . . . and the consequent elimination of God from nature has played a decisive role in the secularization of Western society. Darwinian theory broke man's link with God and set him adrift in a cosmos without purpose or end." Johnson suggested that evolution—and perhaps the entire scientific approach to nature—should be abandoned in favor of a return to religion because of the untold damage it has done to the human values that underpin society.

In the following selections, Jack Hitt, who has studied creationist views, reports on a visit with Kurt Wise, a creationist who dreams of replacing not just the theory of evolution but much of the rest of human knowledge as well. Daniel C. Dennett argues that in his theory of evolution by means of natural selection, Charles Darwin had the single best idea of all time and that in it lies our hope of finding the truest meaning of life.

YES

Jack Hitt

ON EARTH AS IT IS IN HEAVEN

"Everyone come this way. Gather around!" shouted Professor John Whitmore above the roar of a swollen creek. Twenty-five undergraduates from his earth-sciences class at Cedarville College bunched up at the edge of a cliff. The field trip to Indian Mound gorge in southwestern Ohio had begun just after lunch. We'd taken a van past spent cornfields. Yellow wildflowers poked up from the brown stubble after days of rain. The hike in was treacherous. Fresh shoots of poison ivy probed the mired path. Already a few girls in fancy shoes were wearing mud slicks up their backs.

"This stratum is known as the Cedarville Dolomite," Whitmore said as students scratched furiously in their pads, "because it is best exposed in Cedarville. Below it you can see a layer of Springfield Dolomite, which would be best exposed in . . . ?" Professor Whitmore paused, inviting an answer. But group dynamics forbade it. The crowd held still, embarrassed, helpless in silence.

"*Springfield.* Exactly," Whitmore answered. The class descended along a rocky path and stopped to draw sketches. A streak of shale appeared, and a gurgling aquifer trickled from the rock, flowing over a smooth limestone tongue. A sweet-faced boy named Jeff with James Dean sideburns fingered the shale and suddenly gripped a loose shard.

"That's a fossil," Professor Whitmore explained, "a crinoid. A nice one too." The preserved animal appeared to be nothing more than an inch of baling string stuck in the shale. The body was neatly segmented into tiny bulbous sections; all in all, in pretty good shape, given that most geologists would date its age at 400 million years.

"It's an echinoderm," Whitmore explained. "It's related to the starfish and the sand dollar." He turned it over in his hand. "I would say it's about 4,500 years old. It was deposited during Noah's Flood." . . .

* * *

For the last ten years or so, I have been dipping into creationist literature. Back in the 1980s, the science was unintentional vaudeville. Zealous devotees of murky academic pedigree performed amusing tests and published the results in their own periodicals, such as the *Creation Science Research Quarterly.* One experiment, I recall, involved pouring drinking water from a pitcher into a saltwater aquarium to determine scientifically the effect on saltwater life of what creationists call the "960 consecutive hours" (40 days and nights) of rain during the "Noachian Deluge" (Noah's Flood). The "fish stopped swimming at 20.3 ± 1.1 0/00 salinity," the researcher reported in deadpan science-ese. "Obviously," he concluded with pep, "additional research is needed."

In the last few years, though, creationism has been revived and transformed by an influx of scientists, some with Ivy League degrees, striving to verify the truths of their discipline using the scientific method. They are young: John Whitmore, for example, is thirty-three years old. But more importantly, they are committed: Whitmore completed a study of geology at Kent State University without disturbing his literalist belief in the Bible. Like his cutting-edge colleagues, he now spends his time in the laboratory and on field trips carrying out experiments and carefully weaving together the "physical-world data" and the "Scriptural data," the warp and the woof of what its advocates call "neo-creationism." What emerges from this labor is the taut canvas of a worldview both syncretic and baroque.

Neo-creationism is the work of a small group of scientists, an academic subset situated within "young-earth creationism," the most literal strain of creationist thought. Young-earthers read the Bible as a scientific source document and labor to find evidence of a world created by God in six days about 6,000 years ago. All other creationists, closer to the mainstream, are known as "old-agers"—those who accept most current scientific thinking (including the fact that the universe is billions of years old) but blend it with certain assumptions drawn from the Bible. The most accommodationist old-agers, known as "theistic evolutionists," are little more than latter-day deists; they believe that God kickstarted the universe billions of years ago with the Big Bang and has sat back ever since. "Progressive creationists," a bit farther down the spectrum, believe that God intervenes only rarely, once an aeon or so.

Variations of creationist thought have cropped up since Darwin's day, when it first appeared that scientific progress was going to shove God out of His own universe. For the longest time, those defenders of the faith who didn't want to appear downright antediluvian found themselves struggling to keep up with science, in the process often falling into traps of logic. One strategy, for example, was to find a point where scientific theory was having trouble with a mechanical explanation, and then to argue that it was at precisely *that* mysterious place where God could be found—a position that became known, dismissively, as the God of the Gaps theory. The problem was that every time science progressed, God was forced to retreat to a smaller, more humiliating niche. Among some old-agers, the retreat continues: one subset now holds that God does indeed interfere in the universe, but only at the molecular level, a God of the Gaps position wherein the Lord is on the lam among muons and leptons.

Much of contemporary creationist literature is mere antievolutionism. In 1991 Berkeley law professor Phillip Johnson published the best-selling *Darwin on Trial.* His book exposes the flimsy logic underlying some of evolution's stubborn orthodoxies: How *does* an animal evolve an eye or a wing? Why isn't there any good proof for macroevolution—the transition from, say, primate to man? (Microevolution, which is adaptation within species, is accepted even by the strictest creationists.) Why is there always talk of missing links? All good questions, but Johnson doesn't rely on creationist research to pose them; rather, he plunders the internal debates of evolutionists. For example, Johnson cites a particularly gnarly problem for evolutionists: the massive and rapid appearance of numerous phyla about half a billion years ago, a phenomenon known as the Cambrian explosion. The gradualist idea of evolution is shaken by this moment. But the main popularizer of this flaw is, ironically, creationism's public enemy number one, Stephen Jay Gould, Harvard paleontologist and dispenser of evolutionary doctrine in his monthly column in *Natural History* magazine. Gould long ago put his peers on notice of the Cambrian problem, and then offered a solution known as "punctuated equilibrium." Johnson mentions Gould's solution but dwells on Gould's revelation of the problem. Throughout the book, in fact, Johnson holds up evolution's animated, healthy debate as proof against itself and applies the pitiless rules of logic to a world bright with controversy and contradiction. Ultimately, his book advances no positive theory of creationism; he's just a pissed-off Christian. Imagine the reaction if another sophist had raked through the tissuey pages of the Bible with the same angry rigor.

More recently, some old-agers have floated the "intelligent design," or ID, theory. Drawn from the ancient philosophical position known as "argument from design," ID theory is today's gloss on the position that this intricate universe couldn't have "just happened." It rejects evolutionary science with a commonsense variation on probability theory, arguing that the odds of natural selection producing a world as wondrous and magical as ours are about as likely, as the evolutionary critic Fred Hoyle has put it, as "a hurricane blowing through a junkyard and spontaneously having the luck to put together a Boeing 747."

ID is the theory that has been adopted by those agitating for creationism in the high schools. In their sample textbook, called *Of Pandas and People,* published by the Foundation for Thought and Ethics, the writing is calm and cool. Evolution is dismissed in plain English. Although "God" has been banished from the book, the careful reader can find Him hiding out among the charts and graphs, slipping from chapter to chapter under the nom de Dieu "intelligent designer."

In the 1980s, the Supreme Court stifled the movement to teach creationism in the schools by ruling that the practice would violate the separation of church and state. Now creationists are trying to dodge that bullet by relocating the skirmish from the classroom to the textbook. As creationists become increasingly able to argue that their ideas constitute "just another scientific theory," they will be able to make a first-amendment case with Madisonian gusto: if the evolutionists are so certain they are right, then what do they fear from a rigorous clash of ideas? As *Pandas* concludes, quoting no less an authority than John Scopes himself: "[I]f you limit a teacher to only one side of anything the whole country will eventually have only one thought, be one individual. I believe in teaching every aspect of every problem or theory." This strategy will work beautifully in the he said/she said medium of television, casting Stephen Jay Gould in the demonic role of scowling censor.

Meanwhile, creationist beliefs are inching their way into the mainstream. During the Republican primaries, Pat Buchanan told Sam Donaldson that he favored creationism over "godless evolution," and this summer [1996] five Republican state parties wrote platform planks calling for a creationist curriculum. Though the Supreme Court will probably continue to strike down the inclusion of creationism in a formal public-school curriculum, what actually gets taught in a classroom is impossible to monitor from district to district. Once the textbooks are available, the teachers can take over from there. And in many towns, anyway, few parents will complain; 58 percent of Americans believe that it's only fair to teach creationism in the schools.

* * *

These trends involve casting creationism's broadest principles into language that will appeal to a mass audience—essentially rewriting God's Word into hollow sound bites that will go down easily on a *Crossfire* episode. The other wing of creationism, meanwhile, is moving in the opposite direction. Instead of secularizing their ideas and watering down their faith, they clutch the Bible to their lab coats more fervently than ever.

Young-earth creationism is built around a central scriptural truth derived from the Bible's genealogy of patriarchs. By counting up the "begat"s in the book of Numbers, a seventeenth-century Irish bishop named James Ussher—perhaps the very first creation scientist—dated the birth of the universe precisely to the year 4004 B.C. (If you perform your own calculations, you will see that 1996 marks the 6,000th anniversary of God's first week of work.) Strict creationists still believe that Ussher's general method is sound, but they are put off by the crackpot exactitude of his math. They prefer twentieth-century scientific notation, and therefore express creation's date as 4121.0 ± 49.7 B.C. By the same math, Christ was born in 3.5 ± 0.5 B.C. and Noah's Flood concluded in 2363.0 ± 44.7 B.C.

According to a handout in Professor Whitmore's class, the two models—creationism and evolutionism—can be considered side by side. One theory holds that creation occurred 6,000 years ago; the other says the Big Bank exploded between eight and twenty billion years ago. Day One of creation gave us "the heavens and the earth"—or the origin of the solar system occurred five billion years ago. Plant life was created two days later—or life emerged on earth 3.8 billion years ago. On the fifth day, animals appeared—or the Cambrian explosion happened 570 million years go. Man was created that first weekend—or five million years ago, during the late Cenozoic Era, *Homo erectus* developed. Noah's Flood occurred about 4,500 years ago—or modern man stepped forth about 500,000 years ago.

Noah's Flood is central to creationist thinking not only because it is mentioned in the scriptural data but because it provides an explanation for the entire fossil record. Rather than accepting the usual view of trilobites drifting to the bottom of a placid sea to be covered by sediment and then fossilized, the creationist view is that every fossilized animal died in the carnage of the Flood....

* * *

My encounters with the newest texts and young creationists seemed incomplete, as did creationism itself. These scientists were frustratingly content to putter around with their ideas but not carry them through to their radical conclusion. How strange to postulate that dinosaurs lived with Noah or that sin is a mutation in our genetic structure or that the forbidden fruit was an addictive hallucinogen or that a fully mature universe appeared *ex nihilo* 6,000 years ago or that extraterrestrials are angels or that the earth literally stopped rotating for a whole day after Joshua's prayer or that Adam lived to be 930 years old—but then not see the larger implications that if these premises are true, then science is in need of a wholesale revolution.

Then I began to hear creationists mention one of their own with an uncommon deference..., The word was that [this man] intended to gather up creationism's tiny hypotheses and integrate them into a Grand Unified Theory. Creationism, as I found, did have a Stephen Hawking. His name is Kurt Wise, and he conducts his work, appropriately enough, out of Dayton, Tennessee, site of the Scopes Monkey Trial.

On a warm summer morning, I pulled into town and slowed down before the famous courthouse, where a plague commemorates the 1925 trial. A cast-iron paragraph tells the story of the first modern media circus. The last three words form a sentence that lingers in the ear: "Scopes was convicted."

Up a hill is the little-known William Jennings Bryan College, a few buildings beneath some fat, shady oaks. Kurt Wise, who is the director of the college's Origins Research Department, graduated from the University of Chicago with a geology degree, and then received his Ph.D. from Harvard. Wise once served as a teaching assistant for none other than Stephen Jay Gould, a man he still calls "Steve."

Wise has a number of major research projects under way. In the course of two days, he and I hiked and spelunked through his field sites—including a fossil-filled gorge, an abandoned coal mine, and Tennessee's seventh-largest cave. Wise is a tall, thin guy with neat brown hair and a tidy auburn mustache. His broad face is reminiscent of Bill Gates's, and he wears the same glasses. His laugh is explosive and excessive, like an overeager camp counselor, yet it's charged with a breezy confidence that comes easily to a geologist who has the finest credentials in the world.

Unlike his colleagues, Wise harbors no bitterness toward establishment science. Quite the opposite. Whenever our conversation idled, Wise filled up the dead space with anecdotes that all had the same narrative thrust: Wise subduing an opponent in friendly argument. One was about demolishing an old-age creationist's logic at a particular meeting. On another occasion, he warmly recalled his evolutionary sparring partner at the University of Chicago, who admitted upon graduation that Wise's arguments were unanswerable. Another story featured a professor who forced Wise to write a paper on Jacques Monod's book on randomness, *Chance and Necessity.* "Using Monod's own premises, I arranged them logically such that I proved the existence of God!" he said, and exploded with self-pleased laughter. Wise could not resist the classic Harvard kicker to every anecdote: "I got an A."

As we headed to Arby's for lunch, Wise explained that he was remapping a cave and that his work was leading him to a new theory on cave formation. He loaned me a lantern and helmet, and we set off to explore. By early afternoon we had turned off the main highway and were speeding down an old two-lane macadam into a bright, empty valley off the Cumberland Plateau, a bowl of green except for an occasional farmhouse and some ruminants. . . .

Wise was quick with opinions about his chosen specialty, almost all of them surprising. For example, creationists backed a recent bill before the Tennessee legislature that would have criminalized teaching evolution as anything but a "theory." Wise opposed it. He doesn't want politicians mucking around in this issue "until we have an adequate creation-science curriculum." Right now, he said, "we don't."

. . . "Most creation science is garbage," he easily admitted. And he just as easily suggested that it was his job to change that. His goal is fairly immodest: he means to undo the Great Divorce—that time after Galileo when theology and science went their separate ways—by serving as the intellectual engine behind his own cosmic theory. He calls it the Great Synthesis. It makes Stephen Hawking's Grand Unified Theory look as ambitious as a high school litmus test. In his own words, Wise intends to use the instruments of science "to restore the Bible to its place at the center of all human thought."

. . . "My idea is not to attack evolution," he said. "My goal is to develop a theory that explains the data of the universe better than conventional theory but is consistent with Scripture." His major beef with other creationists, he explained, is that they onl[illegible] in picking at the weaknesses [illegible] "It's a small person who is [illegible] attacking a theory. By the time [illegible]ished at Harvard, I realized I could destroy macroevolutionary theory at will." . . .

"I don't want to *challenge* evolution," he said . . . "I intend to replace it." . . .

* * *

At one point in our talks, back at his office, he explained that he is currently writing a book that will attempt to unify all the new theories of neo-creationism—mainly his own, but some currently being developed by other serious neo-creationists as well. (He said that there are only about two dozen such scientists right now, but he hopes to inspire a new generation.) He tossed me his outline, a seventy-two-page, single-spaced arguement laid out in classical style. The grandeur and scope of these pages cannot really be conveyed [here]. It is a rethinking of all of science—every branch—alongside a close reading of the Bible.

Perhaps it is enough to note that Wise hopes to publish his book in a spiral binder so that updates can be popped right in. He is certain that 60 percent of the book will change every six months for the foreseeable future—at least until the scientific research begins to prove Scripture definitely; then it will slow down.

Wise squeaked backward in his chair, spread his long arms and legs into a variation on Leonardo's Man, and laid out his plan. "The Great Synthesis is prenatal, embryonic. First we'll have to develop an epistemology—a philosophy of knowledge—that will tell us how to look at the scriptural and physical-world data. We will need a philosophy of science and a philosophy of philosophy. Then each field of science will be examined in that new light—a new geology, a new paleontology, a new cosmology, a new archaeology, and a new history of the origin of languages, culture, and history. This would include replacing the Linnean classification system."

I asked: "You would need a new taxonomic system?"

"Taxonomy!" he guffawed. "Taxonomy is the paint on the building. I am *rebuilding* the building! But what you name things is critical, because it reflects the concepts that underlie it." Science's ever-branching tree of phylum, order, family, genus, and species, Wise explained, presupposes evolution. Wise believes in discrete "created kinds" (implied in Genesis 1:21), and already a new discipline—"bariminology"—has emerged to reclassify every living thing. "I intend to replace the evolutionary tree with the creationist orchard," Wise said, "separately created, separately planted by God."

... He paused before another bellow nearly propelled my chair backward into the wall. "It's going to be a long process!"

In time I realized why I liked Wise's laughter. It's honest and beguiling, yet charged with self-awareness. Unlike so many of his colleagues, who look out into the world to see a horizon darkened by pagans and pederasts and high school teachers, he is unafraid of Harvard and Chicago and science's reductionist establishment. He has matriculated through the most prestigious quarters of that world and emerged with his faith unscathed. He quite intends to return—not as a peer but as a conquering hero.

Other creationists rage at a world sinking under the weight of its abominations. Wise believes that science can lift that weight and heal a riven world. He is a modern Adam who quite literally plans to rename all creatures great and small. He wants to restore them to their place in the original Grand Unified Theory, the one that's never been any farther away than the nearest bedside table in every motel.

This recovery means going back, back before postmodernism, before modernism, before naturalism, before romanticism, before humanism, before rationalism, before—well, I asked him precisely how much of human knowledge creationism intended to revise. He replied solemnly, "I'd say everything after, oh, about 3200 B.C." Then another peal of laughter broke, ranging up and down the science hall.

At times, as I listened to Wise's plan for the Great Synthesis, I wanted to ask him if what he was doing wasn't sacrilegious, a modern attempt at Babel: piling all those data atop one another until he arrived at an absolute proof of God. He possesses a faith so certain that he feels it can be proved through science. But what kind of faith is it that can be replaced by proof? In a universe of certainty and knowledge, faith is superfluous. If Wise is successful, who would ever need a leap of faith? We could all walk.

And yet, it is not God's approval Wise seeks (he's convinced he already has it). It's approval from other scientists. Harvard isn't that far behind him....

I felt again the warmth of believing that for every inch of infinity there has already been an accounting. Everything has a reason for being where it is. Every fossil situated out of its normal orientation, every mountain peak, every anfractuous cave, every silted delta, every meandering river valley, every sheered tectonic plate studied through the lens of scriptural inerrancy (and tested by scientific theory) reveals its place in a knowable past and future—all linked, I realized at last, by a sense of purpose. That was the source of nostalgia. I had felt it before, in childhood, when everything around me radiated with specific meaning and parental clarity. That, after all, is what all creationists feel that evolution has stolen from them. Yet only Wise—and quite possibly he alone—has dared to follow the honest logic of creationism's view out to its end. He is a man whose pure faith and fine intellect are woven so tightly together that he can find no ordinary perch from which to view the gathering evidence of human knowledge. He would have to go back to where Moses went, to the mountaintop, back to the beginning, and rewrite the world.

As he talked on, all the difficulties of modern science disappeared. Chaos exerted no influence up here. Complexity theory was far away. Random evolution, meaningless mutations, trial and error (mostly error), aimless procreation, the pointless void of space, the cold materialism of Darwin's damn theory bereft of the tiniest significance—all had fled before the soothing unity of synthesis and the restoration of purpose to a fallen world.

I would have liked to sit up there forever.

NO

Daniel C. Dennett

DARWIN'S DANGEROUS IDEA: EVOLUTION AND THE MEANINGS OF LIFE

We used to sing a lot when I was a child, around the campfire at summer camp, at school and Sunday school, or gathered around the piano at home. One of my favorite songs was "Tell Me Why."...

Tell me why the stars do shine,
Tell me why the ivy twines,
Tell me why the sky's so blue.
Then I will tell you just why I love you.

Because God made the stars to shine,
Because God made the ivy twine,
Because God made the sky so blue.
Because God made you, that's why I love you.

This straightforward, sentimental declaration still brings a lump to my throat—so sweet, so innocent, so reassuring a vision of life!

And then along comes Darwin and spoils the picnic. Or does he?... From the moment of the publication of *Origin of Species* in 1859, Charles Darwin's fundamental idea has inspired intense reactions ranging from ferocious condemnation to ecstatic allegiance, sometimes tantamount to religious zeal. Darwin's theory has been abused and misrepresented by friend and foe alike. It has been misappropriated to lend scientific respectability to appalling political and social doctrines. It has been pilloried in caricature by opponents, some of whom would have it compete in our children's schools with "creation science," a pathetic hodgepodge of pious pseudo-science.[1]

Almost no one is indifferent to Darwin, and no one should be. The Darwinian theory is a scientific theory, and a great one, but that is not all it is. The creationists who oppose it so bitterly are right about one thing: Darwin's dangerous idea cuts much deeper into the fabric of our most fundamental beliefs than many of its sophisticated apologists have yet admitted, even to themselves.

The sweet, simple vision of the song, taken literally, is one that most of us have outgrown, however fondly we may recall it. The kindly God who lovingly fashioned each and every one of us (all creatures great and small) and sprinkled the sky with shining stars for our delight—*that* God is, like Santa Claus, a myth of childhood, not anything a sane, undeluded adult could literally believe in. *That* God must either be turned into a symbol for something less concrete or abandoned altogether.

Not all scientists and philosophers are atheists, and many who are believers declare that their idea of God can live in peaceful coexistence with, or even find support from, the Darwinian framework of ideas. Theirs is not an anthropomorphic Handicrafter God, but still a God worthy of worship in their eyes, capable of giving consolation and meaning to their lives. Others ground their highest concerns in entirely secular philosophies, views of the meaning of life that stave off despair without the aid of any concept of a Supreme Being—other than the Universe itself. Something *is* sacred to these thinkers, but they do not call it God; they call it, perhaps, Life, or Love, or Goodness, or Intelligence, or Beauty, or Humanity. What both groups share, in spite of the differences in their deepest creeds, is a conviction that life does have meaning, that goodness matters.

But can *any* version of this attitude of wonder and purpose be sustained in the face of Darwinism? From the outset, there have been those who thought they saw Darwin letting the worst possible cat out of the bag: nihilism. They thought that if Darwin was right, the implication would be that nothing could be sacred. To put it bluntly, nothing could have any point. Is this just an overreaction? What exactly are the implications of Darwin's idea—and, in any case, has it been scientifically proven or is it still "just a theory"?

Perhaps, you may think, we could make a useful division: there are the parts of Darwin's idea that really are established beyond any reasonable doubt, and then there are the speculative extensions of the scientifically irresistible parts. Then—if we were lucky—perhaps the rock-solid scientific facts would have no stunning implications about religion, or human nature, or the meaning of life, while the parts of Darwin's idea that get people all upset could be put into quarantine as highly controversial extensions of, or mere interpretations of, the scientifically irresistible parts. That would be reassuring.

But alas, that is just about backwards. There are vigorous controversies swirling around in evolutionary theory, but those who feel threatened by Darwinism should not take heart from this fact. Most—if not quite all—of the controversies concern issues that are "just science"; no matter which side wins, the outcome will not undo the basic Darwinian idea. That idea, which is about as secure as any in science, really does have far-reaching implications for our vision of what the meaning of life is or could be.

In 1543, Copernicus proposed that the Earth was not the center of the universe but in fact revolved around the Sun. It took over a century for the idea to sink in, a gradual and actually rather painless transformation. (The religious reformer Philipp Melanchthon, a collaborator of Martin Luther, opined that "some Christian prince" should suppress this madman, but aside from a few such salvos, the world was not particularly shaken by Copernicus himself.) The Copernican

Revolution did eventually have its own "shot heard round the world": Galileo's *Dialogue Concerning the Two Chief World Systems,* but it was not published until 1632, when the issue was no longer controversial among scientists. Galileo's projectile provoked an infamous response by the Roman Catholic Church, setting up a shock wave whose reverberations are only now dying out. But in spite of the drama of that epic confrontation, the idea that our planet is not the center of creation has sat rather lightly in people's minds. Every schoolchild today accepts this as the matter of fact it is, without tears or terror.

In due course, the Darwinian Revolution will come to occupy a similarly secure and untroubled place in the minds—and hearts—of every educated person on the globe, but today, more than a century after Darwin's death, we still have not come to terms with its mind-boggling implications. Unlike the Copernican Revolution, which did not engage widespread public attention until the scientific details had been largely sorted out, the Darwinian Revolution has had anxious lay spectators and cheerleaders taking sides from the outset, tugging at the sleeves of the participants and encouraging grandstanding. The scientists themselves have been moved by the same hopes and fears, so it is not surprising that the relatively narrow conflicts among theorists have often been not just blown up out of proportion by their adherents, but seriously distorted in the process. Everybody has seen, dimly, that a lot is at stake.

Moreover, although Darwin's own articulation of his theory was monumental, and its powers were immediately recognized by many of the scientists and other thinkers of his day, there really were large gaps in his theory that have only recently begun to be properly filled in. The biggest gap looks almost comical in retrospect. In all his brilliant musings, Darwin never hit upon the central concept, without which the theory of evolution is hopeless: the concept of a *gene.* Darwin had no proper *unit* of heredity, and so his account of the process of natural selection was plagued with entirely reasonable doubts about whether it would work. Darwin supposed that offspring would always exhibit a sort of blend or average of their parents' features. Wouldn't such "blending inheritance" always simply average out all differences, turning everything into uniform gray? How could diversity survive such relentless averaging? Darwin recognized the seriousness of this challenge, and neither he nor his many ardent supporters succeeded in responding with a description of a convincing and well-documented mechanism of heredity that could combine traits of parents while maintaining an underlying and unchanged identity. The idea they needed was right at hand, uncovered ("formulated" would be too strong) by the monk Gregor Mendel and published in a relatively obscure Austrian journal in 1865, but, in the best-savored irony in the history of science, it lay there unnoticed until its importance was appreciated (at first dimly) around 1900. Its triumphant establishment at the heart of the "Modern Synthesis" (in effect, the synthesis of Mendel and Darwin) was eventually made secure in the 1940s, thanks to the work of Theodosius Dobzhansky, Julian Huxley, Ernst Mayr, and others. It has taken another half-century to iron out most of the wrinkles of that new fabric.

The fundamental core of contemporary Darwinism, the theory of DNA-based reproduction and evolution, is now beyond dispute among scientists. It demonstrates its power every day, contributing crucially to the explanation of planet-sized facts of geology and meteorology, through middle-sized facts of ecology and agronomy, down to the latest microscopic facts of genetic engineering. It unifies all of biology and the history of our planet into a single grand story. Like Gulliver tied down in Lilliput, it is unbudgeable, not because of some one or two huge chains of argument that might —hope against hope—have weak links in them, but because it is securely tied by hundreds of thousands of threads of evidence anchoring it to virtually every other area of human knowledge. New discoveries may conceivably lead to dramatic, even "revolutionary" *shifts* in the Darwinian theory, but the hope that it will be "refuted" by some shattering breakthrough is about as reasonable as the hope that we will return to a geocentric vision and discard Copernicus.

Still, the theory is embroiled in remarkably hot-tempered controversy, and one of the reasons for this incandescence is that these debates about scientific matters are usually distorted by fears that the "wrong" answer would have intolerable moral implications. So great are these fears that they are carefully left unarticulated, displaced from attention by several layers of distracting rebuttal and counter-rebuttal. The disputants are forever changing the subject slightly, conveniently keeping the bogeys in the shadows. It is this misdirection that is mainly responsible for postponing the day when we can all live as comfortably with our new biological perspective as we do with the astronomical perspective Copernicus gave us.

Whenever Darwinism is the topic, the temperature rises, because more is at stake than just the empirical facts about how life on Earth evolved, or the correct logic of the theory that accounts for those facts. One of the precious things that is at stake is a vision of what it means to ask, and answer, the question "Why?" Darwin's new perspective turns several traditional assumptions upside down, undermining our standard ideas about what ought to count as satisfying answers to this ancient and inescapable question. Here science and philosophy get completely intertwined. Scientists sometimes deceive themselves into thinking that philosophical ideas are only, at best, decorations or parasitic commentaries on the hard, objective triumphs of science, and that they themselves are immune to the confusions that philosophers devote their lives to dissolving. But there is no such thing as philosophy-free science; there is only science whose philosophical baggage is taken on board without examination.

The Darwinian Revolution is both a scientific and a philosophical revolution, and neither revolution could have occurred without the other. As we shall see, it was the philosophical prejudices of the scientists, more than their lack of scientific evidence, that prevented them from seeing how the theory could actually work, but those philosophical prejudices that had to be overthrown were too deeply entrenched to be dislodged by mere philosophical brilliance. It took an irresistible parade of hard-won scientific facts to force thinkers to take seriously the weird new outlook that Darwin proposed. Those who are still ill-acquainted with that beautiful procession can be forgiven their continued allegiance to the pre-Darwinian ideas. And the battle is not yet over; even among the scientists, there are pockets of resistance.

Let me lay my cards on the table. If I were to give an award for the single best idea anyone has ever had, I'd give it to Darwin, ahead of Newton and Einstein and everyone else. In a single stroke, the idea of evolution by natural selection unifies the realm of life, meaning, and purpose with the realm of space and time, cause and effect, mechanism and physical law. But it is not just a wonderful scientific idea. It is a dangerous idea. My admiration for Darwin's magnificent idea is unbounded, but I, too, cherish many of the ideas and ideals that it *seems* to challenge, and want to protect them. For instance, I want to protect the campfire song, and what is beautiful and true in it, for my little grandson and his friends, and for their children when they grow up. There are many more magnificent ideas that are also jeopardized, it seems, by Darwin's idea, and they, too, may need protection. The only good way to do this—the only way that has a chance in the long run—is to cut through the smokescreens and look at the idea as unflinchingly, as dispassionately, as possible.

On this occasion, we are not going to settle for "There, there, it will all come out all right." Our examination will take a certain amount of nerve. Feelings may get hurt. Writers on evolution usually steer clear of this apparent clash between science and religion. Fools rush in, Alexander Pope said, where angels fear to tread. Do you want to follow me? Don't you really want to know what survives this confrontation? What if it turns out that the sweet vision—or a better one—survives intact, strengthened and deepened by the encounter? Wouldn't it be a shame to forgo the opportunity for a strengthened, renewed creed, settling instead for a fragile, sickbed faith that you mistakenly supposed must not be disturbed?

There is no future in a sacred myth. Why not? Because of our curiosity. Because, as the song reminds us, *we want to know why.* We may have outgrown the song's answer, but we will never outgrow the question. Whatever we hold precious, we cannot protect it from our curiosity, because being who we are, one of the things we deem precious is the truth. Our love of truth is surely a central element in the meaning we find in our lives. In any case, the idea that we might preserve meaning by kidding ourselves is a more pessimistic, more nihilistic idea than I for one can stomach. If that were the best that could be done, I would conclude that nothing mattered after all....

* * *

At what "point" does a human life begin or end? The Darwinian perspective lets us see with unmistakable clarity why there is no hope at all of *discovering* a telltale mark, a saltation in life's processes, that "counts." We need to draw lines; we need definitions of life and death for many important moral purposes. The layers of pearly dogma that build up in defense around these fundamentally arbitrary attempts are familiar, and in never-ending need of repair. We should abandon the fantasy that either science or religion can uncover some well-hidden fact that tells us exactly where to draw these lines. There is no "natural" way to mark the birth of a human "soul," any more than there is a "natural" way to mark the birth of a species. And, contrary to what many traditions insist, I think we all do share the intuition that there are gradations of value in the ending of human lives. Most human embryos end in spontaneous abortion—fortunately, since these are mostly *terata*, hopeless monsters whose lives are all but impossible. Is this a terrible evil? Are the mothers whose bodies abort these embryos guilty of involuntary manslaughter? Of course not. Which is worse, taking "heroic" measures to keep alive a severely deformed infant, or taking the equally "heroic" (if unsung) step of seeing to it that such an infant dies as quickly and painlessly as possible? I do not suggest that Darwinian thinking gives us answers to such questions; I do suggest that Darwinian thinking helps us see why the traditional hope of solving these problems (finding a moral algorithm) is forlorn. We must cast off the myths that make these old-fashioned solutions seem inevitable. We need to grow up, in other words.

Among the precious artifacts worth preserving are whole cultures themselves. There are still several thousand distinct languages spoken daily on our planet, but the number is dropping fast (Diamond 1992, Hale et al. 1992). When a language goes extinct, this is the same kind of loss as the extinction of a species, and when the culture that was carried by that language dies, this is an even greater loss. But here, once again, we face incommensurabilities and no easy answers.

I began . . . with a song which I myself cherish, and hope will survive "forever." I hope my grandson learns it and passes it on to his grandson, but at the same time I do not myself believe, and do not really want my grandson to believe, the doctrines that are so movingly expressed in that song. They are too simple. They are, in a word, wrong—just as wrong as the ancient Greeks' doctrines about the gods and goddesses on Mount Olympus. Do you believe, literally, in an anthropomorphic God? If not, then you must agree with me that the song is a beautiful, comforting falsehood. Is that simple song nevertheless a valuable meme? I certainly think it is. It is a modest but beautiful part of our heritage, a treasure to be preserved. But we must face the fact that, just as there were times when tigers would not have been viable, times are coming when they will no longer be viable, except in zoos and other preserves, and the same is true of many of the treasures in our cultural heritage.

The Welsh language is kept alive by artificial means, just the way condors are. We cannot preserve *all* the features of the cultural world in which these treasures flourished. We wouldn't want to. It took oppressive political and social systems, rife with many evils, to create the rich soil in which many of our greatest works of art could grow: slavery and despotism ("enlightened" though these sometimes may have been), obscene differences in living standards between the rich and the poor—and a huge amount of ignorance. Ignorance is a necessary condition for many excellent things. The childish joy of seeing what Santa Claus has brought for Christmas is a species of joy that must soon be extinguished in each child by the loss of ignorance. When that child grows up, she can transmit that joy to her own children, but she must also recognize a time when it has outlived its value.

The view I am expressing has clear ancestors. The philosopher George Santayana was a Catholic atheist, if you can imagine such a thing. According to Bertrand Russell (1945, p. 811), William James once denounced Santayana's ideas as "the perfection of rottenness," and one can see why some people would be offended by his brand of aestheticism: a deep appreciation for all the formulae, ceremonies, and trappings of his religious heritage, but lacking the faith. Santayana's position was aptly caricatured: "There is no God and Mary is His Mother." But how many of us are caught in that very dilemma, loving the heritage, firmly convinced of its value, yet unable to sustain any conviction at all in its truth? We are faced with a difficult choice. Because we value it, we are eager to preserve it in a rather precarious and "denatured" state—in churches and cathedrals and synagogues, built to house huge congregations of the devout, and now on the way to being cultural museums. There is really not that much difference between the roles of the Beefeaters who stand picturesque guard at the Tower of London, and the Cardinals who march in their magnificent costumes and meet to elect the next Pope. Both are keeping alive traditions, rituals, liturgies, symbols, that otherwise would fade.

But hasn't there been a tremendous rebirth of fundamentalist faith in all these creeds? Yes, unfortunately, there has been, and I think that there are no forces on this planet more dangerous to us all than the fanaticisms of fundamentalism, of all the species: Protestantism, Catholicism, Judaism, Islam, Hinduism, and Buddhism, as well as countless smaller infections. Is there a conflict between science and religion here? There most certainly is.

Darwin's dangerous idea helps to create a condition in the memosphere that in the long run threatens to be just as toxic to these memes as civilization in general has been toxic to the large wild mammals. Save the Elephants! Yes, of course, but not *by all means.* Not by forcing the people of Africa to live nineteenth-century lives, for instance. This is not an idle comparison. The creation of the great wildlife preserves in Africa has often been accompanied by the dislocation —and ultimate destruction—of human populations. (For a chilling vision of this side effect, see Colin Turnbull 1972 on the fate of the Ik.) Those who think that we should preserve the elephants' pristine environment *at all costs* should contemplate the costs of returning the United States to the pristine conditions in which the buffaloes roam and the deer and the antelope play. We must find an accommodation.

I love the King James Version of the Bible. My own spirit recoils from a God Who is He or She in the same way my heart sinks when I see a lion pacing neurotically back and forth in a small zoo cage. I know, I know, the lion is beautiful but dangerous; if you let the lion roam free, it would kill me; safety demands that it be put in a cage. Safety demands that religions be put in cages, too—when absolutely necessary. We just can't have forced female circumcision, and the second-class status of women in Roman Catholicism and Mormonism, to say nothing of their status in Islam. The recent Supreme Court ruling declaring unconstitutional the Florida law prohibiting the sacrificing of animals in the rituals of the Santeria sect (an Afro-Caribbean religion incorporating elements of Yoruba traditions and Roman Catholicism) is a borderline case, at least for many of us. Such rituals are offensive to many, but the protective mantle of religious tradition secures our tolerance. We are wise to respect these traditions. It is, after all, just part of respect for the biosphere.

Save the Baptists! Yes, of course, but not *by all means.* Not if it means tolerating the deliberate misinforming of children about the natural world. According to a recent poll, 48 percent of the people in the

United States today believe that the book of Genesis is literally true. And 70 percent believe that "creation science" should be taught in school alongside evolution. Some recent writers recommend a policy in which parents would be able to "opt out" of materials they didn't want their children taught. Should evolution be taught in the schools? Should arithmetic be taught? Should history? Misinforming a child is a terrible offense.

A faith, like a species, must evolve or go extinct when the environment changes. It is not a gentle process in either case. We see in every Christian subspecies the battle of memes—should women be ordained? should we go back to the Latin liturgy?—and the same can also be observed in the varieties of Judaism and Islam. We must have a similar mixture of respect and self-protective caution about memes. This is already accepted practice, but we tend to avert our attention from its implications. We preach freedom of religion, but only so far. If your religion advocates slavery, or mutilation of women, or infanticide, or puts a price on Salman Rushdie's head because he has insulted it, then your religion has a feature that cannot be respected. It endangers us all.

It is nice to have grizzly bears and wolves living in the wild. They are no longer a menace; we can peacefully coexist, with a little wisdom. The same policy can be discerned in our political tolerance, in religious freedom. You are free to preserve or create any religious creed you wish, so long as it does not become a public menace. We're all on the Earth together, and we have to learn some accommodation. The Hutterite memes are "clever" not to include any memes about the virtue of destroying outsiders. If they did, we would have to combat them. We tolerate the Hutterites because they harm only themselves—though we may well insist that we have the right to impose some further openness on their schooling of their own children. Other religious membes are not so benign. The message is clear: those who will not accommodate, who will not temper, who insist on keeping only the purest and wildest strain of their heritage alive, we will be obliged, reluctantly, to cage or disarm, and we will do our best to disable the memes they fight for. Slavery is beyond the pale. Child abuse is beyond the pale. Discrimination is beyond the pale. The pronouncing of death sentences on those who blaspheme against a religion (complete with bounties or rewards for those who carry them out) is beyond the pale. It is not civilized, and is owed no more respect in the name of religious freedom than any other incitement to cold-blooded murder.[2] . . .

Long before there was science, or even philosophy, there were religions. They have served many purposes (it would be a mistake of greedy reductionism to look for a single purpose, a single *summum bonum* which they have all directly or indirectly served). They have inspired many people to lead lives that have added immeasurably to the wonders of our world, and they have inspired many more people to lead lives that were, given their circumstances, more meaningful, less painful, than they otherwise could have been. . . .

Religions have brought the comfort of belonging and companionship to many who would otherwise have passed through this life all alone, without glory or adventure. At their best, religions have drawn attention to love, and made it real for people who could not otherwise see it, and ennobled the attitudes and refreshed the spirits of the world-beset. Another thing religions have accomplished, without this being thereby their *raison d'être,* is that they have kept *Homo sapiens* civilized enough, for long enough, for us to have learned how to reflect more systematically and accurately on our position in the universe. There is much more to learn. There is certainly a treasury of ill-appreciated truths embedded in the endangered cultures of the modern world, designs that have accumulated details over eons of idiosyncratic history, and we should take steps to record it, and study it, before it disappears, for, like dinosaur genomes, once it is gone, it will be virtually impossible to recover.

We should not expect this variety of respect to be satisfactory to those who wholeheartedly embody the memes we honor with our attentive—but not worshipful—scholarship. On the contrary, many of them will view anything other than enthusiastic conversion to their own views as a threat, even an intolerable threat. We must not underestimate the suffering such confrontations cause. To watch, to have to participate in, the contraction or evaporation of beloved features of one's heritage is a pain only our species can experience, and surely few pains could be more terrible. But we have no reasonable alternative, and those whose visions dictate that they cannot peacefully coexist with the rest of us will have to quarantine as best we can, minimizing the pain and damage, trying always to leave open a path or two that may come to seem acceptable.

If you want to teach your children that they are the tools of God, you had better not teach them that they are God's rifles, or we will have to stand firmly opposed to you: your doctrine has no glory, no special rights, no intrinsic and inalienable merit. If you insist on teaching your children falsehoods—that the Earth is flat, that "Man" is not a product of evolution by natural selection—then you must expect, at the very least, that those of us who have freedom of speech will feel free to describe your teachings as the spreading of falsehoods, and will attempt to demonstrate this to your children at our earliest opportunity. Our future well-being—the well-being of all of us on the planet—depends on the education of our descendants.

What, then, of all the glories of our religious traditions? They should certainly be preserved, as should the languages, the art, the costumes, the rituals, the monuments. Zoos are now more and more being seen as second-class havens for endangered species, but at least they are havens, and what they preserve is irreplaceable. The same is true of complex memes and their phenotypic expressions. Many a fine New England church, costly to maintain, is in danger of destruction. Shall we deconsecrate these churches and turn them into museums, or retrofit them for some other use? The latter fate is at least to be preferred to their destruction. Many congregations face a cruel choice: their house of worship costs so much to maintain in all its splendor that little of their tithing is left over for the poor. The Catholic Church has faced this problem for centuries, and has maintained a position that is, I think, defensible, but not obviously so: when it spends its treasure to put gold plating on the candlesticks, instead of providing more food and better shelter for the poor of the parish, it has a different vision of what makes life worth living. Our people, it says, benefit more from having a place of splendor in which to worship than from a little more food. Any atheist or agnostic who finds this cost-benefit analysis ludicrous might pause to consider whether to support diverting all charitable and governmental support for museums, symphony orchestras, libraries, and scientific laboratories to efforts to provide more food and better living conditions for the least well off. A human life worth living is not something that can be uncontroversially measured, and that is its glory.

And there's the rub. What will happen, one may well wonder, if religion is preserved in cultural zoos, in libraries, in concerts and demonstrations? It is happening; the tourists flock to watch the Native American tribal dances, and for the onlookers it is folklore, a religious ceremony, certainly, to be treated with respect, but also an example of a meme complex on the verge of extinction, at least in its strong, ambulatory phase; it has become an invalid, barely kept alive by its custodians. Does Darwin's dangerous idea give us anything in exchange for the ideas it calls into question?

... [T]he physicist Paul Davies proclaim[ed] that the reflective power of human minds can be "no trivial detail, no minor by-product of mindless purposeless forces," and [I] suggested that being a by-product of mindless purposeless forces was no disqualification for importance. And I have argued that Darwin has shown us how, in fact, *everything* of importance is just such a product. Spinoza called his highest being God or Nature *(Deus sive Natura)*, expressing a sort of pantheism. There have been many varieties of pantheism, but they usually lack a convincing *explanation* about just how God is distributed in the whole of nature.... Darwin offers us one: it is in the distribution of Design throughout nature, creating, in the Tree of Life, an utterly unique and irreplaceable creation, an actual pattern in the immeasurable reaches of Design Space that could never be exactly duplicated in its many details. What is design work? It is that wonderful wedding of chance and necessity, happening in a trillion places at once, at a trillion different levels. And what miracle caused it? None. It just happened to happen, in the fullness of time. You could even say, in a way, that the Tree of Life created itself. Not in a miraculous, instantaneous whoosh, but slowly, slowly, over billions of years.

Is this Tree of Life a God one could worship? Pray to? Fear? Probably not. But it *did* make the ivy twine and the sky so blue, so perhaps the song I love tells a truth after all. The Tree of Life is neither perfect nor infinite in space or time, but it is actual, and if it is not Anselm's "Being greater than which nothing can be conceived," it is surely a being that is greater than anything any of us will ever conceive of in detail worthy of its detail. Is something sacred? Yes, say I with Nietzsche. I could not pray to it, but I can stand in affirmation of its magnificence. This world is sacred.

NOTES

1. I will not devote any space [here to] cataloguing the deep flaws in creationism, or supporting my peremptory condemnation of it. I take that job to have been admirably done by others.

2. Many, many Muslims agree, and we must not only listen to them, but do what we can to protect and support them, for they are bravely trying, from the inside, to reshape the tradition they cherish into something better, something ethically defensible. *That* is—or, rather, ought to be—the message of multiculturalism, not the patronizing and subtly racist hypertolerance that "respects" vicious

and ignorant doctrines when they are propounded by officials of non-European states and religions. One might start by spreading the word about *For Rushdie* (Braziller, 1994), a collection of essays by Arab and Muslim writers, many critical of Rushdie, but all denouncing the unspeakably immoral "fatwa" death sentence proclaimed by the Ayatollah. Rushdie (1994) has drawn our attention to the 162 Iranian intellectuals who, with great courage, have signed a declaration in support of freedom of expression. Let us all distribute the danger by joining hands with them.

POSTSCRIPT

Should the Theory of Evolution Be Replaced by Creationism?

In October 1996 Pope John Paul II announced that "new knowledge leads us to recognize that the theory of evolution is more than a hypothesis." This endorsement had little noticeable impact on the creationism-evolution debate because creationism is a thing of fundamentalist Protestant sects. The debate between creationists and evolutionists will thus go on for the foreseeable future. At least on occasion the debate will surely involve verbal abuse, as it did in May 1996, when biologists attempting to inform the Ohio House Education Committee of how thoroughly the evidence supports the theory of evolution were heckled, jeered, and shouted down. See Karen Schmidt, "Creationists Evolve New Strategy," *Science* (July 26, 1996).

As Janet Raloff notes, in "When Science and Beliefs Collide," *Science News* (June 8, 1996), harsh reactions to scientists are not surprising, considering that nearly half the U.S. population misunderstands and/or rejects "many of the basic precepts and findings of science." And these reactions are not seen only in churches and before legislative committees but also at academic meetings. As Barbara Ehrenreich and Janet McIntosh note, in "The New Creationism: Biology Under Attack," *The Nation* (June 9, 1997), there is a movement among feminists, cultural anthropologists, social psychologists, and other academics, amounting to a kind of "secular creationism" that insists human beings are not shaped by their biology, unlike all other living things, and shouts down all mention of Darwin, DNA, and even science. This "new creationism... represents a grave misunderstanding of biology and science generally," say Ehrenreich and McIntosh, but it is not about to go away, because secular creationists do not brook contradiction of their cherished beliefs.

Stephen Jay Gould, in "The Persistently Flat Earth," *Natural History* (March 1994), makes the point that irrationality and dogmatism serve the adherents of neither science nor religion well: "The myth of a war between science and religion remains all too current and continues to impede a proper bonding and conciliation between these two utterly different and powerfully important institutions of human life." Those who wish to see more of the creationism-evolution debate can turn to Dean H. Kenyon and Percival Davis, *Of Pandas and People*, 2d ed. (Haughton Publishing, 1993) for a view of biology that credits the intricate complexity of living things to a master intellect or designer. This argument from design is well critiqued by Kenneth R. Miller in "Life's Grand Design," *Technology Review* (March 1994). Also invaluable is Ronald L. Numbers, *The Creationists: The Evolution of Scientific Creationism* (Alfred A. Knopf, 1992).

On the Internet . . .

The Worldwatch Institute

The Worldwatch Institute is dedicated to fostering the evolution of an environmentally sustainable society, one in which human needs are met in ways that do not threaten the health of the natural environment or the prospects of future generations.
http://www.worldwatch.org/

National Oceanic and Atmospheric Administration

The mission of the National Oceanic and Atmospheric Administration (NOAA) is to describe and predict changes in Earth's environment and to conserve and manage wisely U.S. coastal and marine resources to ensure sustainable economic opportunities.
http://www.noaa.gov/

National Renewable Energy Laboratory

The National Renewable Energy Laboratory (NREL) is the leading center for renewable energy research in the United States. *http://www.nrel.gov/*

EPA and Ozone Depletion

This site, sponsored by the U.S. Environmental Protection Agency (EPA), contains information about ozone depletion, ozone-protective regulations in the United States, and other related topics.
http://www.epa.gov/docs/ozone/index.html

Electromagnetic Fields and Human Health

At this site, John E. Moulder, professor of radiation oncology, radiology, and pharmacology/toxicology at the Medical College of Wisconsin, answers frequently asked questions about the relationship between electromagnetic fields (EMFs) and cancer.
http://www.mcw.edu/gcrc/cop/powerlines-cancer-FAQ/toc.html

The Heritage Foundation

The Heritage Foundation is a think tank whose mission is to formulate and promote conservative public policies based on the principles of free enterprise, limited government, individual freedom, traditional American values, and a strong national defense.
http://www.conservative.org/heritage/

PART 2

The Environment

As the damage that human beings do to their environment in the course of obtaining food, wood, ore, fuel, and other resources has become clear, many people have grown concerned. Some of that concern is for the environment—the landscapes and living things with which humanity shares its world. Some of that concern is more for human welfare; it focuses on the ways in which environmental damage threatens human health or even human survival.

Some environmental issues are well known. These include overpopulation and the prospect for an adequate future food supply, global warming, ozone depletion, and the impact of environmentalism on individual freedoms and property rights. Other environmental issues, such as the possible health risks of electromagnetic fields, are less familiar to the general population but are perhaps no less worthy of concern. All have provoked extensive debate over details and degrees of certainty, over what can or should be done to prevent future difficulties, and even over whether or not the issues are real.

- Will Future Generations Have Enough to Eat?
- Should Society Be Concerned About Global Warming?
- Is Ozone Depletion a Genuine Threat?
- Are Electromagnetic Fields Dangerous to Your Health?
- Are Environmental Regulations Too Restrictive?

ISSUE 4

Will Future Generations Have Enough to Eat?

YES: World Bank, from "Food Security for the World," Paper Prepared for the World Food Summit in Rome, Italy (November 13–17, 1996)

NO: Lester R. Brown, from "Can We Raise Grain Yields Fast Enough?" *World Watch* (July/August 1997)

ISSUE SUMMARY

YES: The World Bank argues that the only barriers to meeting the need for food of future generations are political and economic, not physical and biological. If we make the right decisions, agricultural productivity can keep up with population growth.

NO: Lester R. Brown, president of the Worldwatch Institute, argues that the physical and biological barriers are so significant that there is a strong chance that agricultural productivity will fall behind population growth. Politics and economics alone are not enough, he contends; population must be stabilized and soil must be protected.

In 1798 the British economist Thomas Malthus published his *Essay on the Principle of Population.* In it, he pointed with alarm at the way the human population grew geometrically (a hockey-stick curve of increase) while agricultural productivity grew only arithmetically (a straight-line increase). It was obvious, he said, that the population must inevitably outstrip its food supply and experience famine. Contrary to the conventional wisdom of the time, Malthus argued, population growth was not necessarily a good thing. Indeed, it led inexorably to catastrophe. For many years, Malthus was something of a laughingstock. The doom he forecast kept receding into the future as new lands were opened to agriculture, new agricultural technologies appeared, new ways of preserving food limited the waste of spoilage, and the birth rate dropped in the industrialized nations (the "demographic transition"). The food supply kept ahead of population growth and seemed likely —to most observers—to continue to do so. Malthus's ideas were dismissed as irrelevant fantasies.

Yet overall population kept growing. In Malthus's time, there were about 1 billion human beings on Earth. By 1950—when Warren S. Thompson worried that civilization would be endangered by the rapid growth of Asian and Latin American populations during the next five decades ("Population," *Sci-*

entific American, February 1950, pp. 11–15)—there were a little over 2.5 billion. Before the end of the twentieth century, the tally will pass 6 billion. While global agricultural production has also increased, it has not kept up with rising demand, and—because of the loss of topsoil to erosion, the exhaustion of aquifers for irrigation water, and the high price of energy for making fertilizer (among other things)—the prospect of improvement seems to many observers exceedingly slim. The statistics presented in *World Resources 1996–97* (Oxford University Press, 1996), a report of the World Resources Institute in collaboration with the United Nations Environment and Development Programmes, are positively frightening. The Worldwatch Institute's report *State of the World 1997* (W. W. Norton, 1997) is no less so.

Some people are still laughing at Malthus and his forecasts of doom that two centuries never saw come to pass. Among the scoffers are Julian Simon, a "cornucopian" economist who believes that the more people we have on Earth, the more talent we have available for solving problems, and that humans can indeed find ways around all possible resource shortages (see his essay "Life on Earth Is Getting Better, Not Worse," *The Futurist*, August 1983).

But more and more people—including some economists—are coming to realize that Malthus's error lay not in his prediction but in his timing. There is a growing consensus that he was quite correct to say that a growing population must inevitably outrun its food supply. The only question is how long human ingenuity can stave off the day of reckoning.

How long *can* human ingenuity stave off the day of reckoning? The World Resources Institute sets the global human population at about 8.5 billion in 2025. By 2050 the population is expected to hit 10 billion *and to still be rising*; some estimates peg the 2050 population at 12.5 billion. The UN expects that *if* human fertility can be reined in, the population may stabilize in the neighborhood of 11.5 billion by 2150.

Can population really go that high? If it does, can we possibly feed it? There are famines in the world *today*. Won't they grow far, far worse long before we double our numbers—indeed, well before we hit the 10 billion (or more) mark in 2050? Paul R. Ehrlich and Anne H. Ehrlich, in "Ehrlich's Fables," *Technology Review* (January 1997), write, "A new kit of tools to expand food production is required to carry us into the future, yet no such kit appears to be on the horizon."

In the selections that follow, the World Bank asserts that catastrophe is preventable if we make the right political and economic decisions. The solution lies in encouraging international trade, meeting the needs of the poor, and boosting agricultural research and development. Lester R. Brown grants that such measures are necessary, but he sees physical and biological barriers in population, loss of cropland to other uses, loss of irrigation water, and limits to how much crop yields can be increased even with fertilizer and selective breeding.

YES

World Bank

FOOD SECURITY FOR THE WORLD

INTRODUCTION

Today there are 800 million people in the world who are hungry. Many more are at risk from micronutrient deficiencies (vitamin A, iodine, and iron). This is unacceptable. The fundamental challenge for the [World] Bank and the world in the 21st century is to ensure that the hundreds of millions of families living in poverty in rural and urban areas throughout the world have access to enough food to maintain a healthy and active life.

Fostering the growth of *national and global* food supplies is essential for eliminating hunger and reducing poverty. But increased national and global food supplies are not enough. Today, even when the world produces more than 1 kilogram of grain per person per day, people are hungry because they cannot afford to buy the food they need.

To reduce poverty and hunger demands a multi-pronged strategy with a cornerstone of *rural development,* and, in particular, a prosperous smallholder private sector agricultural economy. Today almost 75 percent of the poor in developing countries live in rural areas. Thus it matters where agricultural production takes place and who receives the associated income. Only if more rapid agricultural growth takes place in countries with impoverished rural populations, can rural farm and nonfarm incomes rise sufficiently to enable the rural poor to afford more and better food. And only if the many millions of men and women smallholders participate in agricultural growth will rural poverty be reduced. Rural growth also contributes to reducing *urban poverty.* When agricultural productivity improves, rural wages and employment rise, reducing labor flows to urban areas—leading to wage increases for the unskilled and semi-skilled in cities too. Increased farm productivity also reduces the price of food in urban areas, often a significant component of household expenditures for the urban poor. Reducing poverty and hunger also requires policies and strategies to address undernutrition in cities, where the numbers of poor and hungry are rapidly increasing.

The challenges are technological, economic, institutional and political. The Bank believes that assuring food for all—now and in the future—can be met

From World Bank, "Food Security for the World," Paper Prepared for the World Food Summit in Rome, Italy (November 13–17, 1996).

only if the global community and individual nations commit themselves now to a set of actions. This document outlines how the World Bank proposes to act....

THE THREE DIMENSIONS OF FOOD SECURITY: HOUSEHOLD, NATIONAL, AND GLOBAL

The job of reducing hunger involves (a) increasing global food supplies to meet the demands of a growing world population; (b) reducing poverty to allow people to buy the food they require; and (c) nutrition education programs to provide people with the information they need to eat healthy diets.

Adequate food supplies. Over the next 30 years, food needs in developing countries could nearly double just to meet the demands created by population growth and modest income growth. Countries worldwide will need to raise the productivity of agriculture to meet this challenge. *Research and extension* at both the national and international levels are fundamental to raising agriculture productivity and increasing global and national food supplies. A *fair trading regime* is critical, since only then can countries refrain from costly self-sufficiency policies and specialize in producing the commodities which are most profitable for them.

Poverty reduction. The best way to reduce poverty and hunger is for countries to grow economically. Few countries have significantly reduced poverty without also experiencing economic growth. *For most developing countries, improved agricultural productivity can be the engine of economic growth.* Indeed, growth in food and agricultural output has been the main basis of economic growth, higher per capita incomes, and better diets for most countries. Most of the developing countries that grew rapidly during the 1980s, and achieved the largest improvements in diets, experienced rapid agricultural growth in the preceding years. For example, China's extraordinary annual economic growth rate of 9.5 percent during the 1980s and 1990s was preceded by rural and agricultural policy reforms of the late 1970s and early 1980s. Indonesia and Thailand also experienced strong agricultural growth prior to the period of high nonagricultural growth, which continues today.

Raising the incomes of the rural poor requires the *development of markets and agribusinesses, investment in people,* and *investment in infrastructure.* It demands programs and processes which guarantee *broad participation* of all community members, including the poor. Extensive evidence shows that investments in health, basic education, agricultural extension, and business training for both men and women greatly increase the productivity of the poor. And it demands *investment in infrastructure.*

Health and nutrition programs. Eliminating the consequences of hunger for the 800 million who are undernourished today and the more than 2 billion who receive insufficient amounts of essential vitamins and minerals requires targeted nutrition, health, and food programs. Increasing family income alone does not assure that people consume the right kind of nutrients in the right quantities at the right times to maintain health and productivity. Today, most households could prevent child malnutrition if they used existing resources optimally, making small changes in health and nutrition behavior. Thus, while general poverty, infrastructure, [and] agriculture programs will improve nutrition eventually, direct nutrition actions are likely to have a greater impact in a shorter time.

TRENDS IN HOUSEHOLD, NATIONAL, AND GLOBAL FOOD SECURITY

Developments of the Past Twenty-Five Years

There has been substantial progress. During the last twenty-five years, there has been real progress in improving living standards among people in the developing world. The proportion of the world's people living in poverty has declined, average incomes per person have doubled, infant mortality has fallen by half, and people can expect to live ten years longer than in the 1970s. In addition, global agricultural productivity has risen sharply, per capita calorie supplies have risen by nearly 30 percent, and real food prices have fallen by more than 50 percent. The increase in productivity has allowed consumers to improve their diets in terms of both calories consumed and variety of foods eaten. Between 1961 and 1992, calories available in developing countries rose from about 1,925 per person per day to about 2,540 per person per day, which is higher than the minimum daily requirement of 2,200–2,300 defined by the United Nations Food and Agriculture Organization. This increase in per capita supplies came about although world population nearly doubled over the same period, rising from about 3 billion to about 6 billion. More than 80 percent of people in developing countries now have adequate diets, as compared with 64 percent in 1970. The *number* of undernourished people has also fallen, from about 940 million in 1970 to 800 million in 1996. The improvement in agricultural productivity and the concomitant fall in food prices has been a major factor in improving living standards in many developing countries. The improvements have resulted from actions deliberately taken in the past to raise productivity —especially investments in research, at both the international and national levels.

Not all have benefited from improving agricultural productivity. Despite these impressive achievements, rising populations and unequal participation in growth have left 1.3 billion people in the world struggling to survive on less than one dollar a day, and the number continues to rise. About 15 percent of the world's people or about 20 percent of the population of the developing world are hungry, undernourished or malnourished. There are over 190 million children under the age of 5 who are not receiving the nutrition they need to fully develop mentally and physically, as indicated by low weight for age—40 percent of preschool-age in the developing world.

Who Are the Hungry?

Ironically, almost three-quarters of the poor and hungry are rural people living in places where food is grown. These people include the landless, those living in poor nations, or living in areas with poor agricultural potential or which are environmentally fragile. The remaining one-quarter of poor are unemployed or underemployed urban dwellers, who live on less than a dollar day. Both the absolute numbers and proportion of poor people living in cities is expected to grow rapidly: by early in the next century the number of urban poor will likely exceed the number of rural poor, as people leave rural areas to pursue higher-paying urban and industrial jobs. These people will be at great risk of undernutrition and malnutrition, unless food is abundant and affordable. But for now, poverty remains a predominantly rural issue.

The poor and hungry are distributed unequally across regions and countries of the world. Most of the poor and hungry live in Asia and sub-Saharan Africa. Two-thirds of undernourished people live in Asia. The Indian subcontinent alone contains almost one-half of the world's hungry people. However, Africa has the greatest proportion of people who are undernourished, one-third of the total population. Countries at war are especially likely to have large numbers of poor and hungry people.

What Factors Contribute to Household and National Food Security?

Poverty. Hunger is most prevalent in countries with low per capita incomes. Countries with low per capita incomes tend to have very low agricultural output. For most developing countries, increasing agricultural output—for home consumption and export—is essential to stimulating economic growth generally, and improving the diets of the poor and hungry. This is so because, despite trade, 90 percent of the world's grain is consumed in the country where it is produced. Agricultural growth stimulates economic growth in nonagricultural activities, which results in increased employment and reduced poverty. Fostering rural and agricultural development, especially among smallholders, would make powerful contributions towards increasing household and national food security.

While hungry people are generally poor, poor people are not always hungry. Some countries with low per capita incomes have been able to achieve a relatively high degree of household food security, including China, Indonesia, and Costa Rica. And rich countries with skewed income distributions or inadequate poverty and food programs may have significant numbers of hungry people.

High rates of population growth. The world's most poorly nourished countries are also generally those with the fastest growing populations. Countries with low per capita food supplies and with fast growing populations must pay especial attention to raising agricultural output. Otherwise, demographic growth alone will increase the total number of hungry people. Rural development, particularly measures which raise the health and productivity of women, can lead to sharp reductions in birth rates, and an increase in well-being for women and children alike.

World Food Demand and Supply

Future demand for food will come from population growth and from higher incomes; the latter increases the demand for meat, vegetables, fruits, and of grains for livestock feed. The population of the world is expected to exceed 8 billion by 2025, rising by more than 2.5 billion over the next 30 years. Most of the increase will take place in developing country cities, where urban populations are expected to triple. *Given modest income growth, food needs in developing countries could nearly double over the next 30 years.*

Agricultural growth in the future must come primarily from rising biological yields rather than from area expansion or intensification through irrigation. Why? Because most fertile lands are already under cultivation, and most areas suitable for irrigation have already been exploited. Furthermore, people everywhere are becoming increasingly concerned about the environmental impacts of bringing new land into agricultural production. With population growth and urban expansion, there is rising competition for water and land from urban and industrial users. Doubling the yields of complex farming systems without damaging the environment is an enormous challenge.

What Are the Prospects for Future Global Food Supplies?

There is considerable disagreement about how easy or difficult it will be to meet the challenge. Views range from "there is no problem" to "the Malthusian nightmare is imminent." The predominance of views is towards the "no problem" end of the spectrum, even bordering on complacency.

This is not a new worry. Societies have long been concerned that food supplies could not grow in step with population, leading to wide-spread food shortages and famine. Malthus articulated this view in his famous *Essay on the Principle of Population as it Affects the Future Improvement of Society,* published in 1798, which argues that population grows geometrically, yet food supplies can grow only linearly. While the situation that Malthus envisioned has never materialized, concerns about imminent food shortages have continued to arise. People were worried about global food shortages immediately after World War II, and again in 1965–66, following two bad monsoons in South Asia. Both periods were followed by years of expanded output. Then, in 1972–74, a confluence of production shortage and escalating demand, particularly from the Soviet Union, led to a tripling of grain prices over an eight-month period, again giving rise to predictions of disaster. Farmers responded to the price incentives, and by the early 1980s, the concern was about surpluses, not shortages.

Optimistic scenario. The debate continues today. On one side are the optimists who anticipate that yields in the future can continue to grow fast enough to not only feed the world's growing population, but also to contribute to future declines in the real price of food. Mitchell and Ingco (1993), the Food and Agriculture Organization (1995), and the International Food Policy and Research Institute (1995) all estimate that world grain production can increase by an average rate of 1.5 percent per year, slightly more than the projected population growth rate of 1.4 percent. The International Food Policy and Research Institute projects that by 2020, world cereal production will rise by 56 percent from 1990 levels, while meat output will rise by 74 percent. Per capita *demand* for calories in developing countries will rise by 12.3 percent from 1990 to 2020 and per capita calories *supplies* will rise by 12.8 percent. Increases in output will come primarily from rising yields which are projected to increase by 1.5–1.7 percent per year, although contributions may also be made by modest increases in the area under cultivation and the area that is irrigated. The optimists also argue that world trade in grains can expand significantly, so that better endowed regions of the world can feed those that are less favored.

Among optimists are those who believe that "getting prices right" is all that is needed to ensure sufficient global food supplies: With accurate price signals, farmers will respond by planting idled acreage, by utilizing land under production more intensively, and by making key investments in genetic stock, irrigation, and agricultural chemicals. Other, more moderate, optimists believe the world can meet the challenge—but only if governments provide adequate resources for research, health and education, and rural roads, irrigation, and water and sanitation.

Pessimistic scenario. On the other side are pessimists who debate that the 1990s are the beginning of a new era in which it will be much more difficult to expand food production. They observe that the growth in yields has perceptibly slowed, and argue that fisheries and rangelands have reached their limits of production, water supplies are nearly fully exploited, land under cultivation is rapidly becoming degraded, substantial cropland is being converted for factories, roads, and urbanization, and climate change threatens existing crop production. Meanwhile, the scale of world population growth is unprecedented, putting strains on global systems unlike those ever before encountered.

Why the Deeply Divergent Views?

The optimists and pessimists look at the same "facts" and reach vastly different conclusions. This is so because of differences in four common and critical projection parameters:

1. The rate of increase in biological cereal yields to be expected over the next 15 to 30 years;
2. The amount of land to be added to or lost from agricultural production;
3. The amount of land subject to intensification primarily through irrigation;
4. The impact of environmental degradation on food production capacity.

The optimists assume the yields will continue to grow at a rate close to that of the past. They assume little increase in the amount of land needed for cultivation and minimize the negative impact on production of resource degradation.

The pessimists assume the yields can grow no faster than around 1 percent a year. Therein lies the source of much of the difference in views. A difference of 1 percent in a compound growth rate makes an enormous difference at the end of a 30 year projection period. The pessimists further assume that growing competition from cities and industries for land and water, and environmental degradation will lead to reductions in cultivated and irrigated acreage.

WHAT DO WE DO NOW? A WORLD BANK PERSPECTIVE

Whether optimist or pessimist, everyone agrees that what happens in the future will be heavily conditioned on what we do now. There is absolutely no room for complacency. The achievements of the past 30 years resulted from concentrated investment in research and dissemination in both the Consultative Group on International Agricultural Research (CGIAR) and the national agricultural research systems. To achieve future yield growth requires many actions. Among the key ingredients for success are the following:

- ***Sound and stable macroeconomic and sector policies.*** Governments must assure that foreign exchange, trade, price, and taxation regimes do no discriminate against agriculture. Large farms and large agro-industrial firms should not receive special privileges, or be able to reduce competition in output, input, land or credit markets.
- ***Enabling policy environments.*** Governments should work towards establishing a policy environment in which markets can flourish. This includes clearly defining and protecting property rights, establishing strong financial institutions, and providing public goods such as education, extension, and information.
- ***Rapid technological change in agriculture in the developing and developed world.*** The major reason for productivity growth in agriculture over the last 25 years has been technological change. Technological change will be even more important in the future to achieve the increases in yields essential to double agricultural output. The private sector will have to undertake an increasing share of the necessary research and dissemination. Public sector financing will be needed for areas of limited interest to the private sector, such as genetic resource conservation, common property resource management, integrated pest management, and research on subsistence crops.
- ***Massive increases in the efficiency of irrigation water use.*** Irrigation uses 70 percent of the water used by man, and has contributed greatly to the yield increases of the twentieth century. However, agriculture is increasingly competing for water with urban and industrial users. Efficiency of their water use must be improved through better water policies, clearer water rights, and stronger institutions for allocating water, as well as by technical improvements in water conveyance and use.
- ***Education, health, and nutrition services for both boys and girls.*** Providing education and health services to both girls and boys is one of the key ways to reduce poverty and hunger. There is substantial evidence that education is closely linked to an individual's income, and that it contributes to national economic growth. Education and health services are especially important for women, who have a major role to play in growing crops and in reducing hunger. Better-educated and healthier women are much more productive, and in general feed their families more nutritional diets. Education for girls also lowers fertility rates and improves environmental management.
- ***Dramatic improvements in the management of soils, watersheds, forests, and biodiversity by local and community-based institutions.***
- ***Adequate infrastructure.*** Adequate roads, communications, storage facilities, and electricity allows farmers to produce the highest-valued crops, store them, move them to market, and receive the best price for them. In most countries, rural infrastructure has receive inadequate attention. Furthermore, whatever infrastructure exists is often poorly operated and maintained.
- ***Broad participation.*** Experience shows that programs and projects are much more likely to reflect a community's priorities, reach their goals, and be sustainable when they are designed and executed with a high degree of influence by the people who benefit from them.

- ***Health and nutrition programs.*** Studies have shown, surprisingly, that families experiencing an increase in income may not use the added resources to provide healthier diets. Improving diets often requires nutrition counseling, prenatal nutrition services, and public health interventions. In some places it also requires investments to correct micronutrient deficiencies. These often cost little, but generate large returns. For example, in a country of 50 million people, adding iron, iodine, vitamin A and other vitamins and minerals to food and water supplies would cost about US$25 million per year and yield a return 40 times the cost.
- ***Improved intra-household distribution of food.*** At the household level, access to food can depend on factors such as the age and composition of family members, and the state of their health. In many countries, *female-headed households* with no adult males are especially likely to have insufficient food. Within households, *pregnant and lactating women,* whose need for calories is especially high, may consume less than they require to bear and sustain normal-weight, healthy babies. Indeed about 20 million infants are born with low birth weight each year, most because of maternal malnutrition. This has lifetime consequences for the infants, who are more likely to contract diseases and die prematurely than those born at normal weight even if they receive sufficient food when weaned. *Infants and children* (especially females and children born lower in the birth order) are also less likely than other family members to receive sufficient food. Malnourished children are much more likely to die from infectious diseases than others. They are also more likely to lag in school, grow up to be poor, and be the parents of poorly nourished children themselves. Ensuring future food security thus means providing targeted programs for the vulnerable now.
- ***Programs to help countries cope with price volatility.*** Countries sometimes face hardships due to fluctuations in commodity prices. For example, a number of countries faced serious difficulties in financing food imports over the period between June 1993 and May 1996, when world food prices rose 56 percent. Others had problems during the drought in southern Africa in 1992. Fortunately, there are a variety of instruments available to help countries cope with short-term food supply and price shocks, including balance of payments assistance from the International Monetary Fund, and emergency assistance from the World Food Programme of the United Nations and bilateral donors. Some countries were able to avoid serious crises during these two recent events by using these programs. It is certain that new crises will arise in the future, requiring the coordinated assistance of all actors.

CONCLUSIONS

Thirty years from now there will be 2.5 billion more people to feed in the world, and most of them will live in developing countries. Not to recognize this fundamental reality and not to increase efforts now will have serious consequences, particularly since there is at least a twenty-year time lag between initiating strategic research and significant increases in farmers' fields. It is within our power to assure that everyone has enough to eat. We know what to do, but we must have the determination to do it.

The challenge is worldwide, and both technological and political in nature. The technological challenge is enormous, requiring the development of new, high productivity, environmentally sustainable production systems. It is not more of the same. Private firms must be induced to develop and apply much of the new technology required. However, there are large areas of technology development which are of little interest to the private sector, including subsistence crops, or truly public goods, such as some aspects of natural resources management. It is here that public sector finance is critical at international, national, and local levels. Yet, in many countries, research capacity and funding are stagnating or even declining.

The political challenge has received much less attention: all major regions of the world need to contribute to cost-effective and sustainable food supply growth, including Eastern and Central Europe, Africa, and Latin America which are not doing their share today. *The challenge can be met only if international and domestic policies, institutional frameworks, and public expenditure patterns are conducive to cost-effective and sustainable agricultural development.* Otherwise the required technologies will not be developed and adopted, the supportive infrastructure will not be built and maintained, land and water will not be allocated to their highest-valued uses, and farmers will not have incentives to maintain and improve the natural resources on which their livelihoods depend.

NO

Lester R. Brown

CAN WE RAISE GRAIN YIELDS FAST ENOUGH?

After a half-century of global surpluses of wheat, rice, corn, and other grains, it is easy to be complacent about the food prospect for the twenty-first century. We have come to take for granted the supply of grain that provides half of humanity's food energy when consumed directly and a good portion of the remainder when consumed indirectly in the form of livestock products.

But this complacency can be dangerous. Each year, as population continues to expand, the world's farmers must stretch their production capacity to feed an additional 80 million people. Beyond that, they must now satisfy the needs generated by record rises in affluence. As people make more money, they consume more beef, pork, poultry, milk, eggs, beer, and other grain-intensive products. A kilogram of pork, for example, may require four kilograms of grain to produce, so as people are able to afford more pork their demand for grain increases. And Third World incomes are now rising at record rates. In Asia, where more than half the world's people live, incomes are rising faster than they have on any continent at any time in history. This combination of more people and more consumption per person is putting heavy pressure on the land.

The world's farmers responded heroically to past increases in demand, nearly tripling the grain harvest from 630 million tons in 1950 to 1.8 billion tons in 1990. Most of this expansion came not from plowing a lot more land, but from more than doubling the amount of grain produced on existing farmland. Between 1950 and 1990, the yield per hectare grew at 2.1 percent per year. Augmented by whatever new land could be added to grain production, including that from expanded irrigation in arid regions, this boosted total grain production by an average of nearly 3 percent a year throughout that four-decade run—well ahead of population growth.

Although there were disastrous shortages from time to time during this period—in China, Ethiopia, and Somalia, for example—and although some 800 million people are still hungry and malnourished, *overall* supply has not been a major issue. In the United States, the government paid farmers not to plant part of their land. The steady growth in the harvest, and resultant

From Lester R. Brown, "Can We Raise Grain Yields Fast Enough?" *World Watch* (July/August 1997).

decline in grain prices, created a psychology of surpluses—a psychology that has made it easy for policymakers both to put off the difficult task of stabilizing human population and to take their farmers' capacities to meet future challenges for granted.

The world's total demand for food is likely to nearly double its present level by 2030, and there is little new land available to plow. The key to food security in the years ahead, then, is whether farmers can continue to rapidly raise the productivity of their land, as they have done in the past. However, assessments of the potential for raising land productivity vary widely. In a recent World Bank report, researchers indicated that they expect grain yields to increase at 1.5 to 1.7 percent per year, or "at rates comparable to those in recent years...." With this rosy outlook, the Bank projects a surplus capacity in world agriculture as a whole, accompanied by declining food prices. This Worldwatch Institute analysis comes to a very different conclusion.

The World Bank economists base their projections on simple extrapolation, arguing that "historically, yields have grown along a linear path from 1960 to 1990, and they are projected to continue along the path of past growth." Although extrapolating past yield trends worked well enough in previous decades, it won't work in a world where the yields simply are not continuing to climb rapidly. In contrast to the robust increases of 2.1 percent per year between 1950 and 1990, the rise between 1990 and 1995 averaged only 1 percent a year. Although this period is too short to establish a clear trend, it may offer a strong indication of what the future holds.

Reliance on the World Bank projections by governments is leading to underinvestment in both agriculture and family planning. Funding for agricultural research is being cut by many governments, including that of the United States. At the international level, a striking example is the fate of the Philippines-based International Rice Research Institute (IRRI), which gave Asia the high-yielding rices. In 1996, several donor governments, facing cutbacks in their aid budgets, cut their general support funding of IRRI, the world's premier rice research institute, forcing a heavy cutback in core staff. Similarly, in 1996 the U.S. Congress voted to cut fiscal year 1997 funding of international family planning assistance by 60 percent from 1995 levels, with little consideration of how this would affect the increasingly precarious balance between population and food supply. Fortunately, the new congress voted in early 1997 to restore part of the funding.

The Bank projections breed complacency, not urgency. They permit governments to treat prime cropland like a surplus commodity—one that can be paved over, built on, or otherwise frittered away with impunity. One result can be seen in California's Central Valley, where housing projects are marching up the valley unimpeded, consuming some of the world's finest farmland. In China, this process is taking place on an even larger scale, as the government paves over millions of hectares of cropland so the bicycle can be replaced with the automobile. And in Indonesia, fertile riceland is being converted to golf courses.

The central question now is whether farmers can restore the rapid rise in land productivity. Moreover, that question needs to be addressed in terms that point to realistic possibilities for farmers working under the natural constraints of their own environments—the availability of sunlight, water, and good soil. Rather than analyzing yields on experimental plots or those achieved by the best farmers, this analysis will assess the long-term yield potential under field conditions by individual countries.

THE CENTURY OF SOARING PRODUCTIVITY

The first recorded case in which a country's farmers achieved a sharp increase in output per unit of land—a "yield takeoff"—began more than a century ago in Japan. In 1878, Japanese rice farmers got an average of 1.4 tons of grain per hectare. By 1984, the average yield had more than tripled, to 4.7 tons. Since then, it has plateaued—fluctuating between 4.3 and 4.6 tons in all but three years.... Despite the fact that Japan supports the price paid to its farmers for rice at four times the world level, thereby offering a powerful financial incentive to raise yields higher—and despite its ability to provide the best technology available—it has been unable to improve average yields for more than a decade.

In the United States, the first yield takeoff came more than a half century later, with wheat. During the nearly 80 years between the Civil War and World War II, U.S. wheat yields had fluctuated around 0.9 tons per hectare. As World War II got underway, and demand for U.S. grain rose as production was disrupted abroad, farmers began investing in higher-yielding seeds and in fertilizer. By 1983, yields had climbed to 2.65 tons per hectare, nearly tripling the traditional level. Since then, however, there has been no further rise. Although the wheat yield takeoff in the United States began decades after that of rice in Japan, farmers in the two countries appear to have "hit the wall" at about the same time. Two questions arise: Can scientists restore the historical growth in yields, or does this plateauing in two of the most agriculturally advanced countries signal a future leveling off in other countries, as farmers exhaust the principal means of increasing yields?

THE FACTORS THAT INCREASE YIELDS

The 2.5-fold increase in world grain land productivity since 1950 has come from three sources: genetic advances, agronomic improvements, and some synergies between the two.

On the genetic front, most growth has come from redistributing the share of the plant's photosynthetic product (photosynthate) going to the various plant parts (leaves, stems, roots, and seeds) so that a much larger share goes to the seed—the part we use for food. On the agronomic front (where farmers' practices weigh heavily), advances that help plants realize their full genetic potential include the use of fertilizer, irrigation, the control of plant diseases and predatory insects, and the eradication of weeds.

Scientists estimate that the originally domesticated wheats devoted roughly 20 percent of their photosynthate to the development of seeds; they were stalk-heavy, harvest-light. Through plant breeding, it has been possible to raise the share of photosynthate going into seed—the "harvest index"—in today's high-yielding grain to some 50 to 55 percent. Given the plant's basic requirements of an adequate root system, a strong stem, and sufficient leaves for photosynthesis, scientists believe the physiological limit is around 60 percent.

One of the earliest gains in this area came in the late nineteenth century, when Japanese scientists incorporated a dwarf gene into both rice and wheat plants. Traditional varieties of these grasses were tall and thin, because their ancestors growing in the wild needed to compete with other plants for sunlight. But once farmers began controlling weeds among the domesticated plants, there was no longer a need for tall varieties. As plant breeders shortened both wheat and rice plants, reducing the length of their straw, they also lowered the share of photosynthate going into the straw and increased that going into seed. I. T. Evans, a prominent Australian soil scientist and plant physiologist who has long studied cereal yield gains and potentials, notes that in the high-yielding dwarf wheats, "the gain in grain yield approximately equals the loss in straw weight."

With corn, similarly, varieties grown in the tropics were reduced in height from an average of nearly three meters to less than two. But Don Duvick, for many years the director of research at the Pioneer Hybrid seed company, observes that with the hybrids used in the U.S. corn belt, the key to higher yields is the ability of varieties to "withstand the stress of higher plant densities while still making the same amount of grain per plant." One of the keys to growing more plants per hectare is to replace the horizontally inclined leaves of traditional strains that droop somewhat with more upright leaves, thereby reducing the amount of self-shading.

But while breeders can manipulate the distribution of photosynthate within the plant, the amount produced by a given leaf area remains unchanged from that of the plant's wild ancestors. Although plant breeders have greatly increased the share of the photosynthate going to the seed of the various grains, they have not been able to alter the basic process of photosynthesis itself.

On the agronomic front, the principal means of increasing land productivity have been to expand irrigation, use more fertilizer, and to more effectively control diseases, insects, and weeds. All of these tactics help plants reach more of their full genetic yield potential. Between 1950 and 1990, the amount of irrigated land in the world increased from 94 million to 240 million hectares, or 2.4 percent per year. Between 1990 and 1994, however, the official data show irrigated area increasing by only another 9 million hectares, or 0.9 percent per year. And because governments do not always report the land taken out of irrigation, some analysts doubt that there has been any net growth in irrigated area at all since 1990.

In the United States and China, the world's two largest grain producers, losses are all too visible—and in some instances perhaps irreversible. Texas, a major farming state that has historically relied heavily on irrigation, has lost 14 percent of its irrigated area since 1980 as a result of aquifer depletion. California, Kansas, and Oklahoma, too, are losing irrigation water. In China's Hebei Province, irrigation water is being diverted to cities to satisfy mushrooming urban and industrial demands for water. In the agricultural region around Beijing, farmers have not been allowed to draw water from the reservoirs since 1994, because all the region's water is now needed to satisfy the capital city's growing thirst.

David Seckler, head of the International Irrigation Management Institute in Sri Lanka, believes that world irrigated area actually may have started to shrink. If the future brings little or no growth in irrigated area, the world will have lost a major source of rising land productivity, since the expansion of irrigation also greatly expands the potential for using fertilizer.

Fertilizer helps to ensure that plant growth won't be inhibited by any lack of nutrients. With a ten-fold rise in fertilizer use, from 14 million tons in 1950 to some 140 million in 1990, this has been by far the most important agronomic source of higher land productivity since mid-century. But in the 1990s, use of fertilizer —like that of irrigation—has leveled off in many countries. U.S. farmers, after discovering that there are optimal levels beyond which further applications aren't cost-effective, are using less fertilizer in the mid-1990s than they were in the early 1980s. The leveling off in the United States has been followed by a similar trend in Western Europe and Japan. In the former Soviet Union, fertilizer use fell precipitously after subsidies were removed in 1988 and fertilizer prices climbed to world market levels.

Other agronomic contributions to higher cropland productivity include the more timely planting of crops made possible by mechanization and higher plant populations per hectare, the latter applying particularly to corn. More timely planting boosts yields because in the temperate zones there is typically a brief window of time for seeding, usually measured in days, when optimum yields can be obtained. If planting is delayed, then yields decline with each day of delay.

Advances in plant breeding and agronomy often reinforce each other. The dwarfing of wheat and rice plants not only reduced the amount of photosynthate that went for straw, for instance, but increased the benefit of adding more fertilizer. For example, the traditional tall, thin-strawed wheat varieties grown in India could effectively use only about 40 kilograms of nitrogen per hectare. Applications above that made the plants grow heavier heads of grain, but these would often "lodge," or fall over (especially in storms), leading to crop losses. With the dwarf varieties, however, farmers could boost nitrogen applications up to 120 kilograms per hectare or more, thus greatly increasing the yield, but with little fear of lodging. This synergy between genetics and agronomics helps to explain the doubling or tripling of yields achieved with the first generation of high-yielding wheats and rices that were at the heart of the Green Revolution.

With corn, the greater tolerance for crowding enabled growers to greatly increase the plant population—and hence the number of ears harvested—per hectare. At the same time, herbicides were being developed that would control weeds, eliminating the traditional need to plant corn rows far enough apart to permit mechanical cultivators to pass through the field during the earlier part of the growing season. As a result of these two advances, plant populations have climbed. In Iowa, for example, corn plant densities have nearly tripled since 1930.

For each of the three major grains—wheat, rice, and corn—the major worldwide gains in productivity took place between 1950 and 1990. Since 1990, gains have been much smaller, and the question now facing planners is just how much more can be expected....

FACING BIOLOGICAL REALITY

For individual grains in individual countries, the historic trends show a sobering pattern. In every farming environment, where yields are increased substantially, there comes a time when the increase slows and either levels off or shows signs of doing so. It is equally revealing to look at the global trends. In doing so, we use three-year averages for the decennial or mid-decennial years in order to minimize the effects of weather variations. For example, the yield shown for 1990 is an average of the yield from 1989–91 and that for 1995 is the average for 1994–96.

During the four decades from 1950 to 1990, the world's grain farmers raised the productivity of their land by an unprecedented 2.1 percent per year, but since 1990, there has been a dramatic loss of momentum in this rise. If the former Soviet Union is excluded from the global data for 1990 to 1995, because of the uncharacteristic drop in yields associated with economic reforms and the breakup of the country into its constituent republics, then the rate of yield gain is not... 0.7 percent per year... but 1.1 percent—roughly half that of the preceding 40 years. And while the first half of the 1990s is too short a period to determine a new trend, it does provide a reason for concern. In addition to the plateauing of wheat yields in the United States and Mexico,... those in Canada and Egypt have shown no improvement so far in the 1990s.

Global trends for the three major individual grains, moreover, follow the pattern seen for grain as a whole. Rice production, which was modernized later than wheat and corn, achieved an annual increase in productivity of 2.1 percent per year between 1960 and 1990, but has dropped to 1.0 percent per year since 1990. Wheat yields grew between 1960 and 1990 at an average of 2.6 percent per year, then slowed to 0.1 percent during the 1990s. (If the former Soviet Union is excluded from the global trend after 1990, wheat yields increase by 1.0 percent.) Corn averaged 2.6 percent from 1950 to 1980, then fell to 1.3 percent in the 1980s. The rise in corn yields accelerated slightly during the first half of the 1990s, reaching 1.7 percent, largely because of a belated surge in yields in both China and Brazil.

With this slower rise in grainland productivity thus far during the 1990s, the obvious next question is whether the momentum can be regained through biotechnology. Yet, on that front too, progress is not promising. After two decades of research, biotechnologists have not yet produced a single, high-yielding variety of wheat, rice, or corn. Why haven't some of the leading seed companies put their biotechnologists to work to develop a second generation of varieties that would again double or triple yields, enabling farmers to sustain a rapid rise?

The answer, say plant scientists, is that plant breeders using traditional techniques have largely exploited the genetic potential for increasing the share of photosynthate that goes into seed. Once this share is pushed to its limit, the remaining options tend to be relatively small, clustering around efforts to raise the plant's tolerance of various stresses, such as drought or soil salinity. The one major option left to scientists is to increase the efficiency of the process of photosynthesis itself—something that has thus far remained beyond their reach.

Once plant breeders have pushed genetic yield potential close to the physiological limit, then further advances rely on the expanded use of basic inputs such as the fertilizer and irrigation needed to realize the plant's full genetic potential, or on the fine-tuning of other agronomic practices such as the use of optimum planting densities or more effective pest controls. Beyond this, there will eventually come a point in each country, with each grain, when farmers will not be able to raise yields any further.

U.S. Department of Agriculture plant scientist Thomas R. Sinclair observes that advances in plant physiology now enable scientists to quantify crop yield potentials quite precisely. He notes that "except for a few options, which allow small increases in the yield ceiling, the physiological limit to crop yields may well have been reached under experimental conditions." This means that in those situations where farmers are using the highest yielding varieties that plant breeders can provide, and the agronomic inputs and practices needed to realize their genetic potential, there may be few options left for raising land productivity.

Viewed broadly, one can begin to see an S-shaped growth curve emerging for the historical rise in world grainland productivity. Throughout most of human history, land productivity was static. Then, beginning around 1880, Japan began to raise its rice yield per hectare in a steady, sustained fashion. By the mid-1950s, nearly all the industrial countries were expanding their grain harvest by raising grainland productivity. And by 1970, they had been joined by nearly all the leading grain producers in the developing world.

For the 15 years from 1970 to 1985, yields rose in a steady, sustained fashion in virtually all the grain-producing countries of any size. Then this unique period came to an end, as wheat yields in the United States and Mexico and rice yields in Japan leveled off. If these countries cannot restore the rise in yields and if, as now seems likely, more countries "hit the wall" in the years immediately ahead, it will further slow the rise in world grainland productivity, dropping it well below growth in the world demand for grain.

EIGHT OBSERVATIONS

Except for the general warning by biologists that grain yields would eventually plateau, there does not seem to be any record of specific warnings in the early 1980s that the long rise of rice yields in Japan, or the shorter term rise of wheat yields in the United States or Mexico, were about to level off. Nor is anyone likely to anticipate precisely when, for example, wheat yields will level off in France or China, though this could occur at any time. A review of the last half-century's experience in raising yields does, however, offer certain generalizations.

One, the slower rise in grain yields since 1990 is not the result of something peculiar to individual grains or to individual countries, but rather reflects a systemic difficulty in sustaining the gains that characterized the preceding four decades.

Two, every country that initiated a yield takeoff was able to sustain it for at least a few decades.

Three, most countries that have achieved a yield takeoff have managed at least to double, if not triple or even quadruple, their traditional grain yields. Among those that have quadrupled traditional levels are the United States and China with corn; France, the United Kingdom, and Mexico with wheat; and China with rice.

Four, once plant breeders have essentially exhausted the possibilities for raising the genetic yield potential and farmers are using the most advanced agronomic practices, including irrigation, the yield potential for any particular grain in a given country is determined largely by the physical environment of the country—most importantly by soil moisture, but also by temperature, day length, and solar intensity. These factors are fundamentally unalterable.

Five, all countries are drawing on a common backlog of unused agricultural technology that is gradually diminishing and, for some crops in some countries —such as wheat in the United States and rice in Japan—that has largely disappeared.

Six, as a general matter, the more recently a country has launched a yield takeoff, the faster its yields rise and the shorter the time between yield takeoff and level-off.

Seven, despite the slower rise in yields worldwide in recent years and the plateauing of yields in a few countries, there are still many opportunities for raising grainland productivity in most countries. These are most promising in those countries where there is room for improvement in economic policies affecting agriculture. Although most governments subsidize agriculture, some still have economic policies that discourage investment in agriculture. In these countries, the key to realizing the full genetic yield potential of crops is the restructuring of economic policies to encourage investment in agriculture, such as that now underway in Argentina.

Eight, even with a concerted worldwide effort to increase grain yields, the rise during the last half of this decade could slow still further, dropping below 1 percent per year—far below the 2.1 percent that sustained the world from 1950 to 1990.

RESPONDING TO THE CHALLENGE

This slowdown comes at a time when population growth and rising affluence are combining to drive up the demand for grain at a near-record pace. Even as the rate of world population growth edges down, the absolute number of people on the planet is projected to climb by some 80 million people per year well into the next century. Meanwhile, record numbers of people are shifting to diets rich in grain-intensive livestock products, which means that in some countries per-capita grain consumption is growing even faster than population is.

The resulting disparity between projected growth in demand and supply cannot, of course, exist in the real world. Prices will rise. That won't greatly hurt the affluent, who spend only a small portion of their income for food. But for the world's poor, particularly the 1.3 billion people who live on $1 per day or less, higher grain prices could quickly become life-threatening. Heads of households who cannot afford enough food to keep their families alive may well hold their governments responsible and take to the streets. The result could be unprecedented political instability in Third World cities.

If widespread political instability does materialize, it could affect the earnings of transnational corporations, the performance of stock markets, the earnings of pension funds, and even the stability of the international monetary system. As rising food prices threaten political stability and economic progress in a world economy more integrated than ever before, the problem of the world's poor becomes everyone's problem.

Among other things, the slower rise in world grainland productivity calls for an urgent assessment of the carrying capacity of land and water resources in countries everywhere, but particularly in the low-income countries where they are using all their available land and where the demand for water already exceeds the sustainable yield of aquifers.

In a world where land productivity is rising more slowly, the food security of the next generation depends on quickly slowing population growth. This begins with public education programs on the consequences of continuing rapid increases in population. It means filling the family planning gap by getting family planning services to those women who need them. And it means investing heavily in the education of young females in the Third World to accelerate the shift to smaller families.

Future food security also depends on a sharp increase in investment in agricultural research and in better grain storage facilities. It means boosting the efficiency of water use, largely by moving to water markets. And it means taking strong steps to protect cropland from conversion to nonfarm uses.

To mobilize on all these fronts, political leaders need to know that the rise in cropland productivity is slowing and that the growth in the grain harvest is falling behind the growth of demand. If the foregoing analysis is at all close to the mark, it would be irresponsible for the World Bank not to revise its outmoded supply and demand projections. Otherwise, its projection of surplus capacity and falling grain prices will only reinforce the prevailing complacency, and lead to potentially tragic underinvestment in both food production and population stabilization.

POSTSCRIPT

Will Future Generations Have Enough to Eat?

Janet Raloff, in "Can Grain Yields Keep Pace?" *Science News* (August 16, 1997), notes that many experts believe that although there are genuine difficulties in continuing to produce a food supply that is adequate to feed a growing world population, it can be done. She quotes Gurdev S. Khush, the chief rice breeder at the International Rice Research Institute in Manila, Republic of the Philippines, as saying, "If we manage our resources properly and continue to put money into research, we should be able to meet world food needs for at least the next 30 years."

Roy L. Prosterman, Tim Hanstad, and Li Ping, in "Can China Feed Itself?" *Scientific American* (November 1996), discuss how China is reforming its agricultural system in ways that seem quite similar to what the World Bank is calling for. China's eventual success in feeding its populace, however, will also depend on its parallel attempts to rein in population growth.

Food and population come together in the concept of "carrying capacity," defined very simply as the size of the population that the environment can support, or "carry," indefinitely, through both good years and bad. It is not the size population that can prosper in good times alone, for such a large population must suffer catastrophically when droughts, floods, or blights arrive or the climate warms or cools. It is a long-term concept, where "long-term" means not decades or generations, nor even centuries, but millennia or more.

What is Earth's carrying capacity for human beings? It is surely impossible to set a precise figure on the number of human beings the world can support for the long run. As Joel E. Cohen discusses in *How Many People Can the Earth Support?* (W. W. Norton, 1996), estimates of Earth's carrying capacity range from under 1 billion to over 1 trillion. The precise number depends on our choices of diet, standard of living, level of technology, willingness to share with others at home and abroad, and desire for an intact physical, chemical, and biological environment, as well as on whether or not our morality permits restraint in reproduction and our political or religious ideology permits educating and empowering women. The key, Cohen stresses, is human choice, and the choices are ones we must make within the next 50 years. To support this point, Cohen judiciously analyzes a great many population studies and resource estimates. Again and again he says that there is no one neat answer. But there are clearly limits, and they seem to be in the neighborhood of population sizes that we may see well before the year 2100.

Others are more willing to be definite. Sandra Postel, in the Worldwatch Institute's *State of the World 1994* (W. W. Norton, 1994), says, "As a result of our population size, consumption patterns, and technology choices, we have surpassed the planet's carrying capacity. This is plainly evident by the extent to which we are damaging and depleting natural capital" (including land and water). Later in the same volume, Project Director Lester R. Brown says, "The world is facing a day of reckoning." In *State of the World 1996* (W. W. Norton, 1996), Brown notes that already "the demands of our generation exceed the... sustainable yield of the earth's ecological endowment." In "Facing Food Scarcity," *World Watch* (November–December 1995) and in "Averting a Global Food Crisis," *Technology Review* (November–December 1995), Brown argues that the most obvious things that we can do to increase food availability are only stopgaps, or temporary measures. For instance, putting back into use cropland the United States and Europe have set aside to keep prices high could produce 34 million tons of grain per year and feed 15 months' worth of additional population growth. Land now used to grow tobacco could feed 6 months' worth of new mouths. Growing food on half of the land now used to grow cotton could cover 11 months' population growth. And feeding 10 percent of the grain now used as animal feed to people instead would take care of the population for 28 months.

Can we avert catastrophe? Courtland L. Smith, in "Assessing the Limits to Growth," *BioScience* (July/August 1995), asserts,

> Culture, institutions and technology can mitigate the growth of population and affluence. When culture, institutions, and technology do not change to reduce environmental impact, then population and affluence do create a greater environmental impact. Biologists with a strict-limits-to-growth view of the future and economists who emphasize human capabilities to moderate their impacts can both be right. The question to be resolved in the limits-to-growth debate is how large and how fast must be the change in culture, institutions, and technology.

Also see Charles Mann, "Reseeding the Green Revolution," *Science* (August 22, 1997).

ISSUE 5

Should Society Be Concerned About Global Warming?

YES: Ross Gelbspan, from "The Heat Is On," *Harper's Magazine* (December 1995)

NO: Wilfred Beckerman and Jesse Malkin, from "How Much Does Global Warming Matter?" *The Public Interest* (Winter 1994)

ISSUE SUMMARY

YES: Journalist Ross Gelbspan argues that the evidence for global warming is incontrovertible, despite the disinformation campaign being waged by the fossil fuels industry, and that the effects on human society will be extreme. Action is therefore needed now.

NO: Economists Wilfred Beckerman and Jesse Malkin argue that global warming, if it even occurs, will not be catastrophic and warrants no immediate action; there are other worldwide concerns that are far more pressing.

Scientists have known for a century that carbon dioxide and other "greenhouse gases" (including water vapor, methane, and chlorofluorocarbons) help prevent heat from escaping Earth. In fact, it is this "greenhouse effect" that keeps Earth warm enough to support life. Yet there can be too much of a good thing. Ever since the dawn of the industrial age, humans have been burning vast quantities of fossil fuels, releasing the carbon they contain as carbon dioxide. Because of this, some estimate that by the year 2050, the amount of carbon dioxide in the air will be double what it was in 1850. By 1982 the increase was apparent. Less than a decade later, many researchers were saying that the climate had already begun to warm. Now there is a strong consensus that the global climate will continue to warm. However, there is less agreement on just how much it will warm or what the impact of the warming will be on human (and other) life. See Spencer R. Weart, "The Discovery of the Risk of Global Warming," *Physics Today* (January 1997).

The June 1992 issue of *The Bulletin of the Atomic Scientists* carries two articles on the possible consequences of the greenhouse effect. In "Global Warming: The Worst Case," Jeremy Leggett says that although there are enormous uncertainties, a warmer climate will release more carbon dioxide, which will warm the climate even further. As a result, soil will grow drier; forest fires will occur more frequently; plant pests will thrive; and methane trapped in the world's seabeds will be released and will increase global warming much further—in effect, there will be a "runaway greenhouse effect." Leggett also hints at the possibility that the polar ice caps will melt and raise sea levels by hundreds of feet.

Taking the opposing view, in "Warming Theories Need Warning Label," S. Fred Singer emphasizes the uncertainties in the projections of global warming and their dependence on the accuracy of the computer models that generate them, and he points out that improvements in the models have consistently shrunk the size of the predicted change. There will be no catastrophe, he argues, and money spent to ward off the climate warming would be better spent on "so many pressing—and real—problems in need of resources."

These scientists are not alone on their sides of the debate. In 1991 many scientists testified on "Global Climate Change and Greenhouse Emissions" before the House Subcommittee on Health and the Environment, Committee on Energy and Commerce. Some scientists maintained that the problem was real and potentially serious. See, for instance, Wallace S. Broecker, "Global Warming on Trial," *Natural History* (April 1992), in which the author asserts that past climate coolings have been so immensely disruptive that a cautious approach to the future would require doing all we can to decrease releases of greenhouse gases to ward off potential disaster.

Other scientists asserted that they were not impressed by the data and computer models assembled to date. For instance, Sallie Baliunas, deputy director of the Mount Wilson Observatory and chair of the Science Advisory Board at the George C. Marshall Institute in Washington, D.C., claimed that global warming in the next century will amount to no more than a few tenths of a degree, "indistinguishable from natural fluctuations in temperature." Richard Lindzen, in "Absence of Scientific Basis," *Research and Exploration* (Spring 1993), stated outright that there is no real evidence at all for global warming.

In the following selections, journalist Ross Gelbspan notes that the last few years have witnessed an alarming number of extremes in the weather. This, he asserts, is consistent with analyses that say global warming puts more energy into weather systems and thus drives rainstorms, droughts, floods, and even blizzards. The scientific consensus is therefore stronger today than it was just a few years ago. Those who still question "global warming," says Gelbspan, are serving a campaign of disinformation for the fossil-fuels industry. Oxford University scholar Wilfred Beckerman and writer Jesse Malkin insist that disaster is not imminent and that people can adapt to whatever changes may occur. They say that there are much better ways to spend money, such as on relieving world poverty, than in trying to ward off a nonproblem like global warming.

YES

Ross Gelbspan

THE HEAT IS ON

After my lawn had burned away to straw last summer, and the local papers announced that the season had been one of the driest in the recorded history of New England, I found myself wondering how long we can go on pretending that nothing is amiss with the world's weather. It wasn't just the fifty ducks near my house that had died when falling water levels in a creek exposed them to botulism-infested mud, or the five hundred people dead in the Midwest from an unexpected heat wave that followed the season's second "one-hundred-year flood" in three years. It was also the news from New Orleans (overrun by an extraordinary number of cockroaches and termites after a fifth consecutive winter without a killing frost), from Spain (suffering a fourth year of drought in a region that ordinarily enjoys a rainfall of 84 inches a year), and from London (Britain's meteorological office reporting the driest summer since 1727 and the hottest since 1659).

The reports of changes in the world's climate have been with us for fifteen or twenty years, most urgently since 1988, when Dr. James Hansen, director of NASA's Goddard Institute for Space Studies, declared that the era of global warming was at hand. As a newspaper correspondent who had reported on the United Nations Conferences on the environment in Stockholm in 1972 and in Rio in 1992, I understood something of the ill effects apt to result from the extravagant burning of oil and coal. New record-setting weather extremes seem to have become as commonplace as traffic accidents, and three simple facts have long been known: the distance from the surface of the earth to the far edge of the inner atmosphere is only twelve miles; the annual amount of carbon dioxide forced into that limited space is six billion tons; and the ten hottest years in recorded human history have all occurred since 1980. The facts beg a question that is as simple to ask as it is hard to answer. What do we do with what we know?

The question became more pointed in September, when the 2,500 climate scientists serving on the Intergovernmental Panel on Climate Change [IPCC] issued a new statement on the prospect of forthcoming catastrophe. Never before had the IPCC (called into existence in 1988) come to so unambiguous a conclusion. Always in years past there had been people saying that we didn't

From Ross Gelbspan, "The Heat Is On," *Harper's Magazine*, vol. 291, no. 1747 (December 1995). Some notes omitted.

yet know enough, or that the evidence was problematical, or our system of computer simulation was subject to too many uncertainties. Not this year. The panel flatly announced that the earth had entered a period of climatic instability likely to cause "widespread economic, social and environmental dislocation over the next century." The continuing emission of greenhouse gases would create protracted, crop-destroying droughts in continental interiors, a host of new and recurring diseases, hurricanes of extraordinary malevolence, and rising sea levels that could inundate island nations and low-lying coastal rims on the continents.

I came across the report in the *New York Times* during the same week that the island of St. Thomas was blasted to shambles by one of thirteen hurricanes that roiled the Caribbean this fall. Scientists speak the language of probability. They prefer to avoid making statements that cannot be further corrected, reinterpreted, modified, or proven wrong. If its September announcement was uncharacteristically bold, possibly it was because the IPCC scientists understood that they were addressing their remarks to people profoundly unwilling to hear what they had to say.

That resistance is understandable, given the immensity of the stakes. The energy industries now constitute the largest single enterprise known to mankind. Moreover, they are indivisible from automobile, farming, shipping, air freight, and banking interests, as well as from the governments dependent on oil revenues for their very existence. With annual sales in excess of one trillion dollars and daily sales of more than two billion dollars, the oil industry alone supports the economies of the Middle East and large segments of the economies of Russia, Mexico, Venezuela, Nigeria, Indonesia, Norway, and Great Britain. Begin to enforce restriction on the consumption of oil and coal, and the effects on the global economy—unemployment, depression, social breakdown, and war—might lay waste to what we have come to call civilization. It is no wonder that for the last five or six years many of the world's politicians and most of the world's news media have been promoting the perception that the worries about the weather are overwrought. Ever since the IPCC first set out to devise strategies whereby the nations of the world might reduce their carbon dioxide emissions, and thus ward off a rise in the average global temperature on the order of 4 or 5 degrees Celsius (roughly equal in magnitude to the difference between the last ice age and the current climatic period), the energy industry has been conducting, not unreasonably, a ferocious public relations campaign meant to sell the notion that science, any science, is always a matter of uncertainty. Yet on reading the news from the IPCC, I wondered how the oil company publicists would confront the most recent series of geophysical events and scientific findings. To wit:

- A 48-by-22 mile chunk of the Larsen Ice Shelf in the Antarctic broke off last March, exposing rocks that had been buried for 20,000 years and prompting Rodolfo del Valle of the Argentine Antarctic Institute to tell the Associated Press, "Last November we predicted the [ice shelf] would crack in ten years, but it has happened in barely two months."
- In April, researchers discovered a 70 percent decline in the population of zooplankton off the coast of southern California, raising questions about the

survival of several species of fish that feed on it. Scientists have linked the change to a 1 to 2 degree C increase in the surface water temperature over the last four decades.

- A recent series of articles in *The Lancet*, a British medical journal, linked changes in climate patterns to the spread of infectious diseases around the world. The *Aedes aegypti* mosquito, which spreads dengue fever and yellow fever, has traditionally been unable to survive at altitudes higher than 1,000 meters above sea level. But these mosquitoes are now being reported at 1,150 meters in Costa Rica and at 2,200 meters in Colombia. Ocean warming has triggered algae blooms linked to outbreaks of cholera in India, Bangladesh, and the Pacific coast of South America, where, in 1991, the disease infected more than 400,000 people.
- In a paper published in *Science* in April, David J. Thomson, of the AT&T Bell Laboratories, concluded that the .6 degree C warming of the average global temperature over the past century correlates directly with the buildup of atmospheric carbon dioxide. Separate findings by a team of scientists at the National Oceanic and Atmospheric Administrations's National Climatic Data Center indicate that growing weather extremes in the United States are due, by a probability of 90 percent, to rising levels of greenhouse gases.
- Scientists previously believed that the transitions between ice ages and more moderate climatic periods occur gradually, over centuries. But researchers from the Woods Hole Oceanographic Institution, examining deep ocean sediment and ice core samples, found that these shifts, with their temperature changes of up to 7 degrees C, have occurred within three to four decades—a virtual nanosecond in geological time. Over the last 70,000 years, the earth's climate has snapped into radically different temperature regimes. "Our results suggest that the present climate system is very delicately poised," said researcher Scott Lehman. "Shifts could happen very rapidly if conditions are right, and we cannot predict when that will occur." His cautionary tone is underscored by findings that the end of the last ice age, some 8,000 years ago, was preceded by a series of extreme oscillations in which severe regional deep freezes alternated with warming spikes. As the North Atlantic warmed, Arctic snowmelts and increased rainfall diluted the salt content of the ocean, which, in turn, redirected the ocean's warming current from a northeasterly direction to one that ran nearly due east. Should such an episode occur today, say researchers, "the present climate of Britain and Norway would change suddenly to that of Greenland."

These items (and many like them) would seem to be alarming news—far more important than the candidacy of Colin Powell, or even whether Newt Gingrich believes the government should feed poor children—worthy of a national debate or the sustained attention of Congress. But the signs and portents have been largely ignored, relegated to the environmental press and the oddball margins of the mass media. More often than not, the news about the accelerating retreat of the world's glaciers or the heat- and insect-stressed Canadian forests comes qualified with the observation that the question of global warming never can be conclusively resolved. The confusion is intentional, expensively gift wrapped by the energy industries.

* * *

Capital keeps its nose to the wind. The people who run the world's oil and coal companies know that the march of science, and of political action, may be slowed by disinformation. In the last year and a half, one of the leading oil industry public relations outlets, the Global Climate Coalition, has spent more than a million dollars to downplay the threat of climate change. It expects to spend another $850,000 on the issue next year. Similarly, the National Coal Association spent more than $700,000 on the global climate issue in 1992 and 1993. In 1993 alone, the American Petroleum Institute, just one of fifty-four industry members of the GCC, paid $1.8 million to the public relations firm of Burson-Marsteller partly in an effort to defeat a proposed tax on fossil fuels. For perspective, this is only slightly less than the combined yearly expenditures on global warming of the five major environmental groups that focus on climate issues—about $2.1 million, according to officials of the Environmental Defense Fund, the Natural Resources Defense Council, the Sierra Club, the Union of Concerned Scientists, and the World Wildlife Fund.

For the most part the industry has relied on a small band of skeptics—Dr. Richard S. Lindzen, Dr. Pat Michaels, Dr. Robert Balling, Dr. Sherwood Idso, and Dr. S. Fred Singer, among others—who have proven extraordinarily adept at draining the issue of all sense of crisis. Through their frequent pronouncements in the press and on radio and television, they have helped to create the illusion that the question is hopelessly mired in unknowns. Most damaging has been their influence on decision makers; their contrarian views have allowed conservative Republicans such as Representative Dana Rohrabacher (R., Calif.) to dismiss legitimate research concerns as "liberal claptrap" and have provided the basis for the recent round of budget cuts to those government science programs designed to monitor the health of the planet.

Last May, Minnesota held hearings in St. Paul to determine the environmental cost of coal burning by state power plants. Three of the skeptics—Lindzen, Michaels, and Balling—were hired as expert witnesses to testify on behalf of Western Fuels Association, a $400 million consortium of coal suppliers and coal-fired utilities.[1]

An especially aggressive industry player, Western Fuels was quite candid about its strategy in two annual reports: "[T]here has been a close to universal impulse in the trade association community here in Washington to concede the scientific premise of global warming... while arguing over policy prescriptions that would be the least disruptive to our economy.... We have disagreed, and do disagree, with this strategy." "When [the climate change] controversy first erupted... scientists were found who are skeptical about much of what seemed generally accepted about the potential for climate change." Among them were Michaels, Balling, and S. Fred Singer.

Lindzen, a distinguished professor of meteorology at MIT, testified in St. Paul that the maximum probable warming of the atmosphere in the face of a doubling of carbon dioxide emissions over the next century would amount to no more than a negligible .3 degrees C. Michaels, who teaches climatology at the University of Virginia, stated that he foresaw no increase in the rate of sea level rise—another feared precursor of global warming. Balling, who works on climate issues at Arizona State University, declared that the increase in emissions would boost the average global temperature by no more than one degree.

At first glance, these attacks appear defensible, given their focus on the black holes of uncertainty that mark our current knowledge of the planet's exquisitely interrelated climate system. The skeptics emphasize the inadequacy of a major climate research tool known as a General Circulation Model, and our ignorance of carbon dioxide exchange between the oceans and the atmosphere and of the various roles of clouds. They have repeatedly pointed out that although the world's output of carbon dioxide has exploded since 1940, there has been no corresponding increase in the global temperature. The larger scientific community, by contrast, holds that this is due to the masking effect of low-level sulfur particulates, which exert a temporary cooling effect on the earth, and to a time lag in the oceans' absorption and release of carbon dioxide.

But while the skeptics portray themselves as besieged truth-seekers fending off irresponsible environmental doomsayers, their testimony in St. Paul and elsewhere revealed the source and scope of their funding for the first time. Michaels has received more than $115,000 over the last four years from coal and energy interests. *World Climate Review,* a quarterly he founded that routinely debunks climate concerns, was funded by Western Fuels. Over the last six years, either alone or with colleagues, Balling has received more than $200,000 from coal and oil interests in Great Britain, Germany, and elsewhere. Balling (along with Sherwood Idso) has also taken money from Cyprus Minerals, a mining company that has been a major funder of People for the West—a militantly anti-environmental "Wise Use" group. Lindzen, for his part, charges oil and coal interests $2,500 a day for his consulting services; his 1991 trip to testify before a Senate committee was paid for by Western Fuels, and a speech he wrote, entitled "Global Warming: the Origin and Nature of Alleged Scientific Consensus," was underwritten by OPEC. Singer, who last winter proposed a $95,000 publicity project to "stem the tide towards ever more onerous controls on energy use," has received consulting fees from Exxon, Shell, Unocal, ARCO, and Sun Oil, and has warned them that they face the same threat as the chemical firms that produced chlorofluorocarbons (CFCs), a class of chemicals found to be depleting atmospheric ozone. "It took only five years to go from... a simple freeze of production [of CFCs]," Singer has written, "... to the 1992 decision of a complete production phase-out—all on the basis of quite insubstantial science."[2]

The skeptics assert flatly that their science is untainted by funding. Nevertheless, in this persistent and well-funded campaign of denial they have become interchangeable ornaments on the hood of a high-powered engine of disinformation. Their dissenting opinions are amplified beyond all proportion through the media while the concerns of the dominant majority of the world's scientific establishment are marginalized. By keeping the discussion focused on whether there is a problem in the first place, they have effectively silenced the debate over what to do about it.

Last spring's IPCC conference in Berlin is a good example. Delegations from 170 nations met to negotiate targets and timetables for reducing the world's carbon dioxide emissions. The efforts of the conference ultimately foundered on foot-dragging by the United States and Japan and active resistance from the OPEC nations. Leading the fight for the most dramatic reductions—to 60 percent of 1990 levels—was a coalition of small island nations from the Caribbean and the Pacific that fear being flooded out of existence. They were supported by most western European governments, but China and India, with their vast coal resources, argued that until the United States significantly cuts its own emissions, their obligation to develop their own economies outranked their obligation to the global environment. In the end, OPEC, supported by the United States, Japan, Australia, Canada, and New Zealand, rejected calls to limit emissions, declaring emission limits premature.

* * *

As the natural crisis escalates, so will the forces of institutional and societal denial. If, at the cost of corporate pocket change, industrial giants can control the publicly perceived reality of the condition of the planet and the state of our scientific knowledge, what would they do if their survival were truly put at risk? Billions would be spent on the creation of information and the control of politicians. Glad-handing oil company ads on the op-ed page of the *New York Times* (from a quarter-page pronouncement by Mobil last September 28: "There's a lot of good news out there") would give way to a new stream of selective finding by privatized scientists. Long before the planet itself collapsed, democracy would break apart under the stress of "natural" disasters. It is not difficult to foresee that in an ecological state of emergency our political liberties would be the first casualties.

Thus, the question must be asked: can civilization change the way it operates? For 5,000 years, we have thought of ourselves as dependent children of the earth, flourishing or perishing according to the whims of nature. But with the explosion of the power of our technology and the size of our population, our activities have grown to the proportion of geological forces, affecting the major systems of the planet. Short of the Atlantic washing away half of Florida, the abstract notion that the old anomalies have become the new norm is difficult to grasp. Dr. James McCarthy of Harvard, who has supervised the work of climate scientists from sixty nations, puts it this way: "If the last 150 years had been marked by the kind of climate instability we are now seeing, the world would never have been able to support its present population of 5 billion people." We live in a world of man-size urgencies, measured in hours or days. What unfolds slowly is not, by our lights, urgent, and it will therefore take a collective act of imagination to understand the extremity of the situation we now confront. The lag time in our planet's ecological systems will undoubtedly delay these decisions, and even if the nations of the world were to agree tomorrow on a plan to phase out oil and coal and convert to renewable energies, an equivalent lag time in human affairs would delay its implementation for years. What too many people refuse to understand is that the global economy's existence depends upon the global environment, not the other way around. One cannot negotiate jobs, development, or rates of economic growth with nature.

What of the standard list of palliatives —carbon taxes, more energy-efficient buildings, a revival of public transportation? The ideas are attractive, but the thinking is too small. Even were the United States to halve its own carbon dioxide contribution, this cutback would soon be overwhelmed by the coming development of industry and housing and schools in China and India and Mexico for all their billions of citizens. No solution can work that does not provide ample energy resources for the development of all the world's nations.

So here is an informal proposal—at best a starting point for a conversation—from one man who is not an expert. What if we turned the deserts of the world into electricity farms? Let the Middle East countries keep their oil royalties as solar royalties. What if the world mobilized around a ten-year project to phase out all fossil fuels, to develop renewable energy technologies, to extend those technologies to every corner of the world? What if, to minimize the conflict of so massive a dislocation, the world's energy companies were put in charge of the transition—answering only to an international regulatory body and an enforceable timetable? Grant them the same profit margins for solar electricity and hydrogen fuel they now receive for petroleum and coal. Give them the licenses for all renewable energy technologies. Assure them the same relative position in the world's economy they now enjoy at the end of the project.

Are these ideas mere dream? Perhaps, but here are historical reasons to have hope. Four years ago a significant fraction of humanity overturned its Communist system in a historical blink of an eye. Eight years ago the world's governments joined together in Montreal to regulate CFCs. Technology is not the issue. The atomic bomb was developed in two and a half years. Putting a man on the moon took eleven. Surely, given the same sense of urgency, we can develop new energy systems in ten years. Most of the technology is already available to us or soon will be. We have the knowledge, the energy, and the hunger for jobs to get it done. And we are different in one unmeasurable way from previous generations: ours is the first to be educated about the larger world by the global reach of electronic information.

The leaders of the oil and coal industry, along with their skeptical scientists, relentlessly accuse environmentalists of overstating the climatic threat to destroy capitalism. Must a transformation that is merely technological dislodge the keystone of the economic order? I don't know. But I do know that technology changes the way we conceive of the world. To transform our economy would oblige us to understand the limits of the planet. That understanding alone might seed the culture with a more organic concept of ourselves and our connectedness to the earth. And corporations, it is useful to remember, are not only obstacles on the road to the future. They are also crucibles of technology and organizing engines of production, the modern expression of mankind's drive for creativity. The industrialist is no less human than the poet, and both the climate scientist and the oil company operator inhabit the same planet, suffer the same short life span, harbor the same hopes for their children.

NOTES

1. In 1991, Western Fuels spent an estimated $250,000 to produce and distribute a video entitled "The Greening of Planet Earth," which was shown frequently inside the Bush White House as well as within the governments of OPEC. In near-evangelical tones, the video promises that a new age of agricultural abundance will result from increasing concentrations of carbon dioxide. It portrays a world where vast areas of desert are reclaimed by the carbon dioxide-forced growth of new grasslands, where the earth's diminishing forests are replenished by a nurturing atmosphere. Unfortunately, it overlooks the bugs. Experts note that even a minor elevation in temperature would trigger an explosion in the planet's insect population, leading to potentially significant disruptions in food supplies from crop damage as well as to a surge in insect-borne diseases. It appears that Western Fuels' video fails to tell people what the termites in New Orleans may be trying to tell them now.

2. Contrary to his assertion, however, virtually all relevant researchers say the link between CFCs and ozone depletion is based on unassailably solid scientific evidence. As if to underscore the point, in May the research director of the European Union Commission estimated that last winter's ozone loss will result in about 80,000 additional cases of skin cancer in Europe. This fall, the three scientists who discovered the CFC-ozone link won the Nobel Prize for Chemistry.

NO

Wilfred Beckerman and
Jesse Malkin

HOW MUCH DOES GLOBAL WARMING MATTER?

More than a billion people in developing countries have no access to safe drinking water, and at least twice that many have no access to adequate sanitation. Consequently, between 1 and 1.5 billion people suffer from water-related diseases such as schistosomiasis, hookworm, and diarrhea. Infant mortality attributable to diarrhea is estimated to be about 5 million per year.

But the environmental problems that dominate the media, that are given the most attention by environmentalist pressure groups, and that capture the imagination of the public, are the melodramatic issues. The myth of "scarce resources" is one.... Another is global warming—"the highest-risk environmental problem the world faces today," according to Vice President Al Gore. The public is bombarded by television images showing the earth surrounded by a layer of "greenhouse gases" (GHGs) that allow the sun's energy to penetrate, but block much of the outgoing radiation from the earth's surface. These images are accompanied by dire predictions that we shall all frizzle up and that the world will become a desert—despite concurrent predictions that rainfall will increase and sea levels will rise. Such scenarios of global warming are much more exciting for the viewer than pictures showing that what the world's population needs most are more lavatories and better sewage systems.

* * *

The "consensus" opinion on climate change, as embodied in the 1990 report of the International Panel on Climate Change (IPCC), is that a doubling of equivalent carbon dioxide [CO_2] (an index that summarizes the effect of all man-made GHGs), is likely to occur within the next fifty years if nothing is done to reduce CO_2 emissions. Because of the time lags in the dynamics of climate change—notably those caused by the inertia introduced into the system as a result of the absorption of carbon dioxide by the oceans—the temperature increase associated with this warming commitment would not be realized until approximately 2100. At that point, the global mean surface temperature is predicted to increase by between 2° and 5° Celsius.

This conclusion has not gone unquestioned. To be sure, the scientific work that has gone into climate modeling represents a major intellectual achievement. Yet it is widely recognized that these estimates have a wide margin of error and that there are still great gaps in our understanding of how the climate is determined. The IPCC report itself contains hundreds of pages of misgivings about the potential temperature increase, and many climatologists have expressed skepticism about the reliability of the global climate models that forecast significant warming.

For example, equivalent CO_2 levels have increased by over 40 percent during the past 100 years, yet the climate has not responded in the manner predicted by the models. Consider the following anomalies:

- The amount of global warming that has occurred over the past century—roughly .45°C—is at least a factor of two less than that predicted by the most sophisticated models.
- The Northern Hemisphere, which the models say should have warmed more rapidly than the Southern Hemisphere, is no warmer than it was a half century ago.
- The models say warming should occur as a result of GHG buildup, but most of the warming during the past 100 years occurred prior to World War II—*before* most of the GHGs were emitted.

Clearly a dose of skepticism is warranted. But let us suppose that the skeptics are wrong—suppose the earth's temperature does rise by somewhere between 2°C and 5°C. How damaging is this likely to be?

* * *

There is one simple piece of evidence, which does not require vast computerized models of the world's climate or economy (our understanding of both being extremely limited), and which does at least refute the widespread notion that the human race is some tender plant that can only survive in a narrow band of plus or minus 3°C. This is the present dispersion of the world's population throughout widely different temperature zones. For example, taking the average temperatures in the coldest month in the countries concerned, 32.3 percent of the world's population lives in a band of 0°C to 3°C, whereas 18.8 percent live in a band of 12°C to 15°C, and 14.6 percent live in a band of 24°C to 27°C. Furthermore, across the world as a whole there appears to be no correlation at all between average temperatures and income levels. . . .

Of course, it will be argued that such cross-country comparisons do not adequately take into account the difficulty of adjusting to relatively rapid changes in temperature. There is some truth in this. But as the distinguished economist Thomas Schelling has observed, the sort of rapid climate changes experienced throughout history by vast migratory movements of population were far greater than those predicted to occur during the next century as a result of global warming. The human race has always been a highly adaptable species, and is likely to become increasingly so, since most of its adaptability comes from its accumulation of technical knowledge.

Similar back-of-the-envelope calculations show that, for the United States at least, global warming could hardly have a significant impact on national income. For the sector most likely to be affected is agriculture, which constitutes 2 percent of U.S. gross domestic product (GDP). Most other sectors of economic activity are not likely to be affected at all, and some, such as construction, will probably be favorably affected. So even if the net output of agriculture fell by 50 percent by the end of the next century, this is only a 1 percent reduction in GDP.

Anyway, the net effect on U.S. agriculture is more likely to be negligible. In the northern states, growing periods would be longer and there would be less disruption by frosts. Further, the predicted rise in carbon dioxide—the most important greenhouse gas leading to global warming—is actually good for plant growth. Authoritative estimates put the impact on the net output of U.S. agriculture at somewhere between plus and minus $10 billion. Even the worst end of this range, minus $10 billion, is a trivial part of a U.S. GDP of about $6 trillion.

Similar estimates for the world as a whole (which, as far as we know, were not done on the back of an envelope) also show that agricultural output in some countries will be favorably affected by global warming, whereas others will lose; and that, on balance, the net effect is likely to be negligible. Of course, the effects depend partly on how climate change affects the regional distribution of rainfall. But this is even more difficult to predict than is the global climate change.

All in all, such estimates as have been made of the overall effect of a doubling of the CO_2 concentration on the world economy suggest that world output would be reduced by about 1 or 2 percent. Suppose that these estimates are much too conservative—as they may well be given that the models on which they are based are extremely shaky. Suppose instead that world output would be reduced by 10 percent by the end of the next century below what it otherwise would be.

Well, what would it otherwise be? Over the whole period 1950 through 1985, the annual average compound rate of growth of world output *per head* has been 1.9 percent. Given that the rate of growth of world population is slowing down and that, at the same time, there has been a rapid increase in the proportion of the world's population receiving higher education or engaged in scientific research—the mainsprings of technical progress—there is good reason to believe that this growth rate will be at least maintained, if not increased. But suppose that it is only, say, 1.5 percent a year. This means that by the year 2093 world output per head will be 4.4 times as great as it is now. If global warming cuts world GDP by 10 percent, then instead of it being 4.4 times as great as it is now, it would be only 3.96 times as great. Would this be such a disaster? Would it justify imposing vast costs on the present generation rather than devoting more resources to helping developing countries overcome the environmental problems that they are facing today?

* * *

So far we have limited our discussion to the effect of global warming on agriculture, since this is the most vulnerable sector. But what about sea level rise, the other eco-catastrophe most frequently associated with global warming?

In 1980 scientists predicted that global warming would lead to a sea level rise of as much as 8 meters. In early 1989 the prevailing estimate was down to about 1 meter. By 1990 (as in the IPCC report) the predicted sea level rise was about 65 centimeters, and current authoritative estimates put it as low as about 30 cm by the end of the next century, assuming a 4°C rise in average temperature by then. (If one were to extrapolate from trends in these estimates, the sea level would be predicted to fall, with consequences for many seaside resorts that would be as serious as sea level rises!)

But even if sea levels did rise appreciably, the economic consequences would not be disastrous. A few years ago, when the sea level was still predicted to rise by 1 meter, the U.S. Environmental Protection Agency estimated that it would cost about $100 billion to protect U.S. cities by building sea walls. Applying a 1.5 percent a year annual growth rate to the present U.S. GDP of $6 trillion gives a GDP of $26.2 trillion in 2093; so as a fraction of GDP in the year 2093, the once-and-for-all capital cost of the sea walls would be about .38 percent. As a fraction of cumulative GDP over the whole of the next 100 years—the time during which the work would have to be carried out—the amounts involved are, of course, trivial.

What about the rest of the world? Estimates by William Cline of the Institute for International Economics, also assuming a 1 meter rise in the sea level and that the costs of sea walls for other threatened coastal cities are comparable to those of the U.S., show costs of adaptation, plus the value of land lost in coastal areas, of about $2 trillion. On the above assumption concerning the growth rate, world GDP in the year 2093 will be about $115 trillion—so the one-time capital cost of the sea walls would still be only about 1.7 percent of one year's GDP. As a fraction of cumulative GDP over the whole period it would still be negligible. Given that the latest predictions of the rise in the sea level are about one third of those assumed in these estimates and that a given reduction in the estimated sea level rise implies more than a proportionate reduction in the costs of adaptation or the damage done through land loss, the costs for the world as a whole would be insignificant even if the above estimates are way off.

Now that may be very well for the world as a whole, but it is little consolation to the people of Bangladesh, where 20 percent of the land could be lost under the sea with a 1 meter sea level rise. But leaving aside the falling trend in the estimate of sea level rises, suppose, purely for the sake of illustrating the logic of the choices to be made, that measures to prevent the climate change and the consequent sea level rise would cost the world community $20 trillion—i.e. ten times as much as the cost of protection against the rising sea level. It would clearly be in everyone's interest, including the Bangladeshis', to strike some sort of a deal. For example, instead of incurring $20 trillion in costs to prevent the climate change and the associated sea-level rise, the rich countries would do better to hand over, say, a quarter of the resulting economy—that is, $5 trillion—to the people who would suffer from the sea-level rise. The Bangladeshis would then gain $5 trillion to carry out work costing only $2 trillion, and the rest of the world still saves $15 trillion.

In other words, the course of action that is being urged by environmentalists—to prevent rising sea levels at any cost—would mean that the world is being asked to incur costs of $20 trillion —or whatever the cost would be—to prevent the Bangladeshis from suffering the effects of the sea level rise when there would be a much cheaper way of sparing them from these effects. Developed countries could, for example, help Bangladeshis move away from threatened coastal areas, diversify their economy, or take some other protective action. They might even consider making it easier for Bangladeshis to export goods and emigrate. The point is that adaptation should not be viewed as an impossibility. After all, over half the population of the Netherlands and virtually all the people in Amsterdam live below sea level.

If the estimates of the costs involved in significant reductions of CO_2 emissions referred to above are anywhere near reality, it is clear that the world and the Bangladeshis would be far better off if adaptive policies were taken rather than drastic action to prevent threatened rises in sea levels. In any event, since it is suspected that far more land is being lost to soil erosion than through climate change, there are policies that could be undertaken to reduce land loss without draconian cuts in carbon dioxide emissions.

* * *

But what would it cost to implement CO_2 cuts by the most economical means, and how would one do it? One view, espoused by the Clinton administration among others, is that cuts can be achieved at no cost whatsoever to the public or private sector. The White House insists that its "Climate Change Action Plan," a collection of mostly voluntary public-private partnerships which purports to stabilize greenhouse gas emissions at 1990 levels by 2000, will save the private sector $207 billion in reduced energy costs by 2010.

Most economists, however, believe the administration is relying on astonishingly optimistic assumptions about energy savings. The reason is simple: If energy-efficiency investments really yield such fantastic returns, businesses and individuals would undertake them without the government's push. To accept the administration's savings estimates, one would have to believe that firms are very stupid indeed.

As many disappointed environmentalists have noted, voluntary measures of the sort in Clinton's plan are likely to have only a very limited effect on greenhouse gas emissions. Deep cuts will require stronger government action, namely a carbon tax or a system of internationally tradable emissions permits.

Making a rapid transition to a less carbon-intensive economy would be painful, but just how much so is unclear. The uncertainties about the magnitude of the costs are as great, if not greater, than those embodied in the scientific models to predict climate change. Nobody can predict with much accuracy the future pattern of economic growth over the next century in a manner that enables one to say how much fossil fuel would be burned and hence how much needs to be done to reduce fossil fuel use by any given amount. Nor can anyone say with any degree of reliability how sensitive the consumption of fossil fuels is to different levels of carbon taxes. The impact of carbon taxes depends in large part on the degree to which carbon-free energy sources like nuclear, wind energy, and photovoltaics become cheaper and easier to use than their fossil rivals.

Finally, there is the huge question of what is done with the revenues raised by carbon taxes. If they are refunded to the economy in a way that merely changes the pattern of fuel use rather than exerting a depressive effect on the economy as a whole, the effect could well be favorable. For example, there is a strong case for reducing government subsidies to the coal and oil industries apart from any considerations about global warming. If the savings from the elimination of those subsidies were used to pursue growth-enhancing policies, so much the better. If, on the other hand, carbon taxes were used to fund extra entitlements or low-return government "investment" programs, the effect would more likely be negative. A tax on a basic input to the economy, which is what fossil fuels are, could have a far more serious impact than other taxes such as those that are levied on different items of consumer expenditure, such as cigarettes or beer.

Vast computerized models tend to blind people with science and impress grant-giving institutions. Most of them produce a maximum amount of statistical "results" with a minimum input of ideas or insight. The estimates that they churn out of the cost of measures designed to cut CO_2 emissions are highly uncertain. Inevitably, different models lead to widely different results. But for what they are worth, it seems that the cost of freezing CO_2 emissions at 1990 levels would be somewhere in the region of 2 percent of output per year in advanced industrial countries and perhaps twice that in developing countries, which have fewer financial and technological resources with which to adapt. To put that in perspective, 2 percent of current U.S. GDP is about $120 billion. The large magnitude of the costs is not surprising since, again, it seems obvious—even without the aid of any sophisticated computerized models—that, if carried out quickly, a drastic cut in the use of fossil fuels would cause severe economic disruption.

* * *

The above discussion of the damage that might be done by a doubling of present concentrations of carbon in the atmosphere considers only the costs over a 100-year period. Some economists have pointed out that there is no reason to confine the analysis to a single century. Fossil fuels will probably continue to be used well after the year 2100, so it is likely that there will be greater atmospheric build-up of CO_2 and more global warming than that indicated by the century-long models. William Cline, for example, has shown that over a period of three centuries, atmospheric concentration of CO_2 might increase eight-fold and temperatures might rise 10°C to 13°C or more unless action is taken to reduce carbon dioxide emissions. This, he argues, could lead to a reduction of world GDP on the order of 10 percent.

Of course, the further one projects into the future the more uncertain the already shaky projections become. There have been vast technological changes in energy use in the last century. And these will probably be dwarfed by the changes that will take place in the next three centuries, during which an incomparably greater number of people will be engaged in technological and scientific research all over the world. Hence, nobody can suppose that the world of the 23rd century will bear much resemblance to the world that we know today. It is most unlikely that energy will still be produced on a large scale by the use of dirty and polluting substances such as coal.

One need only look at the past to see how difficult it is to make predictions over hundreds of years. Who could have predicted three centuries ago that sources of power would shift from wind, water, and wood to coal, oil, natural gas, and nuclear energy? Who could have imagined that modern gas-fired, combined-cycle power plants would be about ten times as energy efficient as power stations built at the beginning of the century, or that the thermal efficiency of steam engines would be about forty times that of the earliest engines, or that the most advanced fluorescent lights would be 900 times more energy efficient than the original kerosene lamp?

And progress is now being made in developing viable forms of renewable energy that emit no carbon at all. These include the photovoltaic cells that convert sunlight directly into storable electricity. They are regarded as having great promise for local power generation in developing countries. During the past twenty years the cost of photovoltaic-generated electricity has fallen from about $60 per kilowatt hour to about 30–50 cents, and is expected to fall to about 12–16 cents within the next few years as a result of further efficiency improvements that are already in the pipeline. Industry analysts say costs could go as low as 6 cents per kilowatt hour by 2020, little more than the price of electricity from a coal-fired station.

Similarly, the cost of wind-generated electricity has dropped from 50 cents per kilowatt hour in 1975, to 25 cents in 1980, to 7–9 cents in the best locations today. A new wind turbine under development is expected to bring costs down to 5 cents or less. The U.S. Department of Energy projects that over the next twenty years, costs in moderately windy sites could fall to 3.5 cents per kilowatt hour. Wind power's contribution will ultimately be limited by the number of suitable sites. The German Future Energies Forum, an energy research group, estimates that wind power can meet no more than 15 percent of the world's energy requirements.

Then there is always the possibility of a breakthrough in geothermal, solar thermal, hydrogen fuel, or nuclear fusion technology. It is impossible to determine the ultimate potential of technologies that are barely off the drawing board, but it would be foolishly pessimistic to assume that none of these carbon-free technologies will ever become cost-competitive with fossil fuels.

Even without any significant recourse to the many forms of renewable energy under investigation, there have been major reductions in the ratio of both energy use and carbon emissions to GDP. For example, from 1950 to 1985 the rate of growth of CO_2 emissions in the US was 1.9 percent a year, compared with a growth of GDP of 3.2 percent a year. This means that there was a 1.3 percent a year fall in carbon emissions per unit of GDP over the period 1950–1985. This is accounted for by a 0.8 percent a year decline in the rate of energy use per unit of output and a 0.5 percent a year decline in the rate of CO_2 emissions per unit of energy.

Cline does not make explicit his assumptions concerning these two influences on the growth of CO_2 emissions in his predictions. But based on his projections of GDP growth and CO_2 emissions, it is possible to work out what assumption is implied. It appears that he is assuming that the carbon emissions per unit of GDP will fall by only about 0.7 percent a year in the latter half of the twenty-first century—i.e. only about half the rate of decline in the U.S. between 1950 and 1985. Further, his estimates imply that the ratio of carbon emissions to GDP will fall even more slowly thereafter, and will actually rise during the last 75 years of the period covered in his projections.

Cline also assumes that every carbon dioxide molecule released into the atmosphere stays there for the duration of his three-century model. Calculating the effective lifespan of carbon dioxide is complex because CO_2 molecules are constantly exchanged between the atmosphere, the oceans, and the biosphere. Yet according to the IPCC Scientific Assessment, the "lifetime" associated with an individual CO_2 surge is only 50 to 200 years.

Adjusting Cline's assumptions yields dramatic changes. If we assume (a) that CO_2 molecules have an average lifespan of 100 years in the atmosphere and (b) that the amount of carbon emissions per unit of GDP declines 1.3 percent a year, both the CO_2 build-up and the warming projections become far less ominous. Instead of Cline's assumed eight-fold increase in carbon dioxide, there would be only about a one-and-a-half-fold increase. And the temperature increase resulting from this carbon buildup falls from between 10°C and 13°C to just 2°C. Even allowing for a generous margin of error, it would seem that Cline's projections are terribly pessimistic.

* * *

Assessing carbon abatement measures requires the comparison of costs and benefits over very long periods of time. This raises an important question: How is a benefit enjoyed in many years' time to be compared with a benefit enjoyed immediately? Most people would be unwilling to exchange $1.00 for, say, only $1.02 in one year's time—even if the $1.02 were indexed for inflation.

The need to discount the future at some rate of interest implies that the benefits in two or three centuries' time of abating GHG emissions now would have to be astronomic to justify significant current sacrifices. In fact, for almost any discount rate at all, what happens in the twenty-third century is of almost no importance. For example, if future costs and benefits are discounted at 2 percent a year—which is well below what most people would expect to earn on their investments, even net of tax and inflation—$1 of cost or benefit in 200 years' time would have a present value today of only 1.9 cents. In other words, if the damage done by global warming in 200 years' time were to be $1,000 billion, it would not be worthwhile taking steps to avoid it if the cost of doing so today would be greater than $19 billion. If a more conventional discount rate were used—say, 6 percent—it would not be worthwhile taking steps to avoid $1,000 billion of damage in 200 years' time if the cost of doing so today exceeded only $8.7 million.

People discount the future for two types of reasons. First, many people simply prefer to consume resources now rather than in the future. This "time preference" generally arises from any of three motives, namely (a) the risk that one might not survive to reap the future rewards for sacrifice of consumption today; (b) an expectation that future consumption levels will be higher (combined with the usual assumption of diminishing marginal utility of consumption), and (c) "pure" time preference—e.g., sheer impatience, or lack of imagination.

But even an individual who had no *basic* time preference whatsoever—who had no worry that her investment might not pay off, who did not expect to be richer, and was not impatient—would still not lend $1 now for $1 next year (leaving aside personal motivations, such as helping out a friend). The reason for this is that she would know that any old fool can get a better rate of return on the market. She would lend, therefore, up to the point where *at the margin* the rate of return she could get on her savings and investments would be equal to the relative valuation she places on consumption today as against consumption tomorrow. The same principle applies—with a few adjustments—to society.

In other words, the use of a discount rate in allocating investments today between alternative uses does not reflect any tendency by economists to attach less value, per se, to future *welfare.* On the contrary, a unit of *welfare* in 200 years' time is given the same value as a unit of welfare today. But, if technological progress and economic growth continue, a unit of *consumption* in 200 years' time will give a smaller unit of welfare than it does today. Taking account of the discount rate when choosing between projects is a technique designed to maximize welfare over the whole time period regarded as relevant. To invest in some environmental project that yields only, say, a 1 percent rate of return, when there are alternative projects around that will yield, say, 5 percent, would leave future generations much worse off than they could otherwise have been.

The simple answer, therefore, to the criticism that, by discounting, one is being unfair to future generations is this: If, Heaven forbid, there were some leap in medical science so that we all now expected to live for 300 years, *we would still use the discount rate in our savings and investment decisions.* So, as a result of our doing so today (when we do not expect to survive for centuries), future generations are left no worse off than we would be if it were us who were to be alive in 200 years' time, not them. Hence, it is absurd to argue that we are somehow or other being unfair to them.

In short, the widespread environmentalist attack on the validity of time preference is a vast red herring. What matters is the rate of technological progress and economic growth that one can expect to take place. As we have argued, there are good reasons for expecting this to be as rapid in the future as in the past. Of course, if it is expected to slow down to zero, or become negative—which is not beyond the bounds of possibility—one should adopt a zero or negative discount rate for later years. But a case should be laid out for such a prediction. The "immorality" of time preference has nothing to do with it.

* * *

Global warming may be a problem, but it is no cause for undue alarm or drastic action. There is plenty of time to improve our understanding of the science and scrap policies that encourage economically inefficient uses of fossil fuels. It does not justify diverting vast amounts of time, energy, and funds from more urgent environmental problems, particularly those in developing countries. Nor does it justify a massive diversion of resources from high-yield projects in the private sector. We are not on the edge of an abyss and the human race is not facing destruction from the accumulation of greenhouse gases. There is far less danger of the human race being wiped out on account of the conflict between Man and the Environment than on account of the conflict between Man and Man (or Woman and Woman). Global warming is far more glamorous and telegenic, of course, than the need for better toilets and drains in the Third World. But if we truly care about the welfare of our fellow world citizens, it is these kinds of environmental issues upon which we must focus our attention.

POSTSCRIPT

Should Society Be Concerned About Global Warming?

In 1992 the United Nations Conference on Environment and Development in Rio de Janeiro was held. High on the agenda was the issue of global warming, but despite widespread concern and calls for reductions in carbon dioxide releases, the United States refused to consider rigid deadlines or set quotas. The Bush administration felt that the uncertainties were too great and that the economic costs of cutting back on carbon dioxide emissions might well be greater than the costs of letting the climate warm.

However, James Kasting of Pennsylvania State University and James Walker of the University of Michigan warn that if one looks a little further into the future than the next century, the prospects look much more frightening. By 2100, they say, the amount of carbon dioxide in the atmosphere will reach double its preindustrial level. By the 2200s it could be 7.6 times the preindustrial level. With draconian restrictions, it could be held to only 4 times the preindustrial level. Correspondingly, they predict, global warming in the twenty-first century will be worse than anyone is currently forecasting.

On the other hand, it remains difficult to be at all sure of such predictions. See, for example, David Schneider, "The Rising Seas," *Scientific American* (March 1997) and Thomas R. Karl, Neville Nicholls, and Jonathan Gregory, "The Coming Climate," *Scientific American* (May 1997).

Are there any practical alternatives? In October 1993 the Clinton administration announced a "Climate Change Action Plan" that called for stabilizing carbon dioxide releases by such measures as planting trees (which remove carbon dioxide from the air) and promoting energy efficiency. Unfortunately, population growth and the need to use land for growing food render this solution unlikely.

In a recent experiment, researchers spread high-iron fertilizer across several square miles of the Pacific Ocean to see whether or not it could stimulate the growth of algae and the absorption of carbon dioxide from the air. The process worked, but it would probably be too expensive to apply as a worldwide anti-greenhouse measure.

Gelbspan is by no means alone in his concerns that the 1990s have seen a worldwide pattern of extreme fluctuations in the weather. Already the world's insurance industry is growing very concerned because its losses due to storm and flood damage have been increasing rapidly. See Christopher Flavin, "Storm Warnings: Climate Change Hits the Insurance Industry," *World Watch* (November/December 1994). And according to Flavin and Odil Tunali, in "Getting Warmer: Looking for a Way Out of the Climate Impasse,"

World Watch (March/April 1995), governments are not acting on the problem. In September 1995 the Intergovernmental Panel on Climate Change (IPCC) reported the consensus view that global warming is real, and Thomas Karl of the National Climatic Data Center in Asheville, North Carolina, said, "There's a 90 to 95% chance that we're not being fooled."

In December 1997 the nations that signed the UN Framework Convention on Climate Change in Rio de Janeiro in 1992 will meet again in Kyoto, Japan. Their goal will be to set carbon emissions limits for the industrial nations. As of May 1997, despite broad agreement that global warming is a real and significant threat, the prospect for agreement on what to do about it did not look good (see Richard Monastersky, "Beyond Hot Air," *Science News*, May 24, 1997). However, by August, Steve Olson (in "White House Enlists Science for Public Education Effort," *Science*, August 1, 1997) could say, "Most of the world's nations are expected to sign an agreement setting limits on greenhouse gas emissions [in Kyoto]."

Even the IPCC admits that the data are inadequate to determine with certainty that human activities are to blame for changes in climate and weather. It can be very difficult to discern a clear trend in a system that varies erratically from year to year and from place to place, such as climate. Yet research is making it very clear that looking for such trends is important. Studies of ice laid down millennia ago in Greenland and in glaciers in South America and China are showing that climate can warm or cool dramatically in very short times and that such changes happen frequently. See Richard Monastersky, "The Case of the Global Jitters," *Science News* (March 2, 1996). The concern here is that even though human-induced climate change may seem gradual, it may suddenly become much more abrupt and leave society very little time to adjust.

On the other hand, there are intriguing new data connecting variations in the amount of energy emitted by the sun to variations in Earth's climate. As reported by Richard A. Kerr, in "A New Dawn for Sun-Climate Links?" *Science* (March 8, 1996), "An effort to reconstruct changes in the sun's brightness over the last 400 years has yielded [a pattern that implies] that the sun could have been responsible for as much as half of the warming of the past century." Scientists are still arguing over the data and their interpretation, but the point remains: Human-released greenhouse gases may not be entirely to blame for global warming. Society may still need to be concerned about global warming and its consequences, but that concern need not have quite so large a component of guilt.

Many people are not pleased with the prospect of having to reduce their use of carbon-emitting fossil fuels. Energy companies are leading the fight against any measures that would affect their operations by stressing the uncertainty in negative projections. See David Helvarg, "The Greenhouse Spin," *The Nation* (December 16, 1996).

ISSUE 6

Is Ozone Depletion a Genuine Threat?

YES: Mary H. Cooper, from "Ozone Depletion," *CQ Researcher* (April 3, 1992)

NO: James P. Hogan, from "Ozone Politics: They Call This Science?" *Omni* (June 1993)

ISSUE SUMMARY

YES: Mary H. Cooper, a staff writer for *CQ Researcher,* asserts that scientific findings in recent years indicate that the ozone layer is being depleted and that such depletion threatens the health of Earth's living organisms.

NO: James P. Hogan, a science fiction writer, maintains that reports of the ozone being destroyed by chlorofluorocarbons are politically motivated scare stories unsupported by any valid scientific evidence.

The debate about ozone depletion may offer a good example of how attempts to make life better may actually do great harm. The controversy can be traced back to early improvements on the refrigerator. The fluid in the cooling coils of early refrigerators was ammonia, which is both highly irritating and very toxic. Chlorofluorocarbons (CFCs) replaced ammonia in the 1930s because they were nontoxic, nonirritating, nonflammable, inexpensive, noncorrosive, and stable (that is, they did not have to be replaced periodically). They were ideal for the purpose, and later on they proved to be ideal for other purposes as well. Consequently, CFC production grew, and by 1974 over 800,000 metric tons of CFCs were being produced per year.

Around this time, however, Sherwood Rowland and Mario Molina, chemists at the University of California at Irvine, realized that those safe, stable chlorofluorocarbon molecules were stable enough to rise all the way into the stratosphere before they broke down. When the molecules did break down, the chemists said, they would release their chlorine component precisely where it could destroy stratospheric ozone. Two years later, the first solid evidence of CFC-induced damage to the ozone layer was reported. As a result, in 1978 the U.S. Environmental Protection Agency and Food and Drug Administration banned the use of CFCs in aerosol spray cans. In 1985 a roughly circular zone of greatly diminished ozone concentration—commonly called the "ozone hole"—was discovered over Antarctica. Since then, several key events have occurred that have firmly established ozone depletion as a worldwide concern: In 1987 the United States and 23 other countries signed the Montreal Protocol (promising to cut CFC use drastically); in 1991 the United Nations Environment Programme (UNEP) issued a report concluding that ozone depletion was a worldwide phenomenon; and in 1992 scientists from the National Aeronautics and Space Administration (NASA) reported alarmingly high levels of ozone-destroying chlorine monoxide in the stratosphere over the Northern Hemisphere, confirming the conclusions of the UNEP.

Why is news of the depleted ozone alarming? It has been known for many years that stratospheric ozone absorbs much of the ultraviolet (UV) light that comes to Earth from the sun (or "solar ultraviolet"). The small amount that reaches Earth's surface stimulates the production of vitamin D in human skin, but it also causes sunburn, skin cancer, and cataracts, as well as other problems. An increase in solar UV at ground level would therefore pose a serious health threat to human beings. It would also threaten other living organisms, including crop plants. See David J. Leffell and Douglas E. Brash, "Sunlight and Skin Cancer," *Scientific American* (July 1996).

As yet, there is no proof that ozone depletion is already causing health problems, but according to everything that scientists know or can calculate, it will. Furthermore, since CFCs can remain in the air for years, many believe that society cannot wait for the signs of damage to become incontrovertible before taking action because by then it will be too late to keep the damage from growing still worse for years.

Although the facts strongly suggest a connection between CFCs and the depletion of stratospheric ozone, science cannot prove beyond a doubt that CFCs are destroying the ozone. Despite the evidence, not everyone accepts as true the threat of CFCs and ozone depletion. Critics have denounced the "ozone scare" as a politically motivated scam, a conspiracy to provide monopoly control to companies that have patents on refrigerant alternatives to CFCs, a way to gain public funding for unnecessary research, and so on.

In the following selections, Mary H. Cooper provides a clear account of the discovery and nature of ozone depletion, and she asserts that the hazards of CFCs are real. James P. Hogan argues that any conclusions about the destruction of the ozone by CFCs are based on sloppy scientific procedures and misinterpretations of data, and he maintains that public fears about the ozone are perpetuated for political reasons.

YES

Mary H. Cooper

OZONE DEPLETION

Scientists at the National Aeronautics and Space Administration (NASA) hadn't planned to hold a news conference on February 3 [1992]. But, they decided at the last minute, their preliminary findings about Earth's upper atmosphere were too important to sit on. Earth's protective ozone layer,[1] they announced, was losing ozone much faster than anyone had predicted, exposing humans to higher amounts of harmful radiation.

Even more ominous, they said, it seemed likely that a highly depleted section of the ozone layer, known as a "hole," would develop over the Arctic, exposing populated areas of the Northern Hemisphere. A similar hole had first been observed over Antarctica in 1985.

The scientists based their startling announcement on new data collected over northern New England, eastern Canada and much of Europe and Asia. What their airborne instruments—carried aloft by a satellite and two high-flying planes—detected was the highest concentration of ozone-destroying chlorine monoxide ever measured in the atmosphere.

Chlorine monoxide is a derivative of an important family of synthetic chemicals that are known as chlorofluorocarbons (CFCs). They have enjoyed wide use for decades as coolants in refrigerators and air conditioners, propellants in aerosol spray cans, blowing agents in the manufacture of plastic and rubber foam products and as solvents in the production of electronic equipment.

Once released into the atmosphere, CFCs drift upward until they reach the ozone layer, which begins in the stratosphere. As long as they remain in their original molecular form, CFCs are harmless. But intense ultraviolet radiation can break the CFC molecule apart, producing chlorine monoxide and setting off a series of reactions that destroy ozone.

High levels of chlorine monoxide are alarming enough by themselves. But NASA's scientists found evidence of even more worrisome atmospheric problems: high levels of bromine monoxide. A byproduct of halons, man-made chemicals used in fire extinguishers, bromine monoxide is even more destructive than chlorine monoxide.

Michael Kurylo, NASA's program manager for the study, estimated that the two chemicals could destroy 1 to 2 percent of the ozone layer daily during brief periods of late winter. At that rate, as much as 40 percent of the ozone over populous areas of the Northern Hemisphere could be depleted by early spring, when ozone destruction ends each year. The resulting hole, scientists said, could be almost as serious as the one over Antarctica, where ozone depletion has been known to reach 50 percent.

In addition to high levels of ozone-destroying chlorine monoxide and bromine monoxide, the NASA team found reduced levels of nitrogen oxides, which protect ozone from the other two gases by converting them into harmless compounds before they have time to destroy ozone. The loss of nitrogen oxides, which scientists attribute to high levels of volcanic ash ejected into the stratosphere last summer during the eruption of Mount Pinatubo in the Philippines, diminishes the atmosphere's natural ability to recover from ozone depletion.

"The latest scientific findings indicate pretty clearly that the atmosphere all over the place, and not just in the polar regions, is nearly devoid of some of the constituents that protect ozone against depletion," says Michael Oppenheimer, senior scientist at the Environmental Defense Fund in New York City.

... [R]ecent findings are serious enough that several countries, including the United States, have taken new steps to slow ozone depletion. In 1987, for example, the main producers and consumers of CFCs and halons signed the Montreal Protocol, which mandated phasing out these destructive chemicals by the year 2000, or sooner. The phaseout was subsequently accelerated in 1990, and several signatories to the protocol have since committed themselves to beating the deadline....

How Dangerous Is Ozone Depletion?

Ozone-destroying chemicals are extremely stable, so they last in the atmosphere for many decades. That means that even if production of all CFCs and halons stopped today, the chemicals already in the atmosphere would go on destroying ozone well into the 21st century. And because large quantities of these chemicals are contained in existing air conditioners and refrigerators, from which they continue to escape through malfunction or intentional venting, it may be a century before the ozone layer has built itself back up.

Just how devastating widespread ozone depletion would be is not known. But a 1975 government study on the environmental effects of an all-out nuclear war—which scientists say would destroy much of the ozone layer—provided a chilling glimpse of the aftermath. Ozone depletion of 50 percent, the study postulated, "would cause [skin] blistering after one hour of exposure. This leads to the conclusion that outside daytime work in the Northern Hemisphere would require complete covering by protective clothing.... It would be very difficult to grow many (if any) food crops, and livestock would have to graze at dusk if there were any grass to eat."

The study speculated that a 25 to 30 percent depletion of stratospheric ozone —which NASA's findings indicate already may have occurred over parts of the Northern Hemisphere—would make it "difficult to imagine" how survivors could carry out postwar recovery operations.

Since the ozone hole opened over Antarctica in 1985, scientists have been assessing the impact of increased ultraviolet (UV) radiation on phytoplankton, the micro-organisms that make up the essential first link in the food chain that maintains all animal life in warm southern waters, including whales. Preliminary findings show that phytoplankton populations have dropped by up to 12 percent in areas where surface UV radiation has increased under the Antarctic ozone hole.

This is the first evidence outside the laboratory that links ozone depletion to damage of living organisms on Earth.

Excessive UV radiation is also thought to disrupt photosynthesis, the process by which green plants use the sun's radiant energy to produce carbohydrates. Ozone depletion could thus cause reduced yields in crops such as soybeans and rice, crops that are essential to feeding large parts of the Third World.

Ultraviolet radiation has long been known to cause health problems in animals, including cataracts in humans—the leading cause of blindness. The United Nations Environment Programme (UNEP), which was set up in 1972 to foster international cooperation in protecting the environment, predicts that ozone depletion will cause an additional 1.6 million cases per year.

There are also preliminary reports of widespread blindness among rabbits, sheep, horses and cattle in southern Chile, where high UV radiation exposure resulted from the ozone hole over Antarctica.

UNEP also foresees an annual increase of 300,000 cases of skin cancer, by the year 2000, particularly in Argentina and Australia, which have come under increased UV radiation. UNEP also estimates that a 10 percent depletion of the ozone layer would cause up to 26 percent more basal and squamous-cell skin cancers. The agency cites new evidence that UV radiation may also contribute to cancers of the lip and salivary glands.

Other studies project that a 10 percent increase in UV penetration would cause up to a 9 percent increase in the incidence of the more deadly malignant melanoma among light-skinned people, the group that is most vulnerable to this virulent form of cancer.

Ultraviolet radiation may also undermine the immune system's ability to ward off infectious diseases. This, says Margaret L. Kripke, an immunologist at the University of Texas' M. D. Anderson Cancer Center in Houston, is the biggest unknown health effect of UV radiation. Animal experiments have indicated that UV radiation may reduce lymphocytes' ability to destroy certain microorganisms that enter the body through the skin, such as Leishmania, malaria, schistosoma and the leprosy bacillus.

Although it is not known whether UV radiation actually reduces human resistance to these agents, Kripke testified last fall, "infectious diseases constitute an enormous public health problem worldwide, and any factor that reduces immune defenses... is likely to have a devastating impact on human health."

Kripke's research was particularly ominous for sun worshipers. She found that commercial sunscreen preparations, which protect against sunburn and other damage to the skin from UV radiation, don't block the immunosuppressive effects of UV radiation. Similarly, skin pigmentation, which protects darker-skinned people from skin cancers that are prevalent among Caucasians, doesn't seem to protect the immune system from UV damage....

FIRST SIGNS OF TROUBLE

Even as industry was finding new uses for CFCs in the early 1970s, scientists were beginning to link them to ozone destruction. In 1974, Ralph Cicerone, then at the University of Michigan, and his colleague, Richard S. Stolarski, investigated the possible effects on stratospheric ozone of chlorine released by NASA rockets. They concluded that a single atom of chlorine would destroy many thousands of ozone molecules.

However, because the number of rockets passing through the ozone layer was small, and no other sources of chlorine at that altitude had been identified, their findings did not cause widespread alarm.

Findings reported later that year, however, showed that rocket engines were not the only source of chlorine in the stratosphere. Sherwood Rowland and Mario Molina at the University of California at Irvine decided to study CFCs after they are released into the atmosphere. They found that CFCs are so durable that they do not break down under the forces of solar radiation and precipitation in the lower atmosphere, but continue to float around in their original state for many years, eventually drifting upward into the stratosphere.

"What we did was to ask a question that hadn't been asked before: What is going to happen to the CFCs?" Rowland recalls. "The conclusion we came to was that nothing would happen quickly, but on the time scale of many decades CFCs would go away into the stratosphere and release chlorine atoms and then that the chlorine atoms would attack the ozone.... We concluded that there was danger to the ozone layer and... that we should quit putting CFCs into the atmosphere."

Not surprisingly, Rowland and Molina faced hostile reactions from the producers of CFCs when they published their results in 1974. "The public was probably more likely to believe it than the chemistry community," Rowland says. "Within the chemistry community then and still now there is a feeling that most environmental problems are really just public relations problems, that they are not real problems."

Rowland says the chemicals manufacturers set up the Committee on Atmospheric Science to discredit the two researchers' findings. Indeed, he adds that many critics dismissed their conclusions as "kooky. One of my favorites was an aerosol-propellant company that claimed [our results were] disinformation put out by the KGB."

But their data held up. In 1976, after a nationwide research effort involving NASA and the National Oceanic and Atmospheric Administration (NOAA), the National Academy of Sciences confirmed that CFC gases released into the atmosphere from spray cans were in fact damaging the ozone layer.

Two years later, after consumer boycotts had reduced the market for spray cans by almost two-thirds, the United States banned the use of CFCs as aerosol propellants in spray cans for most uses.

OZONE HOLE DISCOVERED

Although other industrial nations continued to produce and use CFCs for aerosol sprays and other purposes, the international scientific community continued the search for data on ozone depletion launched by Rowland and Molina. During the early 1980s, most research was confined to computer models of the atmosphere. Then, in 1985, British scientists discovered that ozone depletion had become so severe over a vast area of Antarctica that it amounted to a virtual "hole" in the ozone layer.

Still, resistance to the ozone-depletion theory remained so strong that the British team was refused additional government funding to continue their research. Ironically, they obtained backing instead from the U.S. Chemical Manufacturers Association, whose members had the most to lose from confirmation of Rowland and Molina's theory. Because of mounting pressure at home to find substitutes for CFCs, however, the American chemical industry wanted to resolve the issue once and for all before abandoning CFCs.

Meanwhile, Rowland and other scientists were learning more about ozone depletion and why the phenomenon was so strong over the Antarctic.

They discovered that CFCs are concentrated over the South Pole because of strong circular winds known as the "polar vortex," which sweep unimpeded over the flat, barren continent of Antarctica. The vortex gathers the destructive gases from the surrounding atmosphere into a wide funnel over Antarctica, where they remain isolated during the dark, frigid winter months.

Equally important, they found that as CFCs break down, the resulting chlorine monoxide clings to the ice crystals that form clouds in the stratosphere. These ice crystals provide the surfaces needed for the catalytic reaction in which chlorine breaks down ozone.

With the return of sunlight to Antarctica during September and October, the beginning of spring in the Southern Hemisphere, solar radiation acts as a catalyst enabling the chlorine monoxide produced by CFCs to destroy the surrounding ozone layer.

As the days lengthen, the air over Antarctica warms up, breaking up both the stratospheric ice clouds and the polar vortex. The destruction of ozone slows as the chlorine atoms are once again bound into harmless chlorine nitrate and hydrogen chloride molecules, and the hole disappears as the vortex dissipates, allowing ozone from the surrounding regions to fill the void.

The final confirmation of Rowland and Molina's theory linking CFCs to ozone depletion came in 1987, when NASA undertook a series of aerial tests over Antarctica. From inside the ozone hole, the NASA instruments detected high concentrations of chlorine monoxide.

Montreal Protocol Signed

International reaction to the proof that CFCs were destroying the ozone layer was swift. On Sept. 16, 1987, just nine months after formal negotiations began, 24 nations signed the Montreal Protocol on Substances That Deplete the Ozone Layer. The agreement garnered an unprecedented degree of international support for such a sweeping program to protect the environment: The ratifying nations accounted for 99 percent of the world's production of CFCs and 90 percent of their consumption.

The Montreal Protocol called for freezing halon emissions at 1986 levels by 1992; for halving CFC emissions by 1998; and halving CFC production and importation by 1999. To compensate for their low levels of production of ozone-depleting chemicals, developing nations were given an additional 10 years to meet these deadlines. By Jan. 1, 1989, the protocol had been ratified by enough countries to go into effect.

Richard Elliot Benedick, a Foreign Service officer who led the U.S. delegation in negotiating the Montreal Protocol, identifies several reasons for the treaty's overwhelming success. First, international cooperation among scientists allowed for the rapid discovery of CFCs' role in ozone depletion. Public opinion, which was then beginning to focus on environmental issues throughout the industrial world, was also quick to press governments to act. Negotiations were supported by the UNEP.

Benedick also credits the United States for its leading role in gaining support for the treaty. The United States was the first producer of ozone-depleting chemicals to restrict their production, he points out in his account of the negotiations surrounding the protocol. In addition, Congress passed ozone-protection legislation as early as 1977, long before the governments of Western Europe responded at all.

The United States also was primarily responsible for the 1985 Vienna Convention for the Protection of the Ozone Layer, an agreement among the major CFC producers to collect additional data that led up to the Montreal Protocol.

"The U.S. government reflected its concerns over the fate of the ozone layer through stimulating and supporting both American and international scientific research," Benedick wrote. "Then, convinced of the dangers, it undertook extensive diplomatic and scientific initiatives to promote an ozone protection plan to other countries, many of which were initially hostile or indifferent to the idea."

The drafters of the Montreal Protocol also assured its success by making the agreement flexible. As such, it could be rapidly amended to reflect subsequent changes in environmental conditions or new findings. And new findings were soon to test the agreement's flexibility.

The ozone hole over Antarctica continued to appear each September and October after its initial discovery in 1985. In 1988, scientists were encouraged to find that the hole was not as big as before. But the following year, the ozone hole reappeared, covering more than 15 million square miles.

Arctic Expedition Launched

The same year, NASA and NOAA launched an airborne expedition to the Arctic to investigate whether conditions were ripe near the North Pole for another ozone hole. Because the Arctic terrain is not as flat as that of Antarctica—and because temperatures at the North Pole do not fall as low as they do at the South Pole—the polar vortex was found to be weaker in the north. But the scientists did find higher than expected concentrations of chlorine compounds and concluded that an ozone hole could easily develop.

Because more people live at far northern latitudes than in southern Chile, Argentina, Australia and New Zealand, which border the area exposed to UV radiation in the Southern Hemisphere, an ozone hole over the Arctic would pose far greater risks to human health.

Other research revealed a new potential source of ozone depletion in areas far from the polar regions. American chemists Susan Solomon and Dave Hoffman found that sulfate particles spewed into the stratosphere by strong volcanic eruptions could act in much the same way as ice crystals in polar stratospheric clouds by providing surfaces on which chlorine and bromine compounds can destroy ozone more efficiently than when they are floating free.

Studying the impact of volcanic ash in the aftermath of the 1982 eruption of El Chichon in Mexico, Solomon and

HOW OZONE-DEPLETING AGENTS ATTACK THE OZONE LAYER

Beginning in the stratosphere at an altitude of about 15 miles and extending up into the mesosphere, the 25-mile-wide ozone layer protects Earth by blocking out most of the sun's harmful ultraviolet light. Breakdown of ozone by chlorofluorocarbons and other chemicals allows harmful radiation to reach Earth.

1. Oxygen molecules in the stratosphere are transformed into ozone by solar ultraviolet (UV) radiation, which splits the oxygen molecule and releases highly reactive oxygen atoms. The free oxygen atoms then bind to oxygen molecules to form ozone molecules, which also are broken up by UV radiation. This continuous creation and destruction of oxygen and ozone occurs normally in the stratosphere.
2. Once certain chemicals, chiefly chlorofluorocarbons (CFCs), reach the ozone layer, UV radiation bombards the CFC molecule, breaking off an atom of chlorine.
3. The free chlorine atom attacks an ozone molecule, breaking off one of ozone's three oxygen atoms to form one chlorine monoxide molecule and leaving one oxygen molecule.
4. When the chlorine monoxide molecule encounters a free oxygen atom, produced during the natural mixing of oxygen and ozone, the oxygen atom breaks up the chlorine monoxide molecule and binds to its oxygen atom, forming a new oxygen molecule and leaving behind a free chlorine atom.
5. The newly freed chlorine atom can continue to destroy ozone molecules for many years. Oxygen molecules continue to break apart and form ozone, but this natural replenishing process is slowed in the presence of chlorine monoxide.
6. Because oxygen, unlike ozone, does not reflect UV radiation, the sun's potentially harmful UV rays penetrate the depleted areas of the ozone layer and reach Earth's surface.

Hoffman found that ozone concentrations over the middle latitudes were significantly depleted. They concluded that ozone depletion was likely following other major volcanic eruptions.

Although their research was limited to El Chichon, Solomon, a NOAA chemist in Boulder, Colo., says, "We found that similar processes could also take place on the liquid sulfuric acid and water particles that form following major volcanic eruptions."

The implications of Solomon and Hoffman's research are clear. While ice clouds form only over the polar regions, volcanic ash can travel anywhere. If volcanic ash does facilitate ozone depletion even in the absence of ice crystals, an ozone hole could open over any region on Earth.

In the summer of 1990, NASA reported that, globally, the ozone layer had been depleted by 2 to 3 percent over the previous two decades. It was also reported that the ozone layer had already begun to thin over the United States and other populated areas in the middle latitudes.

At the same time, the chemicals industry was quickly bringing into production substitutes for CFCs that are less damaging to the ozone layer. While not completely benign, these hydrochlorofluorocarbons, or HCFCs, were hailed as temporary substitutes for CFCs in many applications, particularly as coolants. Most important, the HCFCs and other substitute chemicals facilitated the rapid phaseout of CFCs.

Montreal Protocol nations were quick to respond to the news that ozone depletion was intensifying. In June 1990, in London, they amended the agreement to accelerate the phaseout of ozone-depleting chemicals. Under the new guidelines, all production and importation of CFCs and halons must stop by the year 2000. Other ozone-depleting agents, such as carbon tetrachloride and methyl chloroform, were added to the list of chemicals to be phased out of production. Developing countries still have an additional 10 years to meet the deadline. As a result of the new deadlines, chlorine pollution was expected by 2075 to fall below levels recorded prior to the first appearance of the ozone hole.

The amendments also addressed the special problems faced by developing nations. Although they produce few ozone-depleting chemicals, India, China and other countries have counted on introducing cheap refrigeration and air conditioning as part of their plans for modernization. They succeeded in convincing the industrial world to set up a fund to help them pay for the more expensive substitutes they will be forced to purchase, as well as information and equipment to help them produce environmentally sound refrigerators and air conditioners themselves.

Also in 1990, Congress passed the far-reaching Clean Air Act Amendments, which call for the complete phaseout of CFCs, halons and carbon tetrachloride by 2000, of methyl chloroform by 2002 and HCFCs by 2030. The law made the United States the first nation to legislate a ban on these chemicals. To reduce emissions of existing stores of ozone-depleting agents, the law called for regulations to require recycling of refrigerants and air-conditioning coolants. Finally, the new law mandated faster elimination of ozone-depleting substances if warranted by new scientific findings of damage to the ozone layer.

VAST AREA AT RISK

No sooner had the ink dried on the revisions to the Montreal Protocol than new information pointed to an even more dire situation. In October 1990, scientists found the lowest ozone levels ever recorded over Antarctica and discovered that the hole had stretched into southern Chile including Punta Arenas, a city of 100,000. There was also further evidence that parts of Australia had been exposed to high levels of UV radiation when bits of the ozone hole broke away as the polar vortex weakened and drifted northward from Antarctica.

On Oct. 22, the UNEP and the World Meteorological Organization announced that ozone depletion had begun to occur at the middle and high latitudes of both the Northern and the Southern Hemispheres in spring, summer and winter.

"Ozone depletion in the middle and high latitudes means that it covers almost all of North America, Europe, the Soviet Union, Australia, New Zealand and a sizable part of Latin America," said UNEP Director Mostafa K. Tolba. "The only area with no indication of change, that is, no visible reduction of ozone, is the tropical belt around the Earth."

European researchers, building on Solomon and Hoffman's volcano-ash findings, are now predicting that last year's eruption of Mount Pinatubo threatens to erode the ozone layer to dangerous levels over much of Europe this spring. Researchers participating in the 17-nation European Arctic Stratospheric Ozone Experiment based in northern Sweden have yet to complete their experiments. But they issued a recommendation in early February that governments in Northern Europe should take more urgent steps to protect the ozone layer.

The most recent signs of severe ozone loss were detected by NASA's Upper Atmosphere Research Satellite (UARS), launched last September [1991] to monitor the ozone layer and measure substances that destroy ozone. On Feb. 3, two months before the current study was scheduled for completion, NASA announced that the satellite had detected high levels of chlorine monoxide over Scandinavia and northern Eurasia, an area that includes London, Moscow and Amsterdam. The levels were comparable to concentrations found in the ozone hole over Antarctica.

NASA predicted that an ozone hole could open over the Northern Hemisphere this spring if chlorine monoxide levels remain high enough. Final results of the study are due in mid-April.

The bad news was not limited to the far north. NASA's satellite observations also showed ozone depletion over the tropics, which the agency suggested was due to plumes of ash from Mount Pinatubo. In addition, the satellite detected areas of low ozone across the western United States. These findings were confirmed by separate measurements taken in Boulder, Colo.

Confirming the satellite data were new findings from the NASA-led Airborne Arctic Stratospheric Expedition, which monitors ozone depletion from two specially equipped aircraft: the ER-2, a converted U-2 spy plane that gathers data at 70,000 feet, and the DC-8-72, a "flying lab" that operates at 41,000 feet. The expedition reported Feb. 3 that it had found even higher levels of chlorine monoxide than the satellite had over eastern Canada and northern New England. The readings—at 1.5 parts per billion by volume—surpass anything ever measured in either polar region.

"These findings have increased our concern that significant ozone loss will occur during any given winter over the Arctic in the next 10 years," scientists announced. "This is based on significant new data with improved instrumentation obtained with broader geographic and seasonal coverage and the knowledge that past release of CFCs will increase chlorine substantially in the stratosphere in the decade to come."

NOTES

1. The ozone layer is a 25-mile-wide band above the Earth with a high but uneven concentration of ozone gas. Starting at an altitude of about 15 miles, it shields humans and other organisms from the most harmful effects of the sun's ultraviolet (UV) radiation.

NO

James P. Hogan

OZONE POLITICS: THEY CALL THIS SCIENCE?

Every age has its peculiar folly: some scheme, project, or fantasy into which it plunges, spurred on by the love of gain, the necessity of excitement, or the mere force of imitation.

—Charles Mackay
Extraordinary Popular Delusions and the Madness of Crowds, 1841

Earlier centuries saw witch-hunting hysteria, the Crusades, gold stampedes, and the South Sea Bubble. Periodically, societies are seized by collective delusions that take on lives of their own, where all facts are swept aside that fail to conform to the expectations of what has become a self-sustaining reality. Today we have the environmentalist mania reaching a crescendo over ozone.

Manmade chlorofluorocarbons, or CFCs, we're told, are eating away the ozone layer that shields us from ultraviolet radiation, and if we don't stop using them now deaths from skin cancer in the United States alone will rise by hundreds of thousands in the next half century. As a result, 80 nations are about to railroad through legislation to ban one of most beneficial substances ever discovered at a cost the public doesn't seem to comprehend but that will be staggering. It could mean having to replace today's refrigeration and air-conditioning equipment with more expensive types running on substitutes that are toxic, corrosive, flammable if sparked, less efficient, and generally reminiscent of the things people heaved sighs of relief to get rid of in the 1930s. And the domestic side will be only a small part. The food industry that we take for granted depends on refrigerated warehouses, trucks, and ships. So do supplies of drugs, medicines, and blood. Whole regions of the sunbelt states have prospered during the last 40 years because of the better living and working environments made possible by air conditioning. And to developing nations that rely completely on modern food-preservation methods, the effects will be devastating.

Now, I'd have to agree that the alternative of seeing the planet seared by lethal levels of radiation would make a pretty good justification for whatever drastic action is necessary to prevent it. The only problem is, there isn't one

From James P. Hogan, "Ozone Politics: They Call This Science?" *Omni,* vol. 15, no. 8 (June 1993).

piece of solid, scientifically validated evidence to support the contention. The decisions being made are political, driven by media-friendly pressure groups wielding a power over public perceptions that is totally out of proportion to any scientific competence they possess. But when you ask the people who do have the competence to know—scientists who have specialized in the study of atmosphere and climate for years—a very different story emerges.

What they're saying, essentially, is that the whole notion of the ozone layer as something fixed and finite, to be eroded away at a faster or slower rate like shoe leather, is all wrong to begin with—it's simply not a depletable resource; that even if it were, the process by which CFCs are supposed to deplete it is highly speculative and has never been observed to take place; and even if it did, the effect would be trivial compared to what happens naturally. In short, there's no good reason for believing that human activity is having any significant effect at all.

To see why, let's start with the basics and take seashores as an analogy. Waves breaking along the coastline continually generate a belt of surf. The surf decomposes again, back into the ocean from where it came. The two processes are linked: Big waves on stormy days create more surf; the more surf there is to decay, the higher the rate at which it does so. The result is a balance between the rates of creation and destruction. Calmer days will see a general thinning of the surf line and possibly "holes" in more sheltered spots —but obviously the surf isn't something that runs out. Its supply is inexhaustible as long as oceans and shores exist.

In the same kind of way, ozone is all the time being created in the upper atmosphere—by sunshine, out of oxygen. A normal molecule of oxygen gas consists of two oxygen atoms joined together. High-energy ultraviolet radiation, known as UV-C, can split one of these molecules apart (a process known as photodissociation) into two free oxygen atoms. These can then attach to another molecule to form a three-atom species, which is ozone—produced mainly in the tropics above a 30-kilometer altitude where the ultraviolet flux is strongest. The ozone sinks and moves poleward to accumulate in lower-level reservoirs extending from 17 to 30 kilometers—the so-called ozone "layer."

Ozone is destroyed by chemical recombination back into normal oxygen—by reaction with nitrogen dioxide (produced in part by high-altitude cosmic rays), through ultraviolet dissociation by the same UV-C that creates ozone, and also by a less energetic band known as UV-B, which isn't absorbed in the higher regions. Every dissociation of an oxygen or ozone molecule absorbs an incoming UV-B photon, and that may be what gives this part of the atmosphere its ultraviolet screening ability.

Its height and thickness are not constant, but adjust automatically to accommodate variations in the incoming ultra-violet flux. When UV is stronger, it penetrates deeper before being absorbed; with weaker UV, penetration is less. Even if all the ozone were to suddenly vanish, there would still be 17 to 30 kilometers of hitherto untouched oxygen-rich atmosphere below, which would become available as a resource for new ozone creation, and the entire screening mechanism would promptly regenerate. As Robert Pease, professor emeritus of physical climatology at the University of California at Riverside, says, "Ozone in the atmosphere is not in finite supply." In other words, as in the case of surf with oceans and shores, it is inexhaustible for as long as sunshine and air continue to exist.

If ozone were depleting, UV intensity at the earth's surface would be increasing. In fact, actual measurements show that it has been decreasing—by as much as 8 percent in some places over the last decade.

Ordinarily, a scientific hypothesis that failed in its most elementary prediction would be dumped right there. But as Dr. Dixy Lee Ray—former governor of Washington state, chairman of the Atomic Energy Commission, and a scientist with the U.S. Bureau of Oceans and the University of Washington—put it: "There are fads in science. Scientists are capable of developing their own strange fixations, just like anyone else." Even though the physics makes it difficult to see how, the notion of something manmade destroying the ozone layer has always fascinated an apocalyptic few who have been seeking possible candidates for more than 40 years. According to Hugh Ellsaesser, guest scientist at the Atmospheric and Geophysical Sciences Division of the Lawrence Livermore National Laboratory, "There has been a small but concerted program to build the possibility of man destroying the ozone layer into a dire threat requiring governmental controls since the time of CIAP [Climatic Impact Assessment Program on the supersonic transport (SST), conducted in the early 1970s]."

In the 1950s, it was A-bomb testing; in the 1960s, the SST; in the 1970s, spacecraft launches and various chemicals from pesticides to fertilizers. All of these claimed threats to the destruction of the ozone layer were later discredited, and for a while, the controversy died out. Then, in 1985 and 1986, banner headlines blared that a huge ozone hole had been discovered in the Antarctic. This, it was proclaimed, confirmed the latest version of the threat.

In 1974, two chemists, Rowland and Molina at the University of California at Irvine, hypothesized that ozone might be attacked by CFCs—which had come into widespread use during the previous 20 years. Basically, they suggested that the same chemical inertness that makes CFCs noncorrosive, nontoxic, and ideal as a refrigerant would enable them to diffuse intact to the upper atmosphere. There, they would be dissociated by high-energy ultraviolet and release free atoms of chlorine. Chlorine will combine with one of the three oxygen atoms of an ozone molecule to produce chlorine monoxide and a normal two-molecule oxygen atom, thereby destroying the ozone molecule. The model becomes more insidious by postulating an additional chain of catalytic reactions via which the chlorine monoxide can be recycled back into free chlorine, hence evoking the specter of a single chlorine atom running amok in the stratosphere, gobbling up ozone molecules like Pac-Man.

Scary, vivid, sensational; perfect for activists seeking a cause, politicians in need of visibility; just what the media revel in. Unfortunately, however, it doesn't fit with a few vital facts. And if you claim to be talking about science, that's kind of important.

First, CFCs don't rise in significant amounts to where they need to be for UV-C photons to break them up. Because ozone absorbs heat directly from the sun's rays, the stratosphere exhibits a reverse temperature structure, or thermal "inversion"—it gets warmer with altitude rather than cooler. As Robert Pease points out, "This barrier greatly inhibits vertical air movements and the interchange of gases across the tropopause [the boundary between the lower atmosphere and the stratosphere], including CFCs. In the stratosphere, CFC gases decline rapidly and drop to only two percent of surface values by thirty kilometers of altitude. At the same time, less than two percent of the UV-C penetrates this deeply." Hence the number of CFC splittings is vastly lower than the original hypothesis assumes—for the same reason there aren't many marriages between Eskimos and Australian Aborigines: They don't mix very much.

For the UV photons that do make it, there are about 136 million oxygen molecules for them to collide with for every CFC—and every such reaction will create ozone, not destroy it. So even if we allow the big CFC molecule three times the chance of a small oxygen molecule of being hit, then 45 million ozone molecules will still be created for every CFC molecule that's broken up. Hardly a convincing disaster scenario, is it?

Ah, but what about the catalytic effect, whereby one chlorine atom can eat up thousands of ozone molecules? Doesn't that change the picture?

Not really. The catalysis argument depends on encounters between chlorine monoxide and free oxygen atoms. But the chances are much higher that a wandering free oxygen atom will find a molecule of normal oxygen rather than one of chlorine monoxide. So once again, probability favors ozone creation over ozone destruction.

At least 192 chemical reactions occur between substances in the upper stratosphere along with 48 different identifiable photochemical processes all linked through complex feedback mechanisms that are only partly understood. Selecting a few reactions brought about in a laboratory and claiming that this is what happens in the stratosphere (where it has never been measured) might be a way of getting to a predetermined conclusion. But it isn't science.

But surely it's been demonstrated! Hasn't a thousand times more chlorine been measured over the Antarctic than models say ought to be there?

Yes. High concentrations of chlorine—or to be exact, chlorine monoxide. But all chlorine atoms look alike. There is absolutely nothing to link the chlorine found over the Antarctic with CFCs from the other end of the world. What the purveyors of that story omitted to mention was that the measuring station at McMurdo Sound is located 15 kilometers downwind from Mount Erebus, an active volcano venting 100 to 200 tons of chlorine every day, and that in 1983 it averaged 1,000 tons per day. Mightn't that just have more to do with it than refrigerators in New York or air conditioners in Atlanta?

World CFC production is currently about 1.1 million tons annually—750,000 tons of which is chlorine. Twenty times as much comes from the passive outgassing of volcanoes. This can rise by a factor of ten with a single large eruption—for example that of Tambora in 1815, which pumped a minimum of 211 million tons straight into the atmosphere. Where are the records of all the cataclysmic effects that should presumably have followed from the consequent ozone depletion?

And on an even greater scale, 300 million tons of chlorine are contained in spray blown off the oceans every year. A single thunderstorm in the Amazon region can transport 200 million tons of air per hour into the atmosphere, containing 3 million tons of water vapor. On average 44,000 thunderstorms occur daily, mostly in the tropics. Even if we concede to the depletion theory and allow this mechanism to transport CFCs also, compared to what gets there naturally, the whiff of chlorine produced by all of human industry (and we're only talking about the leakage from it) is a snowflake in a blizzard.

Despite all that, isn't it still true that a hole has appeared in the last ten years and is getting bigger? What about that, then?

In 1985, a sharp, unpredicted decline was reported in the mean depth of ozone over Halley Bay, Antarctica. Although the phenomenon was limited to altitudes between 12 and 22 kilometers and the interior of a seasonal circulation of the polar jet stream known as the "polar vortex," it was all that the ozone-doomsday pushers needed. Without waiting for any scientific evaluation or consensus, they decided that this was the confirmation that the Rowland-Molina conjecture had been waiting for. The ominous term "ozone hole" was coined by a media machine well rehearsed in environmentalist politics, and anything the scientific community had to say has been drowned out.

Missing from the press and TV accounts, for instance, is that an unexpectedly low value in the Antarctic winter-spring ozone level was reported by the British scientist Gordon Dobson in 1956—when CFCs were barely in use. In a 40-year history of ozone research written in 1968, he notes: "One of the most interesting results... which came out of the IGY [International Geophysical Year] was the discovery of the peculiar annual variation of ozone at Halley Bay." His first thought was that the result might have been due to faulty equipment or operator error. But when such possibilities were eliminated and the same thing happened the following year, he concluded: "It was clear that the winter vortex over the South Pole was maintained late into the spring and that this kept the ozone values low. When it suddenly broke up in November, both the ozone values and the stratosphere temperatures suddenly rose." A year after that, in 1958, a similar drop was reported by French scientists at the Antarctic observatory at Dumont d'Urville—larger than that causing all the hysteria today.

These measurements were on the edge of observational capability, especially in an environment such as the Antarctic, and most scientists regarded them with caution. After the 1985 "discovery," NASA reanalyzed its satellite data and found that it had been routinely throwing out low Antarctic ozone readings as "unreliable."

The real cause is slowly being unraveled, and while some correlation is evident with volcanic eruptions and sunspot cycles, the dominant factor appears to be the extreme Antarctic winter conditions, as Dobson originally suspected. The poleward transportation of ozone from its primary creation zones over the tropics does not penetrate into the polar vortex, where chemical depletion can't be replaced because of the lack of sunshine. Note that this is a localized minimum relative to the surrounding high-latitude reservoir regions, where global ozone is thickest. As Hugh Ellsaesser observes, "The ozone hole... leads only to spring values of ultraviolet flux over Antarctica... a factor of two less than those experienced every summer in North Dakota."

But isn't it getting bigger every year? And aren't the latest readings showing depletion elsewhere, too?

In April, 1991, EPA Administrator William Reilly announced that the ozone layer over North America was thinning twice as fast as expected and produced the figures for soaring deaths from skin cancer. This was based on readings from NASA's Nimbus-7 satellite. I talked to Dr. S. Fred Singer of the Washington-based Science and Environmental Policy Project, who developed the principle of UV backscatter that the ozone monitoring instrument aboard Nimbus-7 employs. "You simply cannot tell from one sunspot cycle," was his comment. "The data are too noisy. Scientists need at least one more cycle of satellite observations before they can establish a trend." In other words the trend exists in the eye of the determined beholder, not in any facts he beholds.

February 1992 saw a repeat performance when a NASA research aircraft detected high values of chlorine monoxide in the northern stratosphere. Not of CFCs; nor was there any evidence that ozone itself was actually being depleted, nor any mention that the Pinatubo volcano was active at the time. Yet almost as if on cue, the U.S. Senate passed an amendment only two days later calling for an accelerated phaseout of CFCs. (It's interesting to note that NASA's budget was under review at the time. After getting its increase, NASA has since conceded that perhaps the fears were premature.)

But apart from all that, yes, world mean-total ozone declined about 5 percent from 1979 to 1986. So what? From 1962 to 1979 it increased by 5½ percent. And since 1986, it has been increasing again (although that part's left out of the story the public gets). On shorter time scales, it changes naturally all the time and from place to place, hence surface ultraviolet intensity is not constant and never was. It varies with latitude—for instance, how far north or south from the equator you are—with the seasons, and with solar activity. And it does so in amounts that are far greater than those causing all the fuss.

The whole doomsday case boils down to claiming that if something isn't done to curb CFCs, ultraviolet radiation will increase by 10 percent over the next 20 years. But from the poles to the equator, it increases naturally by a whopping factor of 50, or 5,000 percent, anyway! —equivalent to 1 percent for every six miles. Or to put it another way, a family moving from New York to Philadelphia would experience the same increase as is predicted by the worst-case depletion scenarios. Alternatively, they could live 1,500 feet higher in elevation—say, by moving to their summer cabin in the Catskills.

Superposed on this is a minimum 25-percent swing from summer to winter, and on top of that, a 10- to 12-year pattern that follows the sunspot cycle. Finally, there are irregular fluctuations caused by the effects of volcanic eruptions, electrical storms, and the like on atmospheric chemistry. Expecting to find some "natural" level that shouldn't be deviated from in all this is like trying to define sea level in a typhoon.

Skin cancer is increasing, nevertheless. Something must be causing it.

An increasing rate of UV-induced skin cancer means that more people are receiving more exposure than they ought to. It doesn't follow that the intensity of ultraviolet is increasing as it would if ozone were being depleted. (In fact, it's decreasing, as we saw earlier.) Other considerations explain the facts far better, such as that sun worship has become a fad among light-skinned people only in the last couple of generations, or the migrations in comparatively recent times of peoples into habitats for which they aren't adapted: for instance, the white population of Australia. (Native Australians have experienced no skin-cancer increase.)

Deaths from drowning increase as you get nearer the equator—not because the water becomes more lethal but because human behavior changes: Not many people go swimming in the Arctic. Nevertheless, when it comes to skin cancer, the National Academy of Sciences [NAS] has decided that only variation of UV matters. And from the measured ozone thinning from poles to equator and the change in zenith angle of the sun they determined that a 1-percent decrease in ozone equates to a 2-percent rise in skin cancer.

How you make a disaster scenario out of this, according to Ellsaesser, is to ignore the decline in surface UV actually measured over the last 15 years, ignore the reversal that shows ozone to have been increasing again since 1986, and extend the 1979–1986 slope as if it were going to continue for the next 40 years. Then, take the above formula as established fact and apply it to the entire U.S. population. Witness: According to the NAS report (1975), approximately 600,000 new cases of skin cancer occur annually. So, by the above, a 1-percent ozone decrease gives 12,000 more skin cancers. Projecting the 5-percent ozone swing from the early 1980s through the next four decades gives 25 percent, hence a 50-percent rise in skin cancer, which works out at 300,000 new cases in the year 2030 A.D., or 7.5 million over the full period. Since the mortality rate is around 2.5 percent, this gives the EPA's "200,000 extra deaths in the United States alone." Voilà: Instant catastrophe.

As if this weren't flaky enough, it's possible that the lethal variety of skin cancer has little to do with UV exposure, anyway. The cancers that are caused by radiation are recognizable by their correlation with latitude and length of exposure to the sun and are relatively easily treated. The malignant melanoma form, which does kill, affects places like the soles of the feet as well as exposed areas, and there is more of it in Sweden than in Spain.

So, what's going on? What are publicly funded institutions that claim to be speaking science doing, waving readings known to be worthless (garbage in, gospel out?), faking data, pushing a cancer scare that contradicts fact, and force-feeding the public a line that basic physics says doesn't make sense? The only thing that comes through at all clearly is a determination to eliminate CFCs at any cost, whatever the facts, regardless of what scientists say.

Would it come as a complete surprise to learn that some very influential concerns stand to make a lot of money out of this? The patents on CFCs have recently run out, so anybody can now manufacture them without having to pay royalties. Sixty percent of the world CFC market is controlled by four companies who are already losing revenues and market share to rapidly growing chemicals industries in the Third World, notably Brazil, South Korea, and Taiwan. Some hold the patents on the only substitutes in sight, which will restore monopoly privileges once again if CFCs are outlawed. Mere coincidence?

Ultraviolet light has many beneficial effects as well as detrimental. For all any one knows, the increase that's being talked about could result in more overall good than harm. But research proposals to explore that side of things are turned down, while doomsayers line up for grants running into hundreds of millions. The race is on between chemicals manufacturers to come up with a better CFC substitute while equipment suppliers will be busy for years. Politicians are posturing as champions of the world, and the media are having a ball.

As Bob Holzknecht, a Florida engineer in the CFC industry for 20 years observes, "Nobody's interested in reality. Everyone who knows anything stands to gain. The public will end up paying through the nose, as always, but the public is unorganized and uninformed."

Good science will be the victim, too, of course. But science has a way of winning in the end. Today's superstitions can spread a million times faster than anything dreamed of by the doom prophets in days of old. But the same technologies which make that possible can also prove equally effective in putting them speedily to rest.

POSTSCRIPT

Is Ozone Depletion a Genuine Threat?

The critics of the "ozone scare" have made very little headway against the evidence. In November 1992 the Montreal Protocol was strengthened, and the deadline for ending production of CFCs was moved up by as much as nine years. By December 92 nations had ratified the treaty.

In March 1993 researchers reported a drop in Northern Hemisphere ozone concentrations that was worse than the one that kicked off the 1992 alarm discussed by Cooper. In April NASA researchers reported record thinning of the ozone layer worldwide. By autumn researchers had added that the 1993 ozone hole over the Antarctic was fully 15 percent worse than the one that had appeared the year before.

Some critics have suggested that the decline in ozone is a perfectly natural response to chlorine injected into the atmosphere (as hydrochloric acid) by volcanic eruptions or that volcanic chlorine could be making the problem worse than it would otherwise be. Separate studies done in May and August 1993, however, indicated that CFCs are the cause of the problem and that volcanic activity would have very little effect on the ozone.

In 1995 F. Sherwood Rowland and Mario J. Molina, with Paul Crutzen, shared the Nobel Prize in Chemistry for discovering the hazards that CFCs pose to the ozone layer. The Nobel committee said that their work contributed to "our salvation from a global environmental problem that could have catastrophic consequences." At the same time, computer models suggested that the ozone hole was not likely to grow any worse—that indeed the problem was on the mend, and over the next 50 years the stratosphere would return to normal.

Some people still refuse to cooperate with the ban on CFCs. In March 1996 the U.S. Customs Service said that smuggling of CFCs, especially Freon (the chief coolant for air conditioners), is now second only to illegal drugs as a smuggling problem.

For more reports on ozone depletion, see Owen B. Toon and Richard P. Turco, "Polar Stratospheric Clouds and Ozone Depletion," *Scientific American* (June 1991) and Don Hinrichsen, "Stratospheric Maintenance: Fixing the Ozone Hole Is a Work in Progress," *Amicus Journal* (Fall 1996). Two articles by critics of the "ozone scare" are Gary Taubes, "The Ozone Backlash," *Science* (June 11, 1993) and Ronald Bailey, "The Hole Story: The Science Behind the Scare," *Reason* (June 1992). For an excellent report on how the consensus has tipped away from criticism and toward the need for action, see Hilary F. French, "Learning from the Ozone Experience," in the Worldwatch Institute's *State of the World 1997* (W. W. Norton, 1997).

ISSUE 7

Are Electromagnetic Fields Dangerous to Your Health?

YES: Paul Brodeur, from *The Great Powerline Coverup: How the Utilities and the Government Are Trying to Hide the Cancer Hazard Posed by Electromagnetic Fields* (Little, Brown, 1993)

NO: Edward W. Campion, from "Power Lines, Cancer, and Fear," *The New England Journal of Medicine* (July 3, 1997)

ISSUE SUMMARY

YES: Writer Paul Brodeur argues that there is an increased risk of developing cancer from being exposed to electromagnetic fields (EMFs) given off by electric power lines and that the risk is significant enough to warrant immediate measures to reduce exposures to the fields.

NO: Physician Edward W. Campion argues that there is no credible evidence that there is any risk of developing cancer from EMF exposure and that it is time to stop wasting research resources on further studies.

Electromagnetic fields (EMFs) are emitted by any device that uses electricity. They weaken rapidly as one gets further from the source, but they can be remarkably strong close to the source. Users of electric blankets, before the blankets were redesigned to minimize EMFs, were among those who were most exposed to EMFs. People who use computers regularly are another highly exposed population. And, since EMF strength depends also on how much electricity is flowing through the source, so are people who live near power lines, especially high-tension, long-distance transmission lines.

Early research shows the difficulties of nailing down any possible side effects of EMF exposure. In 1979 researchers at the University of Colorado Health Center in Denver, Colorado, reported that, in a study of 344 childhood cancer deaths, children whose homes were exposed to higher EMF levels were two to three times more likely to die of leukemia or lymphoma. At the time, however, no one could suggest any mechanism by which EMFs could cause cancer, especially since the body generates its own EMFs of strength similar to those produced in the body by high-tension lines. Some other studies found similar links between EMF exposure and cancer; some did not.

Inconsistency has been the curse of research in this area. Speaking on research into the effects of extremely low frequency (ELF) EMFs on cells in the laboratory (which was performed in an effort to find mechanisms by which EMFs might cause cancer), Larry Cress of the U.S. Food and Drug Administration's Center for Devices and Radiological Health said, "Many researchers have been able to reproduce their effects most, but not all, of the time. And we don't see a dose response, as with some radiation, such as x-ray. Or, one laboratory may see an *increase* in something in a cell when the field is turned on, while another laboratory sees a corresponding *decrease* when the field is turned on." See Dixie Farley, "The ELF in Your Electric Blanket [and Other Appliances]," *FDA Consumer* (December 1992).

In 1992 the Committee on Interagency Radiation Research and Policy Coordination, an arm of the White House's Office of Science and Technology Policy, released *Health Effects of Low Frequency Electric and Magnetic Fields,* a report that concluded, "There is no convincing [published] evidence... to support the contention that exposures to extremely low frequency electric and magnetic fields generated by sources such as household appliances, video terminals, and local powerlines are demonstrable health hazards."

However, at about the same time, Swedish researchers announced that a study of leukemic children showed an association between their disease and the distances of their homes from power lines. The researchers also reported finding that the risk of leukemia increases in adults with exposure to EMFs in the workplace. Critics have objected that the correlations in such studies are weak—that they could easily be due to nothing more than coincidence or that they might reflect exposure to something other than EMFs whose levels nevertheless fluctuate in step with EMF levels (perhaps herbicides used to control the growth of vegetation under power lines or vapors given off by electrical insulation).

In 1996 Jon Palfreman, in "Apocalypse Not," *Technology Review* (April 1996), summarized the controversy and the evidence against any connection between cancer and EMFs. On July 3, 1997, *The New England Journal of Medicine* published a report by Martha S. Linet et al. entitled, "Residential Exposure to Magnetic Fields and Acute Lymphoblastic Leukemia in Children," in which the authors failed to find any support for such a connection.

Yet the associations are there for scientists, as well as for journalists such as Paul Brodeur, to consider. In July 1990 Brodeur published a long article in *The New Yorker* in which he describes clusters of cancer cases that seemed to be linked to EMFs from power lines and reviews both the evidence and the responses of public utility representatives. In a later article, reprinted here, Brodeur summarizes his earlier report, adds further cases, and urges immediate measures to reduce what he feels are dangerous EMF exposures.

In an editorial that accompanied Linet et al.'s 1997 report, Edward W. Campion, an assistant professor of medicine at Harvard Medical School and Massachusetts General Hospital, argues that there is no credible evidence of cancer risk from EMF exposure and that it is time to stop wasting research resources on further studies.

YES

Paul Brodeur

THE GREAT POWERLINE COVERUP

In my Annals of Radiation about the health hazard posed by the sixty-hertz magnetic fields that are given off by high-current and high-voltage power lines (July 9, 1990) I cited evidence suggesting that a cancer cluster had occurred among residents of Meadow Street in Guilford, Connecticut. During the past twenty years, seven tumors—two malignant brain tumors, two cases of meningioma (a rare and generally nonmalignant tumor of the brain), a malignant eye tumor, an ovarian tumor, and a tumor of the tibia—have been recorded among children and adults living on that street, which is only two hundred and fifty yards long and has only nine houses on it. Because all seven tumors developed in people who were living or had lived for significant periods of time in five of six adjacent houses situated near an electric-power substation and next to some main distribution power lines carrying high current from the substation, I suggested that the cancer among the residents of Meadow Street was associated with chronic exposure to the magnetic fields that are given off by such wires. To support that contention, I cited the fact that during the past decade some two dozen epidemiological studies had been conducted and published in the medical literature of the United States and other parts of the world showing that children and workers exposed to power-line magnetic fields were developing cancer—chiefly leukemia, lymphoma, melanoma, brain tumors, and other central-nervous-system cancers—at rates significantly higher than those observed in unexposed people, and the fact that between 1985 and 1989 no fewer than twelve studies had shown more brain tumors than were to be expected among people exposed to electric and magnetic fields at home or at work.

At a public meeting held in the Guilford Public Library on August 20th, David R. Brown, chief of the Connecticut Department of Health Services' Division of Environmental Epidemiology and Occupational Health, and Sandy Geschwind, an epidemiologist with the division, declared that there was no cancer cluster on Meadow Street. To support their contention, they distributed a document entitled "Guilford Cancer Cluster Preliminary Investigation," claiming that "there was not a cluster of the same kind of tumors on Meadow Street," and that from 1968 through 1988 "Guilford as a whole did not experience a higher than expected number of brain cancer or meningioma cases." The document stated, further, that "mapping of these brain tumor and meningioma cases showed that they did not cluster in a particular area but were scattered throughout the town."

At the meeting, Geschwind gave a presentation in which she said that one of the brain cancers on Meadow Street was not a primary tumor but an esophageal cancer that had metastasized. She also said that the malignant eye tumor in question was a melanoma—a type of cancer that she claimed had never been associated with exposure to electromagnetic fields—and she assured her listeners that meningioma had never been associated with exposure to such fields. Toward the end of her presentation, Geschwind displayed a map showing the location of ten meningiomas and nineteen other brain and central-nervous-system tumors listed by the Connecticut Tumor Registry as having occurred in Guilford between 1968 and 1988, and told the hundred or so members of her audience—they included a dozen newspaper and television reporters—that the map proved that there was "absolutely no clustering" in Guilford and that the state investigation showed "no cancer cluster on Meadow Street."

However, the fact that Guilford as a whole—the town now has a population of twenty thousand five hundred, living in seventy-three hundred dwellings—did not experience a higher than expected number of meningiomas and other brain and nervous-system tumors during those twenty-one years does not address the situation on Meadow Street. Second, while there is no reason to doubt Geschwind's assertion that one of the two brain cancers among Meadow Street residents was not a primary tumor, eye melanoma—the one in question was a malignant tumor involving the optic nerve, an extension of the brain—has been found to be "notably high for electrical and electronics workers," who are known to be exposed to strong magnetic fields. The finding appeared in a highly regarded study entitled "Epidemiology of Eye Cancer in Adults in England and Wales, 1962–1977," which was conducted by Dr. A. J. Swerdlow, a physician at the Department of Community Medicine of the University of Glasgow, in Scotland. Swerdlow reported his findings in 1983, in Volume 118, No. 2, of the *American Journal of Epidemiology,* which is published by the Johns Hopkins University School of Hygiene and Public Health, in Baltimore. Moreover, melanoma of the skin is one of three types of cancer listed by scientists of the Environmental Protection Agency in a recent draft report, "An Evaluation of the Potential Carcinogenicity of Electromagnetic Fields," as being prevalent among workers in electrical and electronic occupations, and thus associated with exposure to magnetic fields.

The conclusion of Brown and Geschwind that there is no cancer cluster among people who have lived on Meadow Street seems disingenuous, to say the least. As Geschwind noted, the Connecticut Tumor Registry recorded ten cases of meningioma and nineteen other primary tumors of the brain and central nervous system among Guilford residents between 1968 and 1988—a span in which the average population of the town was seventeen thousand five hundred. Thus the meningioma rate in Guilford is consistent with the Connecticut statewide incidence, of 2.6 cases per hundred thousand people per year—I was in error when I gave it in my article as one case per hundred thousand—and the incidence of other brain and central-nervous-system tumors in Guilford is also close to the number that would normally be expected. The fact that three of the twenty-nine primary brain and central-nervous-system tumors that occurred in Guilford during those twenty-one years developed among a handful of people who lived in four of five adjacent houses on Meadow Street that are situated near a substation and very close to a pair of high-current distribution lines, called feeders, together with the fact that a malignant eye tumor, involving a tract of brain tissue, occurred in a woman who had lived in a sixth adjacent dwelling, next to a third feeder line, surely suggests that there is a cancer cluster of some significance on Meadow Street.

Finally, and somewhat ironically, further evidence of cancer clustering associated with exposure to power-line magnetic fields can be found in the very map that Geschwind displayed in an effort to persuade the people of Guilford that no cancer cluster existed there. Among those listening to her presentation was Robert Hemstock, a Guilford resident, who, in January of this year, first sounded the alarm about a cluster on Meadow Street. When Geschwind held up the map, Hemstock noticed that three of the twenty-nine cancers on it appeared to have occurred along the route of a feeder line that carried high current from the Meadow Street substation to other towns during the nineteen-sixties, seventies, and early eighties, when the substation was being operated by its owner, the Connecticut Light & Power Company, as a bulk-supply station for large-load areas in Madison and Clinton—neighboring towns with a total population of about twenty thousand during that period. He also noticed that an unusually large proportion of the other brain tumors on the map appeared to have occurred among people living along the routes of other primary distribution lines emanating from the substation.

After the meeting, Hemstock shared his observation with Don Michak, a reporter for the Manchester Enfield *Journal Inquirer,* who on August 23rd asked the Department of Health Services for a copy of the map. As it happened, Brown had displayed the map the day before at a Rotary Club meeting in Guilford, and told the Rotarians that he saw no need for the department to make any further inquiry into the incidence of cancer on Meadow Street. However, Health Services officials refused to release the map to Michak, on the ground that to do so might violate the confidentiality of cancer victims by revealing their addresses. The *Journal Inquirer* reported this development in an article by Michak on September 6th, and on September 10th it published an editorial pointing out that if the withheld map showed that the distribution of cancer cases in Guilford corresponded to the Meadow Street substation and to a power line running north from it "the public's concern might be overwhelming not only in Guilford but throughout Connecticut and even nationally." The editorial went on to question Health Services' rationale for secrecy, declaring that the map "is just a matter of dots superimposed on a map of Guilford; it apparently doesn't include names and addresses," and that "anyone seeking to use the map to find people who have or had cancer would have to knock on doors in the area of the dots on the map and ask such people to identify themselves." After observing that "the health department undermined its own rationale by displaying the map at the public hearing in Guilford in the first place," the editorial concluded by stating that if the department failed to make the map available "the public will have to assume that the department wants to protect something else more than it wants to protect public health."

In September, a reporter for the New Haven *Register* obtained a copy of the map from an assistant to the Guilford health officer. (The assistant later said that she had given it out by mistake.) The *Register* reporter also went to the Connecticut Light & Power Company's office in Madison and obtained a company map of the routes of existing high-current and high-voltage distribution lines in Guilford. On October 3rd, the *Register* published its own map—one combining the locations of the brain tumors and other central-nervous-system tumors with the routes of Connecticut Light & Power's distribution lines. It clearly showed that Hemstock's observation was correct—that an inordinately high number of the meningiomas and other brain and central-nervous-system tumors that had occurred in Guilford over the twenty-one-year period between 1968 and 1988 had developed in people living close to primary distribution wires.

This correlation notwithstanding, Brown and Geschwind denied that the map furnished any evidence of a link between the occurrence of such tumors and proximity to power lines in Guilford. "You can't use the map to show that kind of association," Geschwind told the *Register.* She added that such tumors could be found on streets near main distribution power lines because those streets were densely populated, and heavily populated areas would have proportionally higher cancer rates.

To the contrary, anyone who knows the addresses of the twenty-nine brain-and-other-central-nervous-system-tumor victims in Guilford, and follows the routes of the feeder lines and primary distribution wires leading from the Meadow Street substation, will find not only that there is a strong correlation between the occurrence of these tumors and living close to high-current or high-voltage wires but also that most of the tumors have not occurred in areas of notably dense population. The feeder that carried high current from the substation to Madison and Clinton was abandoned a few years ago; it ran across Meadow Street from the substation and proceeded east for about a mile and a half, to a point near the junction of Stone House Lane, South Union Street, and Sawpit Road. (Up to that point, the poles and the wires of the line remain in place, but they have been removed from the rest of the route—across an uninhabited salt marsh and the East River, which is the eastern boundary of Guilford, to a substation on Garnet Park Road, in Madison.) This feeder line ran for a mile and a half through Guilford, and it passed close —within a hundred and fifty feet or so —to only twelve houses. One of the ten meningiomas and two of the nineteen other brain and central-nervous-system tumors listed by the Tumor Registry as having occurred in Guilford between 1968 and 1988 afflicted people living in three of those twelve dwellings. All three are situated within about forty feet of the high-current wires. Moreover, a former Meadow Street resident who developed eye cancer at the age of forty-four, and has since died of it, lived for fourteen years in one of the twelve houses close to the abandoned feeder line. It is at 56 Meadow, and is situated only about thirty feet from the wires....

All told, seven of the ten meningiomas and ten of the nineteen other brain and central-nervous-system tumors—that is, seventeen of the total of twenty-nine—have afflicted people living near high-current or high-voltage power lines in Guilford. The total combined length of the lines is about forty-five miles, and along this distance some seven hundred and twenty-two out of a total of eight hundred and six houses are situated within a hundred and fifty feet of the wires. It seems obvious that in a town of seventy-three hundred dwellings the occurrence of this proportion of meningiomas and other brain and central-nervous-system tumors in residents of just over eight hundred dwellings strung out along some forty-five miles of roadway cannot be ascribed to heavy population—as the Connecticut Department of Health Services has done. It also seems obvious that people living in houses close to high-current wires and high-voltage transmission lines in Guilford are especially susceptible to developing meningiomas and other brain tumors. Particularly disturbing in this regard is the fact that in March of 1989—too late to be counted among the twenty-nine tumors listed by the Registry on the map that the Connecticut Department of Health Services displayed to reassure the townspeople of Guilford—a seventeen-year-old girl living in a house close to one of the high-current feeder lines was found to be suffering from an astrocytoma, the same type of malignant brain tumor that has afflicted a seventeen-year-old girl living near the same line on Meadow Street.

Instead of continuing to extend the presumption of benignity to power-line magnetic fields, the Connecticut Department of Health Services could require its Division of Environmental Epidemiology and Occupational Health to conduct a thorough study of the apparent strong association between the occurrence of meningiomas and other brain and central-nervous-system tumors, on the one hand, and, on the other, chronic exposure to the magnetic fields given off by high-current and high-voltage power lines in Guilford. Moreover, since Connecticut is one of the few states that have collected data on the occurrence of such tumors over a significant period, the department has a unique opportunity to perform an important service for public health nationwide by conducting a detailed investigation of the seventeen hundred and three meningiomas and the four thousand one hundred and two other brain and central-nervous-system tumors that have been diagnosed among Connecticut residents over the twenty-one years between 1968 and 1988, in order to determine whether, as is clearly the case in Guilford, a disproportionately high percentage of them have developed in people living close to wires giving off strong magnetic fields. If such an association should prevail throughout the state, meningioma and other brain tumors would have to be considered marker diseases for exposure to power-line magnetic fields.

* * *

Later in my article I described a cluster of seven brain cancers that had been reported to have occurred over the past fifteen or twenty years among the residents of Trading Ford and Dukeville—two small communities near Salisbury, in Rowan County, North Carolina—who had either worked at a nearby power-generating plant, owned by the Duke Power Company, or lived in a company village, Dukeville, that was situated close to the plant and adjacent to a large substation and some high-voltage transmission lines giving off strong magnetic fields. I suggested that officials of the North Carolina Department of Environment, Health, and Natural Resources' environmental-epidemiology section were remiss in not having investigated this brain-cancer cluster during the eight and a half months since it was reported in the Salisbury *Post* on July 12 and 18, 1989, especially since one of the officials, Dr. Peter D. Morris, had made a point of stating that such a cluster might be significant if all the cancer victims had worked in the same plant twenty years earlier. I also suggested that the health experience of the three hundred or so people who lived in the company village or worked at the plant, or did both, should be thoroughly investigated, because, in addition to the seven of those people who had died of brain cancer, four others, who simply lived near the plant or the high-voltage transmission lines radiating from it, had died of the disease, and because a preliminary inquiry revealed that there had also been at least eight deaths from leukemia, lymphoma, and other cancers among these people.

In a recent letter to the editor of *The New Yorker* three officials of the environmental-epidemiology section stated that they had evaluated the seven cases of brain cancer, in order to "determine whether or not they should be included in our study of brain cancer in Rowan County from 1980 through 1989," and had found that "two of the seven cases had metastatic brain cancer, a different type of tumor originating in another part of the body and later spreading to the brain." They went on to say that four of the remaining cases were excluded from their study because the diagnoses of two of them were made prior to 1979, an unconfirmed diagnosis of another was made prior to 1979, and one of the victims lived outside Rowan County at the time of diagnosis.

In the final report of their study, which is entitled "Rowan County Brain Cancer Investigation," the North Carolina health officials state that Rowan County did not have a significantly greater incidence of malignant brain cancer between 1980 and 1989 than each of the five surrounding counties. During a press conference at the Rowan County Health Department on October 25th, Dr. Morris told the Salisbury *Post* that brain cancer in the Trading Ford–Dukeville area during the ten-year period "was not studied as a separate cluster."

The rationale of the North Carolina health officials is as faulty as that of their counterparts in Connecticut, because they not only have failed to address the brain-cancer situation in Trading Ford and Dukeville in its entirety but also have submerged the small part they did address in the larger study of Rowan County. In order to understand how flawed their investigation has been, one must remember that the power plant, which was built in 1926, was partly shut down during the nineteen-fifties and sixties, and the eighty-six houses in Dukeville, which were built between 1926 and 1945, were moved elsewhere in 1955. Thus, in addition to the one case of primary brain cancer among Trading Ford and Dukeville residents that the North Carolina officials included in their study, and the four cases of brain cancer that they saw fit to exclude, other people who were exposed to the electric and magnetic fields from the plant, its substation, and its high-voltage transmission lines by virtue of working at the plant or living in the company village during the nineteen-thirties, forties, and fifties may well have developed the disease and died of it before 1979. By deciding not to include brain cancers diagnosed among residents of the Trading Ford–Dukeville area before 1979, the North Carolina health officials decided not to investigate the health experience of people who worked at or lived near the Duke Power Company plant—a decision that makes about as much epidemiological sense as a decision to study the incidence of gray hair in a given population after excluding all those persons in the study group who became gray more than ten years earlier.

* * *

Still later in my article I wrote that cancer among the student population of the Montecito Union School—an elementary school with four hundred pupils in Montecito, California—was "at least a hundred times what might have been expected." This was an error. The incidence of cancer at the school is considerably less than that, though far greater than it should be. Between 1981 and 1988, six cases of cancer are known to have occurred among children who attended the Montecito Union School: two children developed leukemia; three children developed lymphoma; and one child developed testicular cancer. As I wrote in my article, cancer is a rare event in children, occurring annually in about one of ten thousand children per year under the age of fifteen. However, as several readers have pointed out, the child-years at risk should be calculated at eight times four hundred students per year; that comes to thirty-two hundred child-years at risk. Six cases of cancer out of thirty-two hundred child-years translates to 18.75 cases per ten thousand children per year. According to the National Cancer Institute's Surveillance, Epidemiology and End Results (SEER) data, the all-sites cancer rate for white children of both sexes, aged five to nine, between 1983 and 1987 in the San Francisco–Oakland area (the closest metropolitan area to Santa Barbara for which SEER data exist) was 11.9 cases per hundred thousand children per year. Thus the cancer rate over those eight years at the Montecito Union School—18.75 cases per ten thousand—is more than fifteen times the expected rate.

In their assessment of this cancer cluster officials of California's Department of Health Services' environmental-epidemiology-and-toxicology branch have maintained that magnetic-field levels at the school—which is situated within forty feet of a sixty-six-thousand-volt feeder line originating at an adjacent substation—were not unusually high, and that there was no evidence that they posed a health hazard. The fact is, however, that magnetic-field levels measured at the school's kindergarten patio by Enertech, an engineering consulting firm in the Bay Area, were between four and six milligauss; that is, approximately twice the levels that have been associated with a doubling of the expected rate of childhood cancer in three epidemiological studies cited by staff scientists of the Environmental Protection Agency as providing the strongest evidence that there may be a causal relationship between certain forms of childhood cancer —chiefly leukemia, nervous-system cancer, and lymphoma—and exposure to power-line magnetic fields. (Incidentally, on February 26th of this year I measured the magnetic fields at the kindergarten patio of the Montecito Union School, and found them to be about the same as those reported by Enertech.) It is also a fact that the magnetic-field levels at the kindergarten patio are at least equal to, and, for the most part, greater than, the exposure levels of forty-five hundred New York Telephone Company cable splicers, in whom cancer of all types—particularly leukemia—has been found to be higher than expected.

California health officials decided not to include the case of testicular cancer, which occurred in a second-grader, in their assessment of the cancer hazard at the Montecito Union School, and that decision seems arbitrary, in the light of the fact that cancer of all types was elevated in the childhood-cancer studies cited by the E.P.A. and also in the study of the telephone-cable splicers. It seems all the more arbitrary in the light of SEER data that estimate the chances of a seven- or eight-year-old child's developing testicular cancer to be nearly zero in one hundred thousand children per year. Also disturbing is the fact that since the publication of my article a teacher's aide with several years' experience in the kindergarten of the Montecito Union School has developed a brain tumor. This occurrence, together with the fact that four cases of leukemia have been reported among children who attended the Montecito Union School in the late nineteen-fifties, should encourage the California officials to conduct a full-scale investigation of the health experience of all the children who have attended this school during the past thirty-five years, just as the cancer clusters that have been found among the residents of Guilford and Dukeville should occasion in-depth investigations of the health experience of all the people who have lived near high-voltage and high-current power lines in those communities over a similarly appropriate period.

While these studies are in progress, interim preventive measures should be undertaken to reduce the magnetic-field exposure of children in hundreds of schools and day-care centers across the nation which have been built perilously close to high-voltage and high-current power lines. That can be accomplished by rerouting such lines, or burying them in a manner that will prevent hazardous magnetic-field emissions. Needless to say, such measures should be supported by the parents of schoolchildren, by members of parent-teacher associations, and by officials of school districts, of city and state health departments, and of the federal Environmental Protection Agency.

NO

Edward W. Campion

POWER LINES, CANCER, AND FEAR

Over the past 18 years, there has been considerable interest in the possible link between electromagnetic fields and cancer, especially leukemia. The story of this highly publicized research has been marked by mystery, contradiction, and confusion. When something as ubiquitous and misunderstood as extremely-low-frequency electromagnetic fields is accused of causing cancer in children, people's reactions may be driven more by passion than by reason.

Each year in this country about 2000 children are given a diagnosis of acute lymphoblastic leukemia (ALL), the most common childhood cancer. Despite the remarkable advances in treatment, ALL still carries a 30 percent mortality. Other than exposure to ionizing radiation, its cause remains a mystery. ALL is more common among whites and children of higher socioeconomic class, and for unclear reasons the incidence of ALL has increased by about 20 percent in the past two decades. [1,2] During the past 50 years, per capita use of electricity has increased more than 10 times. Some investigators have claimed that living close to major power lines causes cancer, particularly leukemia in children.

... Linet et al.[3] report the results of a major study showing that the risk of ALL does not increase with increasing electromagnetic-field levels in children's homes. This study has several strengths. It was large, including 629 children with leukemia and 619 controls, and it included measurements of electromagnetic fields, made by technicians blinded to the case or control status of the subjects, both in the houses where the children had lived and, in 41 percent of cases, in the homes in which their mothers resided while pregnant. Linet et al. also found no relation between the risk of ALL and residential wire-code classifications, again determined by technicians blinded to the children's health status. The wire-code classifications are important, because several of the earlier positive studies relied on these proxy indicators rather than on actual measurements of electromagnetic fields.

This whole saga began when two Denver researchers, puzzled by small clusters of cancer in children, came to believe that living in close proximity to high-voltage power lines was a cause of leukemia.[4] The analysis they published in 1979 was crude and relied on distances from homes to power lines and on wiring configurations rather than on direct measures of exposure to electromagnetic fields. They found that the risk of childhood leukemia was more than doubled among children living near such power lines, a finding that led to more studies and more concern. Soon activists and the media began to spread the word that electromagnetic fields cause cancer.

The hypothesized cause was exposure to extremely-low-frequency magnetic fields generated by the electrical current in power lines. Physicists understand these invisible fields well, but most physicians, parents, and patients do not. The movement of any electrical charge creates a magnetic field that can be measured.[5] Even the 60-Hz residential electric current (50 Hz in Europe) creates a very weak oscillating field, which, like all magnetic fields, penetrates living tissue. These low-frequency electromagnetic fields are known as nonionizing radiation, since the amount of energy in them is far below that required to break molecular bonds such as those in DNA.

One ironic fact about low-frequency electromagnetic fields is that we live and worry about them within the Earth's static magnetic field of 50 μT, which is hundreds of times greater than the oscillating magnetic field produced by 110/220-V current in houses (0.01 to 0.05 μT).[5,6] Even directly under high-voltage transmission lines, the magnetic field is only about 3 to 10 μT, which is less than that in an electric railway car and much weaker than the magnetic field close to my head when I use an electric razor (about 60 μT).

Although most physicists find it inconceivable that power-line electromagnetic fields could pose a hazard to health, dozens of epidemiologic studies have reported weak positive associations between proximity to high-voltage power lines and the risk of cancer.[6,7] The negative or equivocal studies did not end the controversy. Fear of leukemia is a powerful force, and the media response amplified the perception of electromagnetic fields as a health hazard. In 1989 *The New Yorker* published three articles by journalist Paul Brodeur that described in mesmerizing detail how maverick researchers had discovered a cause of cancer that the establishment refused to accept.[8–10] Like many of the epidemiologic studies themselves, these widely quoted articles described biologic mechanisms of action for electromagnetic fields that were hypothetical, even fanciful. Brodeur went so far as to claim that the search for the truth about the hazards of electromagnetic fields was threatened most by the "obfuscation of industry, the mendacity of the military, and the corruption of ethics that industrial and military money could purchase from various members of the medical and scientific community."[8] Suspicion spread to many other wavelengths on the nonionizing electromagnetic spectrum, producing fears about occupational exposure to electricity as well as exposure to microwave appliances, radar, video-display terminals, and even cellular telephones. Dozens of studies looked for associations with brain cancer, miscarriages, fetal-growth retardation, lymphoma, breast cancer, breast cancer in men, lung cancer, all cancers, immunologic abnormalities, and even changes in the behavior of animals.

When people hear that a scientific study has implicated something new as a cause of cancer, they get worried. They get even more worried when the exposure is called radiation and comes from dangerous-looking high-voltage power lines controlled by government and industry, which some distrust deeply. Such exposure seems eerie when people hear that electromagnetic fields penetrate their homes, their bodies, their children. The worried citizens took action. Frightened people, including parents of children with leukemia, undertook their own epidemiologic studies and fought to get high-power transmission lines moved away from their children. Congress responded with large direct appropriations for wider research on the effects of electromagnetic fields. After a large apparently positive study in Sweden,[7] the Swedish government came close to mandating the relocation of schools to at least 1000 meters from large power lines. But cooler heads prevailed once it became clear that the absolute incremental risk was small at most, the conclusions were based on a tiny fraction of all Swedish children with leukemia, and the increase in risk was found only in relation to some estimates of magnetic fields, not to the actual fields measured in children's homes.

Serious limitations have been pointed out in nearly all the studies of power lines and cancer.[11,12] These limitations include unblinded assessment of exposure, difficulty in making direct measurements of the constantly varying electromagnetic fields, inconsistencies between the measured levels and the estimates of exposure based on wiring configurations, recall bias with respect to exposure, post hoc definitions of exposure categories, and huge numbers of comparisons with selective emphasis on those that were positive. Both study participation and residential wire-code categories may be confounded by socioeconomic factors. Often the number of cases of ALL in the high-exposure categories has been very small, and controls may not have been truly comparable. Moreover, all these epidemiologic studies have been conducted in pursuit of a cause of cancer for which there is no plausible biologic basis. There is no convincing evidence that exposure to electromagnetic fields causes cancer in animals,[6] and electromagnetic fields have no reproducible biologic effects at all, except at strengths that are far beyond those ever found in people's homes.

In recent years, several commissions and expert panels have concluded that there is no convincing evidence that high-voltage power lines are a health hazard or a cause of cancer.[6,13] And the weight of the better epidemiologic studies, including that by Linet et al., now supports the same conclusion. It is sad that hundreds of millions of dollars have gone into studies that never had much promise of finding a way to prevent the tragedy of cancer in children. The many inconclusive and inconsistent studies have generated worry and fear and have given peace of mind to no one. The 18 years of research have produced considerable paranoia, but little insight and no prevention. It is time to stop wasting our research resources. We should redirect them to research that will be able to discover the true biologic causes of the leukemic clones that threaten the lives of children.

REFERENCES

1. Pui C-H. Childhood leukemias. N Engl J Med 1995; 332: 1618–30.
2. Ries LAG, Miller BA, Hankey BF, Kosary CL, Harras A, Edwards BK, eds. SEER cancer statistics review, 1973–1991: tables and graphs. Bethesda, Md.: National Cancer Institute, 1994. (NIH publication no. 94-2789.)
3. Linet MS, Hatch EE, Kleinerman RA, et al. Residential exposure to magnetic fields and acute lymphoblastic leukemia in children. N Engl J Med 1997; 337: 1–7.

4. Wertheimer N, Leeper E. Electrical wiring configurations and childhood cancer. Am J Epidemiol 1979; 109: 273–84.
5. Hitchcock RT, Patterson RM. Radio-frequency and ELF electromagnetic energies: a handbook for health professionals. New York: Van Nostrand Reinhold, 1995.
6. National Research Council. Possible health effects of exposure to residential electric and magnetic fields. Washington, D.C.: National Academy Press, 1997.
7. Feychting M, Ahlbom A. Magnetic fields and cancer in children residing near Swedish high-voltage power lines. Am J Epidemiol 1993; 138: 467–81.
8. Brodeur P. Annals of radiation: the hazards of electromagnetic fields. I. Power lines. The New Yorker. June 12, 1989: 51–88.
9. *Idem.* Annals of radiation: the haz magnetic fields. II. Something is h New Yorker. June 19, 1989: 47–73.
10. *Idem.* Annals of radiation: the hazards of electromagnetic fields. III. Video-display terminals. The New Yorker. June 26, 1989: 39–68.
11. Savitz DA, Pearce NE, Poole C. Methodological issues in the epidemiology of electromagnetic fields and cancer. Epidemiol Rev 1989; 11: 59–78.
12. Poole C, Trichopoulos D. Extremely low-frequency electric and magnetic fields and cancer. Cancer Causes Control 1991; 2: 267–76.
13. Oak Ridge Associated Universities Panel. Health effects of low-frequency electric and magnetic fields. Washington, D.C.: Government Printing Office, 1992: V-1-V-18. (Publication no. 029-000-00443-9.)

POSTSCRIPT

Are Electromagnetic Fields Dangerous to Your Health?

Is the EMF scare nothing more than media hype, as suggested by Sid Deutsch in "Electromagnetic Field Cancer Scares," *Skeptical Inquirer* (Winter 1994)? Or do EMFs pose a genuine hazard? If they do, the threat is not yet clear beyond a doubt. However, society cannot always wait for certainty. Gordon L. Hester, in "Electric and Magnetic Fields: Managing an Uncertain Risk," *Environment* (January/February 1992), states that just the possibility of a health hazard from EMFs is sufficient to justify more research into the problem. The guiding principle, says Hester, is " 'prudent avoidance,' which was originally intended to mean that people should avoid fields 'when this can be done with modest amounts of money and trouble.' " H. Keith Florig, in "Containing the Costs of the EMF Problem," *Science* (July 24, 1992), makes a similar point in his discussion of the expenses that utilities, manufacturers, and others are incurring to reduce EMF exposures in the absence of solid evidence that there is a hazard but in the presence of public concern and lawsuits.

And the concern does remain despite the lack of evidence for any real hazard. Researchers Hans Wieser, Michael Fuller, and Jon Paul Dobson reported at the May 1993 meeting of the American Geophysical Union that magnetic fields can affect brain activity, suggesting that the body does respond to EMFs from electrical apparatus. On January 27, 1995, *Science* reported that "the U.S. Navy's 90-kilometer-long Extremely Low Frequency (ELF) antenna, set up [in a Michigan forest] in 1986 to communicate with submarines, is invigorating neighboring plant life"—apparently stimulating tree and algal growth in a way that has led some scientists to consider how ELF EMFs might stimulate the growth of cancer cells.

In August 1995 *Science* published a brief item on a draft report by the National Council on Radiation Protection and Measurements, saying that "some health effects linked to EMFs—such as cancer and immune deficiencies—appear real and warrant steps to reduce EMF exposure," such as not building new housing under high-voltage transmission lines. The final report was promised for 1996. However, in September 1995, *Science* reported a pair of studies that had attempted to confirm work indicating that EMFs could stimulate a cancer gene. Both failed to find any effect. And in November 1996, the National Research Council, in their report *Possible Health Effects of Exposure to Residential Electric and Magnetic Fields,* came down firmly on the "no conclusive and consistent evidence" of hazard side of the question. (See Jocelyn Kaiser, "Panel Finds EMFs Pose No Threat," *Science,* November 8, 1996.)

When the study by Martha S. Linet et al. appeared, Gary Taubes, in "Magnetic Field-Cancer Link: Will It Rest in Peace?" *Science* (July 4, 1997), said, "It could be the obituary [for the EMF-cancer scare]." "Yet many people will refuse to believe it," says Lawrence Fisher, a toxicologist at Michigan State University in East Lansing, "not on any scientific basis, but because of their emotional involvement with the disease." To such people, the jury will forever be out. To them, the question is, What should society do in the face of weak, uncertain, and even contradictory data? Can we afford to conclude that there is no hazard? Or must we redesign equipment and relocate power lines and homes with no other justification than our fear that there *might* be a real hazard? Many scientists and politicians argue that even if there is no genuine medical risk from exposure to EMFs, there is a genuine impact in terms of public anxiety. It is therefore appropriate, they say, to fund further research and to take whatever relatively inexpensive steps to minimize exposure are possible. Failure to do so increases public anxiety and distrust of government and science.

It is worth noting that the EMF scare had a precedent in the late 1800s. See Joseph P. Sullivan, "Fearing Electricity: Overhead Wire Panic in New York City," *IEEE Technology and Society Magazine* (Fall 1995).

ISSUE 8

Are Environmental Regulations Too Restrictive?

YES: John Shanahan, from "Environment," in Stuart M. Butler and Kim R. Holmes, eds., *Issues '96: The Candidate's Briefing Book* (Heritage Foundation, 1996)

NO: Paul R. Ehrlich and Anne H. Ehrlich, from "Brownlash: The New Environmental Anti-Science," *The Humanist* (November/December 1996)

ISSUE SUMMARY

YES: John Shanahan, vice president of the Alexis de Tocqueville Institution in Arlington, Virginia, argues that many government environmental policies are unreasonable and infringe on basic economic freedoms. He maintains that although environmental problems do exist, there is no real environmental crisis.

NO: Environmental scientists Paul R. Ehrlich and Anne H. Ehrlich argue that many objections to environmental protections are self-serving and based in bad or misused science.

Concern for the environment in America is not much more than a century old. In 1785 Thomas Jefferson invented the idea (if not the wording) of NIMBY ("Not In My Back Yard") when he wrote, "Let our workshops remain in Europe." He thought that an American factory system would have undesirable social, moral, and aesthetic effects. Clearly, he was alone in that thought, for America developed its industrial base very quickly. The workshop builders flourished, and the effects that concerned Jefferson did indeed come to pass. The first national park, Yosemite, resulted from legislation signed by President Abraham Lincoln in 1864. Yellowstone was approved in 1872. Both were responses to an awareness that if the areas' unique features were not protected, they would be destroyed by ranchers, miners, loggers, and market hunters, as had already happened elsewhere.

By the 1960s people were beginning to realize that other activities, such as the use of pesticides, also threatened treasured features of the environment, such as songbirds (see Rachel Carson, *Silent Spring*, Houghton Mifflin, 1962), as well as human health. The result was government regulation of pesticides, air pollution, water pollution, and much, much more. Lead has been removed from gasoline and paint, chlorofluorocarbons from aerosol deodorants and refrigerants, and phosphates from laundry detergents. Developers have been told they cannot fill in swamps and other wetlands. Loggers have been forbidden to log in many areas. And commercial fishing seasons have been limited or eliminated entirely.

The economic impact of environmental regulation has not been as great as it might have been if Jefferson had had his way in 1785, but in each case someone's economic benefit has been interfered with. In other cases—such as when users of off-road vehicles have been barred from driving on the nesting grounds of rare shorebirds—the freedom to do as one wishes has been interfered with. In nations such as China, which instituted a "one child per couple" population control policy in 1979, the freedom in question is the freedom to have as many children as one wishes. As the environmental regulations have proliferated, so has the interference with freedoms that people once took for granted. And so have the objections. Conservative politicians and lobbyists for industry, recreation, and home-owner groups struggle to block or weaken every new environmental regulation and to repeal old regulations, often in the name of individual freedom and property rights. Environmentalists counter that freedom must be tempered by responsibility; individual freedom and property rights must have limits, or we will destroy what lets us and our children live on Earth.

The issue is not just America's; it is the world's. Environmentalists are active everywhere, identifying problems, promoting a sense of crisis, and saying what must be done, what behaviors must be controlled, and what freedoms must be limited. They have been successful enough to rouse fears among some far-right political conservatives of a liberal-environmentalist conspiracy to take over the world and impose an antifreedom world government. Similar fears may be shared by more moderate conservatives, but they are rarely voiced explicitly.

In the following selections, John Shanahan, writing for the Heritage Foundation's 1996 *Candidate's Briefing Book* (designed to help conservative candidates get elected), argues that many government environmental policies are unreasonably restrictive, infringe on basic economic freedoms, and are based on bad science. Although environmental problems do exist, he contends, there is no real environmental crisis. Environmental scientists Paul R. Ehrlich and Anne H. Ehrlich argue that many objections to environmental protections are self-serving. The antienvironmentalists, they say, deny the facts in favor of religious and political ideologies. The Ehrlichs assert that the environmental crisis is very real.

YES John Shanahan

ENVIRONMENT

THE ISSUES

Americans want a clean, healthy environment. They also want a strong economy. But environmental protection is enormously expensive, costs jobs, and stifles economic opportunity. On the other hand, before government stepped in, robust economic activity such as manufacturing led to a deteriorating and unhealthy environment. The challenge is how to achieve both a strong economy and a healthy environment. After all, what Americans actually want is a high overall quality of life.

Three decades ago, as people perceived that their quality of life was beginning to deteriorate, they began to support aggressive policies to reduce pollution. These policies frequently failed to live up to their sponsors' claims; they also became increasingly and unnecessarily expensive. But the environment did improve, especially in the early years. Now, however, Americans are becoming aware that many of these policies are unreasonable and that, even when they work, they result only in small improvements at a heavy cost in jobs and freedom. Americans also are beginning to recognize that there often is no sound scientific basis for assertions of environmental harm or risk to the public. The pendulum finally has begun to swing the other way.

Conservatives, like Americans generally, have no wish to return to the days of black smoke billowing out of smokestacks. But they do believe common sense can be brought to bear in dealing with the environment: that it is possible to protect the environment without sacrificing the freedoms for which America stands. Conservative candidates and legislators therefore should stress the following themes:

> **Examples of regulatory abuse.** It is important to show that "good intentions" often are accompanied by oppressive, senseless regulations.
>
> **An ethic of conservation.** Candidates need to explain that conserving or efficiently using natural resources is not in dispute. The debate is over how best to do this: through markets or through government controls.

Economic freedom. Candidates need to point out that many government "solutions" to environment problems conflict with basic economic freedoms.

Property-based solutions. Candidates need to explain that environmental objectives can be achieved best not by issuing thousands of pages of rules that people will try to circumvent, but by capitalizing on the incentives associated with owning property.

Sound science. Candidates need to argue that we need policies based on sound science, not "tabloid science."

Priority setting. Candidates must explain that not all problems are of the same importance or urgency, and that regulating all risks equally means fewer lives are saved for the dollars spent than would be saved if priorities were set.

THE FACTS

While pollution levels have fallen dramatically since 1970, most reductions were achieved early and at relatively low cost. From 1970 to 1990, total emission levels fell 33.8 percent. Over the same period, lead levels in the air fell 96.5 percent, and carbon monoxide levels in the air fell 40.7 percent. But reductions have slowed dramatically....

Unworkable Regulations

Environmental regulation does more than just cost too much. Candidates also should use the growing litany of horror stories to demonstrate how ill-conceived environmental regulations, while delivering little benefit, lead to unintended consequences for businesses especially small businesses, which are disproportionately minority-owned and minority-run.

- Larry Mason's family owned a sawmill employing 40 workers in Beaver, Washington. In the mid-1980s, based on harvest assurances from the U.S. Forest Service and loan guarantees from the Small Business Administration, the family invested $1 million in its business. Then, says Mason, "in 1990, the spotted owl injunctions closed our mill, made my equipment worthless, and my expertise obsolete. The same government that encouraged me to take on business debt then took away my ability to repay."
- While the Clean Water Act (CWA) requires a waste treatment facility to submit a simple form stating that a fence restricts access by the public, the Resource Conservation and Recovery Act (RCRA) requires an additional 25 pages detailing the fence design, the location of the posts and gates, a cross section of the wire mesh, and other minor technical matters. RCRA is so wasteful that one plant, whose CWA permit application was only 17 pages long, had to file a seven-foot stack of supporting documents with its applications.
- Ronald Cahill, a disabled Wilmington, Massachusetts, dry cleaner, purchased expensive dry-cleaning equipment to comply with EPA regulations governing the use of trichlorotrifluoroethane (CFC-113). But the EPA levied a tax on all chlorofluorocarbons (CFCs), making CFC-113 hard to find and extremely expensive. In 1995, Cahill's business went under. Washington, says Cahill, "has put me out of business with excessive taxes and regulations."

Regulatory abuses like these usually are a direct result of the way government bureaucracies attack environmental problems. Typically, these agencies regulate without regard to the cost imposed on individuals and businesses. Yet it makes no sense to issue a regulation for which the burden far outweighs any benefit that might be conferred. In fact, it often is unclear whether there will be any benefit at all because the science on which many regulations are based is so poor.

Also, instead of setting realistic performance standards and giving businesses the freedom to develop innovative ways of meeting them, agencies typically rely on inflexible command-and-control regulations that, for example, specify what technologies companies must use. Since businesses differ in their operating structures, this one-size-fits-all approach rarely leads to cost-effective solutions compared to more flexible and dependable performance standards. Moreover, by eliminating the incentive for companies to seek out these cost-effective solutions, it stifles innovative technologies or techniques that reduce costs. In the end, of course, the consumer is the one who pays.

Perhaps the most troublesome aspect of current environmental policy is the fact that bureaucrats and liberal lawmakers generally consider regulation the only option. Creative solutions shown to be less expensive, more effective, and more respectful of human liberty are rejected out of hand. Instead of setting up a system of incentives to lure businesses into operating with environmental impact in mind, the system relies on punishment regardless of whether this accomplishes the desired goal or creates unintended consequences.

Rejecting Property Rights

Regulations have become increasingly unfair. The Environmental Protection Agency (EPA), Department of the Interior (DOI), Army Corps of Engineers, and other federal agencies operate on the premise that property should be used to satisfy government's needs and objectives without regard to who owns the property or the financial burden imposed on them. It is this mentality that leads government reflexively to reject the creative solutions advanced by free-market advocates, including incentive-based approaches to protecting endangered species. By ignoring property rights, establishment environmentalists, bureaucrats, and liberal legislators also ignore the benefits to be derived from free trade and free markets.

The most unfair and burdensome hardship inflicted by government "regulatory takings" is that property owners are not compensated for their losses. For instance, if an elderly husband and wife spend a large portion of their retirement savings to buy land on which to build their dream home and that land subsequently is designated a wetland, they lose the value of their property as well as their savings. They are stuck with property they cannot use and the government does nothing to reimburse them for their loss. Unfortunately, tales of financial hardship caused by government designation of land as wetland or endangered-species habitat have become common. For instance:

- Bill Stamp's family in Exeter, Rhode Island, has been blocked from farming or developing its 70 acres of land for 11 years, yet has been assessed taxes at rates determined by the land's industrial value up to $72,000 annually. As a result, this fifth-generation farm family may lose its life savings. The government, however, appears unmoved.

Stamp relates what one Army Corps of Engineers enforcement officer told him: "We know that this is rape, pillage, and plunder of your farm, but this is our job."

- A small church in Waldorf, Maryland, was told by the Army Corps of Engineers that one-third of its land, on which it planned to build a parking lot, was a wetland and could not be used. Part of this so-called wetland is a bone-dry hillside which almost never collects water. Says Reverend Murray Southwell of the Freewill Baptists, "this obvious misinterpretation of wetland law made it necessary for us to purchase an additional lot [for $45,000, which] has been a heavy financial burden on this small missions church."
- Developer Buzz Oates wants to develop less than 4 percent of the Sutter Basin in Sacramento, California, where an estimated 1,000 giant garter snakes live. But the federal government mandated that he pay a "mitigation" fee of nearly $3.8 million for the 40 or fewer snakes he might disturb: $93,950 per snake. Says Oates, in an age of "depleted [fiscal] resources and deteriorating school infrastructure, this is a very tough pill to swallow."

Hundreds of such stories have surfaced over the past few years, and many analysts suspect that far more are never made public. According to Bob Adams, Project Director for Environmental and Regulatory Affairs at the National Center for Public Policy Research, "the stories we have compiled are just the tip of the iceberg, but many people are simply too scared to come forward or feel powerless against the government."

Ironically, federal agencies and the Clinton Administration argue that it would cost too much money to compensate landowners. Leon Panetta, then Director of the Office of Management and Budget, told the House Committee on Public Works and Transportation's Subcommittee on Water Resources and the Environment on May 26, 1994, that paying compensation for wetlands regulation would be "an unnecessary and unwise use of taxpayer dollars" and a drain on the federal budget.

Property owners counter that regulatory takings are a drain on the family budget. Nancie Marzulla, President of Defenders of Property Rights, points out that "what people don't realize is that these landowners typically are not wealthy and powerful corporations, but normal Americans schoolteachers and elderly couples whose lives are destroyed by stretched interpretations of a single environmental law." Moreover, the federal government already owns about one out of every three acres in the country (with even more owned by state and local governments). If the federal government can afford to maintain one-third of the nation's land, it should be able to pay landowners for regulatory confiscation of their property. If not, maybe it should consider selling the least ecologically sensitive land from its vast holdings to pay for the land it wants.

Lost Opportunities, Lost Lives

Ask the average American how much a human life is worth, and the answer likely will be that "no amount is too much." This is how Congress and federal agencies justify imposing sometimes staggering costs on businesses to reduce the risks of death by infinitesimal amounts. What policymakers fail to understand is that wasting resources in this way means not being able to use them in other ways that might well produce better results and save even more lives.

If lawmakers ever did consider which environmental policies actually save the most lives, they would scrap many existing rules, freeing up resources to be used in other ways. This commonsense approach would lead to regulation that is very different, in its scope and fundamental assumptions, from that which burdens America today....

What America Thinks About the Environment

When asked by the media, pollsters, or politicians, Americans routinely answer that they want a clean and healthy environment. Indeed, the majority of Americans consider themselves "environmentalists." This does not translate, however, into automatic acceptance of the environmental lobby's agenda. Conservative candidates need to make this clear to discourage voters from supporting policies they do not believe in simply because they are portrayed as "pro-environment."

The dichotomy in public opinion shows up in polling data. When respondents are asked general or theoretical questions that involve little personal sacrifice, or that do not identify those burdened, government intervention fares well. In one poll, for instance, 60 percent of respondents agreed that we must protect the environment even if it costs jobs in the community. In another, 72 percent of respondents said they would pay somewhat higher taxes if the money was used to protect the environment and prevent water and air pollution.

On the other hand, when respondents are asked questions that are more specific, that involve greater sacrifice, or that identify the people losing jobs, government intervention is less popular. When respondents are asked to pay much higher taxes to protect the environment, support drops by almost half. By the same token, only one-third would be willing to accept cuts in their standard of living. When asked to pick between spotted owls and Northwest workers who stand to lose their jobs because of efforts to protect the owls, respondents choose jobs by a margin of 3 to 2....

Perhaps the most refreshing change in attitudes in recent years is the recognition that the country can have economic growth and environmental protection simultaneously. Vice President Al Gore has made the point that economic growth and environmental protection are not incompatible. This is true, but only if America's environmental laws are structured correctly to encourage responsible behavior as part of the business decision-making process. Gore advocates stringent command-and-control regulations that are inconsistent with growth and lead to little real gains in environmental protection.

Whenever this question comes up, Americans must be told that the way to promote both environmental protection and economic growth is to allow them to work hand in hand. The government must stop regarding them as mutually exclusive and stop pitting economic freedom against the environment. Laws must be based on, and work with, a free market. Only then can Americans maximize their economic and environmental quality of life.

The Need for Common Sense

Given Americans' ambivalence on the question of environmental protection, it is all the more important for conservatives to approach the issue in a commonsense way. People must understand that environmental protection need not come at the expense of jobs, but will cost jobs if the socialist model of centralized control for protecting the environment is not set aside. It doesn't work. Rather, the country should adopt a reasonable, commonsense approach to environmental protection that is based on:

Freedom With Responsibility. Conservatives traditionally have stressed economic growth while ignoring the importance of environmental problems. Thus, they have fought environmentalists step by step and have lost step by step. The reason, while unpleasant, is not complicated. Environmentalists have had the moral high ground, even though they typically have not provided the most beneficial solutions. In short, conservatives have been on the wrong side of an emotional issue.

Two lessons demonstrate why:

- **First,** leftists and the public at large understand that publicly owned goods, free of constraints on usage, will be depleted over time. Garrett Hardin, Professor Emeritus of Human Ecology at the University of California, in his seminal 1968 work *The Tragedy of the Commons* showed that when a good is publicly owned, or "owned" in common, no one has an incentive to conserve or to manage it. In fact, there is a perverse incentive to use the good inefficiently to deplete it. This fact is at the heart of most environmental problems, such as air and water pollution and species extinction.
- **Second,** if there are incentives to conserve resources, people will conserve out of self-interest. People with a vested interest in providing environmental benefits through property ownership or other positive incentives will provide them voluntarily, without coercion.

... "Freedom with responsibility for one's actions" should be the conservative message. Responsibility restrains wasteful behavior. Ironically, the old environmentalist slogan "Make the polluter pay" is consistent with this message. But when they say this, conservatives and liberals mean different things. As Al Cobb, then Director of Environment and Energy at the National Policy Forum, has said, "What the environmental lobby means by that phrase is that corporate polluters should be punished severely for any pollution whatsoever. What conservatives mean, however, is that polluters should bear the full cost of environmental degradation, but no more." At the same time, individuals and corporations also should be rewarded for conservation and other environmentally sound practices.

Conservation Through Property Rights. The free market reflects the conservation ethic better than any command-and-control regulation from Washington. A free market can occur, however, only when private citizens engage in trade, and people can trade only what they own: some form of property. Thus, property is the cornerstone of a free market. If property rights are insecure or publicly owned, a market cannot function effectively. Some critics misleadingly call this "market failure," but it is really a failure to use markets and their main engine: property rights. As a result, both environmental protection and personal liberty suffer. A resource that is not owned will deteriorate or be depleted because neither protection of nor damage to that resource is part of the individual's usual decision-making process. Others, however, are still forced to bear the consequences.

Conservative candidates should concentrate on explaining the innovative ways in which property rights can be used to protect the environment. The most efficient method and the most protective of individual rights and freedoms is to enlist self-interest in the service of environmental protection.

Consider [two] examples of how the principle of property rights-based environmentalism works:...

- In Scotland and England, the popularity of fishing has burgeoned in recent decades. Property rights to fishing sites have developed as the building block for markets to provide access to prime fishing spots. As a result, many private, voluntary associations have been formed to purchase fishing rights access. In Scotland, "virtually every inch of every major river and most minor ones is privately owned or leased...." Owners of fishing rights on various stretches of the rivers charge others for the right to fish. These rivers are not overfished because it is not in the owner's best interest to allow the fish population to be depleted. Because he wants to continue charging fishermen for the foreseeable future, the owner conserves his fish stock, allowing them to reproduce, and prevents pollution from entering his stretch of the river. If a municipality pollutes the water upstream, the owner of the fishing rights can sue for an injunction. Everyone wins, including fishermen looking for quality fishing with some privacy.
- One group's approach to wetland protection has shown the power of property rights to achieve environmental goals. Ducks Unlimited, a group consisting of hunters and non-hunters alike, is dedicated to enhancing duck populations. To do this, it has purchased property or conservation easements with privately raised funds. Unlike other groups (for example, the National Wildlife Federation) that began as organizations of hunters and outdoorsmen but later lost much of their original focus and joined forces with the more extreme elements of the environmental lobby, Ducks Unlimited still focuses on protecting duck habitat. In the last 58 years, it has raised and invested $750 million to conserve 17 million acres in Canada alone, an effort which benefits other wildlife as well as ducks. In 1994, it restored or created about 50,000 acres of wetlands. Since Ducks Unlimited itself pays for the habitat it protects, in many ways it embodies the essence of the conservative message: that the market should be allowed to determine the best and highest use of a good or resource in this case, duck habitat.

Unfortunately, property rights are under attack from the environmental lobby. The Fifth Amendment to the U.S. Constitution states, "nor shall private property be taken for public use, without just compensation," but this has been interpreted as protection primarily against the physical taking of property. Most infringements, however, involve federal decrees that deny owners the right to use their property as they see fit, for example, to continue farming. Since the courts have been unclear on the degree of protection property owners should have from such intrusions, legislative protection is needed.

Sound Science, Not Tabloid Science. Before issuing regulations to protect health, regulators should ask whether the science behind a measure justifies the often enormous expenditures involved. Unfortunately, however, the federal government often acts in response to strong environmentalist-generated public pressure without adequate scientific justification....

- In 1992, the National Aeronautics and Space Administration (NASA) reported that hole in the Earth's protective ozone layer might open up over North America that spring. This hypothetical hole, which would have been in addition to the annual Antarctic hole, would be caused by chlorofluourocarbons (CFCs), a refrigerant. After widespread media coverage on the threat of CFCs, the White House moved a production ban, scheduled for the year 2000, up to 1996, raising the cost of the ban by tens of billions of dollars. Unfortunately, NASA held its press conference before it had finished the study or subjected it to even cursory peer review. The hypothesized ozone hole over North America never materialized. Nor could it have. According to Patrick Michaels of the Climatology Department at the university of Virginia, "The only way you could produce an ozone hole in the high latitudes of the Northern Hemisphere that resembles what occurs in the Southern Hemisphere (where the ozone hole occurs) would be to flatten our mountains and submerge our continents. Then you would have airflow patterns similar to those that occur in the Southern Hemisphere, and are the ones that are required to create an ozone hole." One would think NASA would know this as well. Now, although no information other than a thoroughly discredited hypothesis justifies dramatically stepping up the phaseout, the country is redirecting its limited economic resources at an extra cost of hundreds of dollars per household because of the ban, which is now in place.

Instead of merely responding to tabloid claims or politically motivated studies by federal agencies and environmental organizations trying to justify their budgets, regulations should be based on credible scientific findings open to public scrutiny. For instance, agencies should use consistent methodologies to determine risks. Currently, they use different methods. Thus, for example, risk assessments by different agencies may turn up different answers as to whether a chemical at a particular dose level causes cancer. Theoretically, exposure to some level of a chemical could be found to be both deadly and perfectly safe.

Government assessments also should reveal the assumptions and uncertainties in their analyses. Typically, because of missing data, most studies use certain assumptions to estimate these uncertainties. These assumptions, sometimes unreasonably gloomy, usually determine the conclusion reached. For instance, sometimes an estimate of the likely risk from some chemical is multiplied thousands, or even millions, of times just to be "conservative." Yet the analyses used to justify these enormously expensive regulations often are obscure as to their assumptions. Moreover, the reports rarely reveal the level of uncertainty involved in arriving at their conclusions.

Whenever regulations that address risks are considered, each agency should be required to conduct risk assessments if only to aid in intelligent decision-making that are consistent, that are transparent to public scrutiny, and that fully detail their assumptions and levels of uncertainty. Moreover, each study should be reviewed before a regulation is published to ensure that scientific guidelines are strictly followed. If federal agencies cannot meet even this very limited standard, it is unconscionable for them to impose costly standards on others.

The Need to Set Priorities. The economy has a limited capacity to absorb environmental regulations. Simply put, the country cannot afford to eliminate every risk. Thus, there is a trade-off: Attempting to regulate one risk out of existence may mean that another risk (or other risks) will have to be tolerated. In most cases, the cure is worse than the disease. Misguided and excessive regulation can cost lives, so it is critical that regulators recognize the costs of their actions. Spending enormous amounts of money to eradicate small or even hypothetical risks means that those dollars cannot be used in other productive ways public or private that might be of greater benefit to the nation.

... [I]t is essential that policymakers develop a priority list of environmental problems, based on the extent of the possible risk each appears to pose and the cost of reducing that risk to acceptable levels. With such a list, policymakers can know just how much protection is being bought for every dollar spent. Americans finally will get the maximum environmental "bang for the buck." Conversely, the federal government will be able to achieve environmental objectives at the lowest cost, and thus with the fewest "pink slips" for American workers.

Is There an Environmental Crisis?

Is there an environmental crisis? The answer is a resounding "No." Certainly the country and planet have environmental problems that need to be addressed. But overall, the environment has been improving. Unfortunately, the public is subjected only to the "Chicken Little" version of the situation, and reports of environmental progress and refutations of environmental alarmists are rarely covered in the press.

In his 1995 book *A Moment on the Earth,* which details many of the improvements that have taken place in the last three decades, *Newsweek* editor Gregg Easterbrook notes that reports of positive environmental developments, such as significantly lower air pollution in major U.S. cities, are buried inside the newspapers. Negative news, meanwhile, gets front-page attention, and the news that is reported often contains numerous misleading "facts."

The truth is that threats to the environment have lessened considerably. Lead has been almost eliminated. Even in Los Angeles, the most polluted city in the country, levels of Volatile Organic Compounds (VOCs) have fallen by more than half since 1970. In other formerly polluted cities, such as Atlanta, the air is now considered relatively clean as VOCs are down by almost two-thirds and Nitrous Oxide is down 15 percent.

In area after area so-called global warming, endangered species, wetlands, pesticides, hazardous waste, and automotive fuel economy, for example, the problem is the same: only rarely are the facts heard by the American people.

NO Paul R. Ehrlich and Anne H. Ehrlich

BROWNLASH: THE NEW ENVIRONMENTAL ANTI-SCIENCE

Humanity is now facing a sort of slow-motion environmental Dunkirk. It remains to be seen whether civilization can avoid the perilous trap it has set for itself. Unlike the troops crowding the beach at Dunkirk, civilization's fate is in its own hands; no miraculous last-minute rescue is in the cards. Although progress has certainly been made in addressing the human predicament, far more is needed. Even if humanity manages to extricate itself, it is likely that environmental events will be defining ones for our grandchildren's generation—and those events could dwarf World War II in magnitude.

Sadly, much of the progress that has been made in defining, understanding, and seeking solutions to the human predicament over the past 30 years is now being undermined by an environmental backlash. We call these attempts to minimize the seriousness of environmental problems the *brownlash* because they help to fuel a backlash against "green" policies. While it assumes a variety of forms, the brownlash appears most clearly as an outpouring of seemingly authoritative opinions in books, articles, and media appearances that greatly distort what is or isn't known by environmental scientists. Taken together, despite the variety of its forms, sources, and issues addressed, the brownlash has produced what amounts to a body of anti-science—a twisting of the findings of empirical science—to bolster a predetermined worldview and to support a political agenda. By virtue of relentless repetition, this flood of anti-environmental sentiment has acquired an unfortunate aura of credibility.

It should be noted that the brownlash is not by any means a coordinated effort. Rather, it seems to be generated by a diversity of individuals and organizations. Some of its promoters have links to right-wing ideology and political groups. And some are well-intentioned individuals, including writers and public figures, who for one reason or another have bought into the notion that environmental regulation has become oppressive and needs to be severely weakened. But the most extreme—and most dangerous—elements are those who, while claiming to represent a scientific viewpoint, misstate scientific findings to support their view that the U.S. government has gone overboard with regulation, especially (but not exclusively) for environmental protection, and that subtle, long-term problems like global warming are nothing to worry about. The words and sentiments of the brownlash are profoundly troubling to us and many of our colleagues. Not only are the underlying agendas seldom revealed but, more important, the confusion and distraction created among the public and policymakers by brownlash pronouncements interfere with and prolong the already difficult search for realistic and equitable solutions to the human predicament.

Anti-science as promoted by the brownlash is not a unique phenomenon in our society; the largely successful efforts of creationists to keep Americans ignorant of evolution is another example, which is perhaps not entirely unrelated. Both feature a denial of facts and circumstances that don't fit religious or other traditional beliefs; policies built on either could lead our society into serious trouble.

Fortunately, in the case of environmental science, most of the public is fairly well informed about environmental problems and remains committed to environmental protection. When polled, 65 percent of Americans today say they are willing to pay good money for environmental quality. But support for environmental quality is sometimes said to be superficial; while almost everyone is in favor of a sound environment—clean air, clean water, toxic site cleanups, national parks, and so on—many don't feel that environmental deterioration, especially on a regional or global level, is a crucial issue in their own lives. In part this is testimony to the success of environmental protection in the United States. But it is also the case that most people lack an appreciation of the deeper but generally less visible, slowly developing global problems. Thus they don't perceive population growth, global warming, the loss of biodiversity, depletion of groundwater, or exposure to chemicals in plastics and pesticides as a personal threat at the same level as crime in their neighborhood, loss of a job, or a substantial rise in taxes.

So anti-science rhetoric has been particularly effective in promoting a series of erroneous notions, including:

- Environmental scientists ignore the abundant good news about the environment.
- Population growth does not cause environmental damage and may even be beneficial.
- Humanity is on the verge of abolishing hunger; food scarcity is a local or regional problem and not indicative of overpopulation.
- Natural resources are superabundant, if not infinite.
- There is no extinction crisis, and so most efforts to preserve species are both uneconomic and unnecessary.
- Global warming and acid rain are not serious threats to humanity.
- Stratospheric ozone depletion is a hoax.
- The risks posed by toxic substances are vastly exaggerated.
- Environmental regulation is wrecking the economy.

How has the brownlash managed to persuade a significant segment of the public that the state of the environment and the directions and rates in which it is changing are not causes for great concern? Even many individuals who are sensitive to local environmental problems have found brownlash distortions of global issues convincing. Part of the answer lies in the overall lack of scientific knowledge among United States citizens. Most Americans readily grasp the issues surrounding something familiar and tangible like a local dump site, but they have considerably more difficulty with issues involving genetic variation or the dynamics of the atmosphere. Thus it is relatively easy to rally support against a proposed landfill and infinitely more difficult to impose a carbon tax that might help offset global warming.

Also, individuals not trained to recognize the hallmarks of change have difficulty perceiving and appreciating the gradual deterioration of civilization's life-support systems. This is why record-breaking temperatures and violent storms receive so much attention while a gradual increase in annual global temperatures—measured in fractions of a degree over decades—is not considered newsworthy. Threatened pandas are featured on television, while the constant and critical losses of insect populations, which are key elements of our life-support systems, pass unnoticed. People who have no meaningful way to grasp regional and global environmental problems cannot easily tell what information is distorted, when, and to what degree.

Decision-makers, too, have a tendency to focus mostly on the more obvious and immediate environmental problems —usually described as "pollution"— rather than on the deterioration of natural ecosystems upon whose continued functioning global civilization depends. Indeed, most people still don't realize that humanity has become a truly global force, interfering in a very real and direct way in many of the planet's natural cycles.

For example, human activity puts ten times as much oil into the oceans as comes from natural seeps, has multiplied the natural flow of cadmium into the atmosphere eightfold, has doubled the rate of nitrogen fixation, and is responsible for about half the concentration of methane (a potent greenhouse gas) and more than a quarter of the carbon dioxide (also a greenhouse gas) in the atmosphere today—all added since the industrial revolution, most notably in the past half-century. Human beings now use or co-opt some 40 percent of the food available to all land animals and about 45 percent of the available freshwater flows.

Another factor that plays into brownlash thinking is the not uncommon belief that environmental quality is improving, not declining. In some ways it is, but the claim of uniform improvement simply does not stand up to close scientific scrutiny. Nor does the claim that the human condition in general is improving everywhere. The degradation of ecosystem services (the conditions and processes through which natural ecosystems support and fulfill human life) is a crucial issue that is largely ignored by the brownlash. Unfortunately, the superficial progress achieved to date has made it easy to label ecologists doomsayers for continuing to press for change. At the same time, the public often seems unaware of the success of actions taken at the instigation of the environmental movement. People can easily see the disadvantages of environmental regulations but not the despoliation that would exist without them. Especially resentful are those whose personal or corporate ox is being gored when they are forced to sustain financial losses because of a sensible (or occasionally senseless) application of regulations.

Of course, it is natural for many people to feel personally threatened by efforts to preserve a healthy environment. Consider a car salesperson who makes a bigger commission selling a large car than a small one, an executive of a petrochemical company that is liable for damage done by toxic chemicals released into the environment, a logger whose job is jeopardized by enforcement of the Endangered Species Act, a rancher whose way of life may be threatened by higher grazing fees on public lands, a farmer about to lose the farm because of environmentalists' attacks on subsidies for irrigation water, or a developer who wants to continue building subdivisions and is sick and tired of dealing with inconsistent building codes or U.S. Fish and Wildlife Service bureaucrats. In such situations, resentment of some of the rules, regulations, and recommendations designed to enhance human well-being and protect life-support systems is understandable.

Unfortunately, many of these dissatisfied individuals and companies have been recruited into the self-styled "wise-use" movement, which has attracted a surprisingly diverse coalition of people, including representatives of extractive and polluting industries who are motivated by corporate interests as well as private property rights activists and right-wing ideologues. Although some of these individuals simply believe that environmental regulations unfairly distribute the costs of environmental protection, some others are doubtless motivated more by a greedy desire for unrestrained economic expansion.

At a minimum, the wise-use movement firmly opposes most government efforts to maintain environmental quality in the belief that environmental regulation creates unnecessary and burdensome bureaucratic hurdles which stifle economic growth. Wise-use advocates see little or no need for constraints on the exploitation of resources for short-term economic benefits and argue that such exploitation can be accelerated with no adverse long-term consequences. Thus they espouse unrestricted drilling in the Arctic National Wildlife Refuge, logging in national forests, mining in protected areas or next door to national parks, and full compensation for any loss of actual or potential property value resulting from environmental restrictions.

In promoting the view that immediate economic interests are best served by continuing business as usual, the wise-use movement works to stir up discontent among everyday citizens who, rightly or wrongly, feel abused by environmental regulations. This tactic is described in detail in David Helvarg's book, *The War Against the Greens:*

> To date the Wise Use/Property Rights backlash has been a bracing if dangerous reminder to environmentalists that power concedes nothing without a demand and that no social movement, be it ethnic, civil, or environmental, can rest on its past laurels.... If the anti-enviros' links to the Farm Bureau, Heritage Foundation, NRA, logging companies, resource trade associations, multinational gold-mining companies, [and] ORV manufacturers... proves anything, it's that large industrial lobbies and transnational corporations have learned to play the grassroots game.

Wise-use proponents are not always candid about their motivations and intentions. Many of the organizations representing them masquerade as groups seemingly attentive to environmental quality. Adopting a strategy biologists call "aggressive mimicry," they often give themselves names resembling those of genuine environmental or scientific public-interest groups: National Wetland Coalition, Friends of Eagle Mountain, the Sahara Club, the Alliance for Environment and Resources, the Abundant Wildlife Society of North America, the Global Climate Coalition, the National Wilderness Institute, and the American Council on Science and Health. In keeping with aggressive mimicry, these organizations often actively work *against* the interests implied in their names—a practice sometimes called *greenscamming.*

One such group, calling itself Northwesterners for More Fish, seeks to limit federal protection of endangered fish species so the activities of utilities, aluminum companies, and timber outfits utilizing the region's rivers are not hindered. Armed with a $2.6 million budget, the group aims to discredit environmentalists who say industry is destroying the fish habitats of the Columbia and other rivers, threatening the Northwest's valuable salmon fishery, among others.

Representative George Miller, referring to the wise-use movement's support of welfare ranching, overlogging, and government giveaways of mining rights, stated: "What you have... is a lot of special interests who are trying to generate some ideological movement to try and disguise what it is individually they want in the name of their own profits, their own greed in terms of the use and abuse of federal lands."

Wise-use sentiments have been adopted by a number of deeply conservative legislators, many of whom have received campaign contributions from these organizations. One member of the House of Representatives recently succeeded in gaining passage of a bill that limited the annual budget for the Mojave National Preserve, the newest addition to the National Parks System, to one dollar—thus guaranteeing that the park would have no money for upkeep or for enforcement of park regulations.

These same conservative legislators are determined to slash funding for scientific research, especially on such subjects as endangered species, ozone depletion, and global warming, and have legislated for substantial cutbacks in funds for the National Science Foundation, the U.S. Geological Survey, the National Aeronautics and Space Administration, and the Environmental Protection Agency. Many of them and their supporters see science as self-indulgent, at odds with economic interests, and inextricably linked to regulatory excesses.

The scientific justifications and philosophical underpinnings for the positions of the wise-use movement are largely provided by the brownlash. Prominent promoters of the wise-use viewpoint on a number of issues include such conservative think tanks as the Cato Institute and the Heritage Foundation. Both organizations help generate and disseminate erroneous brownlash ideas and information. Adam Myerson, editor of the Heritage Foundation's journal *Policy Review,* pretty much summed up the brownlash perspective by saying: "Leading scientists have done major work disputing the current henny-pennyism about global warming, acid rain, and other purported environmental catastrophes." In reality, however, most "leading" scientists support what Myerson calls henny-pennyism; the scientists he refers to are a small group largely outside the mainstream of scientific thinking.

In recent years, a flood of books and articles has advanced the notion that all is well with the environment, giving credence to this anti-scientific "What, me worry?" outlook. Brownlash writers often pepper their works with code phrases such as *sound science* and *balance* —words that suggest objectivity while in fact having little connection to what is presented. *Sound science* usually means science that is interpreted to support the brownlash view. *Balance* generally means giving undue prominence to the opinions of one or a handful of contrarian scientists who are at odds with the consensus of the scientific community at large.

Of course, while pro-environmental groups and environmental scientists in general may sometimes be dead wrong (as can anybody confronting environmental complexity), they ordinarily are not acting on behalf of narrow economic interests. Yet one of the remarkable triumphs of the wise-use movement and its allies in the past decade has been their ability to define public-interest organizations, in the eyes of many legislators, as "special interests"—not different in kind from the American Tobacco Institute, the Western Fuels Association, or other organizations that represent business groups.

But we believe there is a very real difference in kind. Most environmental organizations are funded mainly by membership donations; corporate funding is at most a minor factor for public-interest advocacy groups. There are no monetary profits to be gained other than attracting a bigger membership. Environmental scientists have even less to gain; they usually are dependent upon university or research institute salaries and research funds from peer-reviewed government grants or sometimes (especially in new or controversial areas where government funds are largely unavailable) from private foundations.

One reason the brownlash messages hold so much appeal to many people, we think, is the fear of further change. Even though the American frontier closed a century ago, many Americans seem to believe they still live in what the great economist Kenneth Boulding once called a "cowboy economy." They still think they can figuratively throw their garbage over the backyard fence with impunity. They regard the environmentally protected public land as "wasted" and think it should be available for their self-beneficial appropriation. They believe that private property rights are absolute (despite a rich economic and legal literature showing they never have been). They do not understand, as Pace University law professor John Humbach wrote in 1993, that "the Constitution does not guarantee that land speculators will win their bets."

The anti-science brownlash provides a rationalization for the short-term economic interests of these groups: old-growth forests are decadent and should be harvested; extinction is natural, so there's no harm in overharvesting economically important animals; there is abundant undisturbed habitat, so human beings have a right to develop land anywhere and in any way they choose; global warming is a hoax or even will benefit agriculture, so there's no need to limit the burning of fossil fuels; and so on. Anti-science basically claims we can keep the good old days by doing business as usual. But the problem is we can't.

Thus the brownlash helps create public confusion about the character and magnitude of environmental problems, taking advantage of the lack of consensus among individuals and social groups on the urgency of enhancing environmental protection. A widely shared social consensus, such as the United States saw during World War II, will be essential if we are to maintain environmental quality while meeting the nation's other needs. By emphasizing dissent, the brownlash works against the formation of any such consensus; instead it has helped thwart the development of a spirit of cooperation mixed with concern for society as a whole. In our opinion, the brownlash fuels conflict by claiming the environmental problems are overblown or nonexistent and that unbridled economic development will propel the world to new levels of prosperity with little or no risk to the natural systems that support society. As a result, environmental groups and wise-use proponents are increasingly polarized.

Unfortunately, some of that polarization has led to ugly confrontations and activities that are not condoned by the brownlash or by most environmentalists, including us. As David Helvarg stated, "Along with the growth of Wise Use/Property Rights, the last six years have seen a startling increase in intimidation, vandalism, and violence directed against grassroots environmental activists." And while confrontations and threats have been generated by both sides—most notably (but by no means exclusively) over the northern spotted owl protection plan—the level of intimidation engaged in by wise-use proponents is disturbing, to say the least....

Fortunately, despite all the efforts of the brownlash to discourage it, environmental concern in the United States is widespread. Thus a public-opinion survey in 1995 indicated that slightly over half of all Americans felt that environmental problems in the United States were "very serious." Indeed, 85 percent were concerned "a fair amount" and 38 percent "a great deal" about the environment. Fifty-eight percent would choose protecting the environment over economic growth, and 65 percent said they would be willing to pay higher prices so that industry could protect the environment better. Responses in other rich nations have been similar, and people in developing nations have shown, if anything, even greater environmental concerns. These responses suggest that the notion that caring about the environment is a luxury of the rich is a myth. Furthermore, our impression is that young people care especially strongly about environmental quality—a good omen if true.

Nor is environmental concern exclusive to Democrats and "liberals." There is a strong Republican and conservative tradition of environmental protection dating back to Teddy Roosevelt and even earlier. Many of our most important environmental laws were passed with bipartisan support during the Nixon and Ford administrations. Recently, some conservative environmentalists have been speaking out against brownlash rhetoric. And public concern is rising about the efforts to cripple environmental laws and regulations posed by right-wing leaders in Congress, thinly disguised as "deregulation" and "necessary budget-cutting." In January 1996, a Republican pollster, Linda Divall, warned that "our party is out of sync with mainstream American opinion when it comes to the environment."

Indeed, some interests that might be expected to sympathize with the wise-use movement have moved beyond such reactionary views. Many leaders in corporations such as paper companies and chemical manufacturers, whose activities are directly harmful to the environment, are concerned about their firms' environmental impacts and are shifting to less damaging practices. Our friends in the ranching community in western Colorado indicate their concern to us every summer. They want to preserve a way of life and a high-quality environment —and are as worried about the progressive suburbanization of the area as are the scientists at the Rocky Mountain Biological Laboratory. Indeed, they have actively participated in discussions with environmentalists and officials of the Department of the Interior to set grazing fees at levels that wouldn't force them out of business but also wouldn't subsidize overgrazing and land abuse.

Loggers, ranchers, miners, petrochemical workers, fishers, and professors all live on the same planet, and all of us must cooperate to preserve a sound environment for our descendants. The environmental problems of the planet can be solved only in a spirit of cooperation, not one of conflict. Ways must be found to allocate fairly both the benefits and the costs of environmental quality.

POSTSCRIPT

Are Environmental Regulations Too Restrictive?

In the original, much longer version of his essay, which is available at the Heritage Foundation's Internet library (`http://www.conservative.org/heritage/library/`), Shanahan spends a fair amount of time on particular environmental issues, such as global warming, ozone depletion, wetlands, pesticides, and hazardous wastes. The arguments will seem familiar to anyone who has paid attention to what conservative Republicans have been saying in their attempts to dismantle environmental protections.

The Ehrlichs have discussed particular issues as well. In "Ehrlich's Fables," *Technology Review* (January 1997), they write about "a sampling of the myths, or fables, that the promoters of 'sound science' and 'balance' are promulgating about issues relating to population and food, the atmosphere and climate, toxic substances, and economics and the environment."

Which side is right? The debate can be seen in the arguments over whether or not "Habitat Conservation Plans," devised as a way to moderate the impact of Endangered Species Act regulations on landowners, actually help endangered species. See Joselyn Kaiser, "When a Habitat Is Not a Home," *Science* (June 13, 1997). Perhaps the answer depends on what one values most: human beings or the environment; present prosperity or future survival; past tradition or new understanding.

In the Worldwatch Institute's *State of the World 1994* (W. W. Norton, 1994), Lester R. Brown says, "As the nineties unfold, the world is facing a day of reckoning. Many knew that this time would come, that at some point the cumulative effects of environmental degradation and the limits of the earth's natural systems would start to restrict economic expansion." In *State of the World 1996* (W. W. Norton, 1996), he states that that point is rushing upon us too fast for political systems to cope, due in large part to the addition of nearly 100 million people to the world's population each year and the consequent rapid increase in humanity's demand for resources. Catastrophe can be avoided, he contends, but not if we insist on doing things the way we always have, "not if we keep sleep-walking through history."

Since the 1992 Earth Summit (the UN Conference on Environment and Development) in Rio de Janeiro, the world's environmental problems have actually gotten worse. But according to Christopher Flavin, in "The Legacy of Rio," in *State of the World 1997* (W. W. Norton, 1997), the answer does not lie in some centralized world government but "in an eclectic mix of international agreements, sensible government policies, efficient use of private resources, and bold initiatives by grassroots organizations and local governments.

On the Internet . . .

National Aeronautics and Space Administration
At this site, you can find out the latest information on the Pathfinder mission—including Mars surface photos—as well as other NASA projects. *http://www.nasa.gov/*

Center for Mars Exploration
This is the starting point for an exploration of the history of Mars, with links to the Whole Mars Catalog and information about the Mars Pathfinder and Mars Global Surveyor missions. *http://cmex-www.arc.nasa.gov/*

The SETI Institute
The SETI Institute serves as a home for scientific research in the general field of life in the universe, with an emphasis on the search for extraterrestrial intelligence (SETI).
http://www.seti-inst.edu/

The SETI League
The SETI League, Inc., is dedicated to the electromagnetic (radio) search for extraterrestrial intelligence.
http://seti1.setileague.org/homepg.htm

National Center for Genome Resources
The National Center for Genome Resources (NCGR) is a not-for-profit organization created to design, develop, support, and deliver resources in support of public and private genome research. This site includes links to the Genome Sequence DataBase; Bugs 'n' Stuff, the NCGR's microbial genome site; and the Genetics and Public Issues program's Web site.
http://www.ncgr.org/

Office of Health and Environmental Research Biology Information Center
The Department of Energy's Office of Health and Environmental Research's Biology Information Center is a listing of Internet-accessible resources for biology and other related areas of research or development.
http://www.er.doe.gov/production/ober/bioinfo_center.html

PART 3

The Cutting Edge of Technology

Many interesting controversies arise in connection with technologies that are so new that they often sound more like science fiction than fact. Some examples are technologies that allow the exploration of outer space, the search for extraterrestrial intelligence, and genetic engineering. Such advances offer capabilities undreamed of in earlier ages, and they raise genuine, important questions about what it is to be a human being, the limits on human freedom in a technological age, and the place of humanity in the broader universe. They also raise questions of how we should respond: Should we accept the new devices and abilities offered by scientists and engineers? Or should we reject them?

- **Should the Goals of the U.S. Space Program Include Manned Exploration of Space?**
- **Is It Worthwhile to Continue the Search for Extraterrestrial Life?**
- **Should Genetic Engineering Be Banned?**

ISSUE 9

Should the Goals of the U.S. Space Program Include Manned Exploration of Space?

YES: Robert Zubrin, from "Mars on a Shoestring," in Robert Zubrin and Richard Wagner, *The Case for Mars: The Plan to Settle the Red Planet and Why We Must* (Free Press, 1996)

NO: John Merchant, from "A New Direction in Space," *IEEE Technology and Society Magazine* (Winter 1994)

ISSUE SUMMARY

YES: Engineer Robert Zubrin argues that it is within our capabilities to establish a human presence on Mars and that doing so would benefit economies on Earth.

NO: John Merchant, a retired staff engineer at Loral Infrared and Imaging Systems, argues that it will be much cheaper to develop electronic senses and remotely operated machines that humans can use to explore other worlds.

The dream of conquering space has a long history. The pioneers of rocketry —the Russian Konstantin Tsiolkovsky (1857–1935) and the American Robert H. Goddard (1882–1945)—both dreamed of exploring other worlds, although neither lived long enough to see the first artificial satellite, the Soviet *Sputnik,* go up in 1957. That success sparked a race between America and the Soviet Union to be the first to achieve each step in the progression of space exploration. The next steps were to put dogs (the Soviet Laika was the first), monkeys, chimps, and finally human beings into orbit. Communications, weather, and spy satellites were then designed and launched. And on July 20, 1969, the U.S. Apollo program landed the first men on the moon (see Buzz Aldrin and Malcolm McConnell, *Men from Earth,* Bantam Books, 1989).

There were a few more Apollo moon landings, but not many. The United States had achieved its main political goal of beating the Soviets to the moon and, in the minds of the government, demonstrating American superiority. Thereafter, the United States was content to send automated spacecraft (computer-operated robots) off to observe Venus, Mars, and the rings of Saturn; to land on Mars and study its soil; and even to carry recordings of Earth's sights and sounds past the distant edge of the solar system, perhaps to be retrieved in the distant future by intelligent life from some other world.

(Those recordings are attached to the Voyager spacecraft, launched in 1977; if you wish a copy, it was advertised in February 1994 as a combination of CD, CD-ROM, and book, *Murmurs of Earth: The Voyager Interstellar Record*, available from Time-Warner Interactive Group, 2210 Olive Avenue, Burbank, CA 91506.) Humans have not left near-Earth orbit for two decades, even though space technology has continued to develop. The results of this development include communications satellites, space shuttles, space stations, and independent robotic explorers such as the Mariners and Vikings and—landing on Mars on July 4, 1997—the Pathfinder craft (now the Sagan Memorial Station) and its tiny robot rover, Sojourner.

Why has human space exploration gone no further? One reason is that robots are now extremely capable and much cheaper; see David Callahan, "A Fork in the Road to Space," *Technology Review* (August/September 1993). Although some robot spacecraft have failed partially or completely, there have been many grand successes that have added enormously to humanity's knowledge of Earth and other planets. Another reason for the reduction in human space travel seems to be the fear that astronauts will die in space. This point was emphasized by the explosion of the space shuttle *Challenger* in January 1986, which killed seven astronauts and froze the entire shuttle program for over two and a half years. Still another is money: Lifting robotic explorers into space is expensive, but lifting people into space—along with all the food, water, air, and other supplies necessary to keep them alive for the duration of a mission—is much more expensive. And there are many people in government and elsewhere who feel that there are many better ways to spend the money on Earth.

Engineer Robert Zubrin believes that we could put humans and their supplies on Mars within a decade for much less money than government studies have estimated. In the following selection, he sketches his plans for how to do it and how to pay the bill. Engineer John Merchant argues in the second selection that there is a third choice besides sending people or independent robots into space, one that takes advantage of modern computer technology: instead of people, he advocates sending their eyes, ears, and hands to space in electronic (or virtual) form.

YES

Robert Zubrin

MARS ON A SHOESTRING

The planet Mars is a world of spectacular mountains three times as tall as Mount Everest, canyons three times as deep and five times as long as the Grand Canyon, vast ice fields, and thousands of kilometers of mysterious dry riverbeds. The planet's unexplored surface may hold unimagined riches and resources for future humanity, as well as answers to some of the deepest philosophical questions that thinking men and women have pondered for millennia.

Mars became an even more tantalizing destination in August 1996 when NASA scientists announced that an Antarctic meteorite—apparently from Mars—contained organic molecules and formations suggestive of microbes. If these are the remains of life, they may well be evidence of only the most modest representatives of an ancient Martian biosphere whose more interesting and complex manifestations are still preserved in fossil beds on the planet. To find them, though, will take more than robotic eyes and remote control. In fact, all that Mars holds will remain beyond our grasp until men and women—agile, autonomous, intuitive beings—walk upon its surface.

The usual proposals for launching a human mission to Mars, be they from the 1950s or the 1990s, call for enormous spaceships hauling to Mars all the supplies and propellant required for a two- to three-year round-trip mission. The size of these spacecraft means that they must be assembled in earth orbit —they're simply too large to launch from the earth's surface in one piece. Thus a virtual "parallel universe" of gigantic orbiting dry docks, hangars, cryogenic fuel depots, checkout points, and crew quarters must also be placed in orbit to enable assembly of the spaceships and storage of the vast quantities of propellant.

Such a mission to Mars would be exceedingly costly and would have to incorporate orbital construction and other technologies that won't be available for another 30 years. One such plan, known as the "90-Day Report," developed in response to President Bush's 1989 call for a Space Exploration Initiative, produced a cost estimate of $450 billion. Sticker shock in Congress doomed Bush's program and has deterred most people from seriously considering a humans-to-Mars program ever since. Continuing this trend, President

From Robert Zubrin, "Mars on a Shoestring," in Robert Zubrin and Richard Wagner, *The Case for Mars: The Plan to Settle the Red Planet and Why We Must* (Free Press, 1996).

Clinton announced in September [1996] that he would postpone any such mission until its cost could be justified.

Yet landing humans on Mars requires neither miraculous new technologies nor vast sums of money. We don't need to build futuristic "Battlestar Galactica" spaceships to go to Mars. We can reach the Red Planet within a decade with relatively small spacecraft launched directly to Mars by boosters embodying the same technology that carried astronauts to the moon more than a quarter-century ago. We simply need to use some common sense to travel light and "live off the land," just as was done by nearly every successful program of terrestrial exploration in history.

Consider what would have happened if Lewis and Clark had decided to bring all the food, water, and fodder needed for their transcontinental journey. They would have required hundreds of wagons, horses, and drivers, creating a logistics nightmare that would have sent the costs of the expedition beyond the resources of the America of Jefferson's time. Is it any wonder that Mars plans that don't make use of local resources manage to ring up $450 billion price tags? Living off the land is not just the way the West was won, it's the way the earth was won, and it's also the way Mars can be won.

Starting in the spring of 1990, I led a team of engineers and researchers at Martin Marietta Astronautics (now called Lockheed Martin Astronautics) in Denver in developing a live-off-the-land plan to pioneer Mars. Called Mars Direct, the plan discards unnecessary, expensive, and time-consuming detours: no need to assemble spaceships in low earth orbit; no need to refuel in space; no need for spaceship hangars at an enlarged space station; and no need for drawn-out development of lunar bases as a prelude to Mars exploration. Avoiding these detours saves perhaps 20 years and avoids the ballooning administrative costs that tend to afflict extended government programs.

If NASA were to undertake Mars Direct, the cost of developing the required hardware would amount to roughly $20 billion, with each individual Mars mission costing about $2 billion once the ships and equipment were in production. Spent over a period of 10 years, $20 billion would represent only about 7 percent of the combined U.S. budgets for military and civilian space exploration. This money could also drive our economy forward, much as spending $70 billion (in today's dollars) on science and technology in the Apollo program contributed to rapid U.S. economic growth during the 1960s.

MARTIAN CHRONICLES

As clearly as can be envisioned from today's vantage point, here's how the Mars Direct plan would work:

August 2005. A new multistage "Ares" rocket fashioned from preexisting parts rests on the launch pad at Cape Canaveral, its dewy metal skin steaming in the morning sunlight. The booster reminds some people of the old Saturn V, the rocket that carried astronauts to the shores of the Sea of Tranquillity. The Ares booster has about the same heavy-lift capacity as a Saturn V, but at its heart are the workhorses of the past 20 years, all drawn from the space shuttle: four main engines and two solid-rocket boosters.

The engines ignite. Flame and smoke signal the start of a new space age as the Ares 1 hurtles skyward. High above earth's atmosphere the upper stage separates from the spent booster, fires its single hydrogen- and oxygen-burning engine, and hurls a crewless 45-tonne payload to Mars (one tonne, or metric ton, is about 2,200 pounds). This payload is the earth return vehicle—the ERV.

As its name implies, the ERV is designed to carry a crew of astronauts back from Mars to a splashdown in earth's waters. On its journey to Mars the ERV carries a small nuclear reactor mounted atop a lightweight truck, fuel-production equipment—an automated chemical processing unit and a set of compressors—and a few scientific rovers. The ERV's crew cabin includes a life support system, food, and other necessities to sustain a four-member crew on an eight-month journey back to earth. Though its two propulsion stages will consume some 96 tonnes of methane and oxygen on the return flight, the ERV heads for Mars carrying just 6 tonnes of liquid hydrogen, the feedstock for producing propellant on the planet's surface.

February 2006. Traveling at an average speed of about 27 kilometers per second, the ERV reaches Mars after a six-month trip. Upon arrival, the ERV uses its aeroshell—a blunt, mushroom-shaped shield—to slow itself down by plowing through the upper reaches of the thin Martian atmosphere. As the craft brakes, it falls into orbit, where it remains for a few days to allow the flight controllers to perform a final system checkout. Then, at dawn on a day free of dust clouds, with well-defined shadows and low winds at the landing site, the craft is steered down into the atmosphere for final entry. Using its aeroshell again, the ERV decelerates to subsonic speeds until a parachute, and finally a set of small rockets, can ease the vehicle to the surface.

Once settled on the rust-colored soil of Mars, the ERV gets to the business at hand: making fuel for the return flight out of thin air. A door pops open on the side of the squat ERV landing stage and the truck carrying the nuclear reactor trundles out. Using a small TV camera on board, mission controllers at NASA's Johnson Space Center in Houston slowly drive the truck a few hundred meters away from the landing site. Once the controllers maneuver the vehicle to an appropriate spot, a winch lifts the reactor from the truck's bed and lowers it into a small crater or other natural depression. The reactor kicks in and begins to energize the chemical processing unit—connected by a cable—with 100 kilowatts of electricity.

Martian air is 95 percent carbon dioxide gas. The chemical plant combines the carbon dioxide with the hydrogen, producing methane, which the ship will store for later use as rocket fuel, and water. This "methanation" reaction is a simple, straightforward chemical process that has been practiced in industry since the 1890s. The reaction eliminates the need to store liquid hydrogen on the Martian surface: as the liquid hydrogen hauled from earth is used up, a second unit of the chemical plant splits the water produced by methanation into hydrogen and oxygen. The oxygen is stored as rocket propellant, while the hydrogen is recycled back into the chemical plant to make more methane and water. To obtain the ideal ratio of oxygen to fuel, a third unit splits Martian carbon dioxide into oxygen, which is stored, and carbon monoxide, which it vents as waste. After just six months, the chemical plant has used Mars's most freely available resource, its air, to leverage the 6 tonnes of liquid hydrogen brought from earth into 108 tonnes of methane and oxygen. This is enough to power the ERV's flight home and to provide 12 tonnes of fuel for ground vehicles.

September 2006. Thirteen months after launch, a fully fueled earth return vehicle sits on the Mars surface, awaiting the arrival of a human crew. Having monitored every step of the chemical production process, engineers in Houston give the go-ahead for the next step in the mission: sending out small robots from the ERV to photograph and conduct seismic exploration of the surrounding terrain. Back on earth, the crew of the first human expedition can then pick an ideal landing spot. One of the ERV robots ambles over to the site and plants a radar transponder to help guide the crew to a safe touchdown when it arrives.

October 2007. The Ares 3 launch vehicle, capped by a spacecraft called the *Beagle,* after the ship that carried Charles Darwin on his historic voyage, towers over the flatlands of the Cape, moments away from launch. Just a few weeks ago a similar booster, Ares 2, climbed into the skies over Florida. Carrying a backup ERV payload similar to the first, Ares 2 hurtles through space even as crowds gather to watch the launch of the ship that will convey the first four humans to Mars.

The main component of the *Beagle* is a habitation module that looks like a huge drum. The module stands about 5 meters high and measures about 8 meters in diameter. Consisting of two decks, each with 2.5 meters (about 8 feet) of headroom and a floor area of 100 square meters (about 1,000 square feet), it is large enough to comfortably house its crew. The "hab," as everybody calls it, contains a closed-loop life support system that recycles oxygen and water (similar to systems planned for the space station); whole food (irradiated for longer life and then canned or frozen) to last three years, plus a large supply of dehydrated emergency rations; and a ground car, pressurized so it provides a shirtsleeve environment and powered by an internal combustion engine that runs on methane and oxygen. Fully loaded, the hab weighs 25 tonnes.

The crew consists of a biogeochemist, a geologist/paleontologist, a flight engineer (who is also a top-notch pilot), and a jack-of-all-trades. This last crew member, who serves as mission commander, is primarily a flight engineer but can provide common medical treatment and understands the broad means and objectives of the scientific investigations.

On board the *Beagle,* the crew prepares for a journey that will return them home in two and a half years—about the same amount of time it took explorers centuries earlier to circumnavigate the globe. After the lower-stage engines erupt and the Ares 3 lifts off the pad, the upper stage fires its own engines, driving the hab to trans-Mars cruise velocity. Four humans are on their way to Mars.

April 2008. On the 180th day of flight, the hab reaches Mars and aerobrakes into orbit. The crew intends to set the *Beagle* down at the landing site near the ERV that flew out to Mars in 2005. In the unlikely event that the *Beagle* misses the site, the crew has three backup options. First, it has aboard the hab a fueled, pressurized rover with a one-way range of nearly 1,000 kilometers. So long as the crew is within that distance of the landing site, it can still get to the ERV by driving overland. If some disaster causes the *Beagle* to miss the mark by more than 1,000 kilometers, the second backup can be brought into play. This is the ERV launched by Ares 2, which, since it has traveled on a slower trajectory than the *Beagle,* is now following the crew to Mars. Even if the crew lands the hab on the wrong side of the planet, this second ERV can be maneuvered to land near them. Finally, as a third-level backup, the crew lands with supplies for three years—if worse comes to worst, the four can just tough it out on Mars until additional supplies and another ERV can be sent out in 2009.

The landing, however, is right on target. Though they have studied images of the landing site in detail, nothing can prepare the crew for the sight of the Martian landscape stretching before them. The soils are rust colored, littered with sharp-edged rocks, large and small. In the distance are small hills and dunes. The scene is akin to the deserts of America's Southwest, save for the skies, which are a constant ruddy salmon color. Although there's an immense amount to be done just after touchdown, the crew members take a moment to savor the knowledge that no human has ever gazed out on this vista.

With the *Beagle* safely down, the Ares 2 ERV lands some 800 kilometers away (distant enough to open a new site for exploration, but close enough to serve as a backup to the first ERV), where it begins filling itself with propellant. It will be used as the ERV for the second human expedition, which will arrive in 2010, along with another ERV that will open up Mars landing site number three.

The crew of the *Beagle* will spend 500 days on the Martian surface. Unlike conventional Mars-mission plans based upon orbiting mother ships with small landing parties, Mars Direct places all the crew on the surface, where they can explore and learn how to live in the Martian environment. No one is left in orbit, vulnerable to the hazards of cosmic rays and zero-gravity living, so there is no strong motive for a quick departure.

With such a substantial amount of surface time, the crew can pursue projects that will vastly expand our knowledge of the planet and pave the way toward future exploration and, eventually, human settlements. Geologic surveys will start piecing together the planet's climatic history, perhaps revealing how and when it lost its warm, wet climate. Such surveys will also prospect for useful minerals and other resources. Mars has undergone the same sorts of volcanic and hydrologic processes that produced a multitude of mineral ores on earth, and virtually every element of significant interest to industry is known to exist on the Red Planet.

Above all, astronauts will seek out easily extractable deposits of water ice or, better yet, subsurface bodies of geothermally heated water. Mars is known to possess veritable oceans of water frozen into its soil as permafrost. If discovered in a usable form, water would free future Mars missions from the need to import hydrogen from earth for making rocket propellant, and would make possible large-scale greenhouse agriculture once a permanent Mars base was established. An inflatable greenhouse will be brought along for experimentation. The exploration that will seize the attention of earth, though, will be the search for Martian life.

Images of Mars taken from orbit show dry riverbeds, indicating that Mars once had flowing water on its surface—that it was once a place potentially friendly to life. The best geologic evidence indicates that this warm and wet period lasted through the planet's first billion years, considerably longer than it took life to emerge on earth. Current theories hold that the evolution of living from nonliving matter is an orderly process with a high probability of occurring whenever conditions are favorable. If this is true, then chances are life should have evolved on Mars.

When NASA announced its discovery of possible microbe fossils last August, the agency termed the evidence compelling but not conclusive. Further investigations by the first human explorers could have a profound impact on our understanding of our place in the cosmos. Conclusive evidence of Martian life, present or fossilized, would strongly suggest that life abounds in the universe. On the other hand, if we find that Mars, despite its once clement climate, never produced any life, we would have to accept the possibility that life on earth is a fluke. We could be virtually alone in the universe.

The search for life will be intensive, and there are many different places to look. The planet's dry riverbeds and lake beds may have been the last redoubts of a retreating Martian biosphere, and thus might contain fossils. Water ice sheets covering the planet's north pole could hold well-preserved remains of actual organisms. Geologically heated ground water beneath the surface—if it exists—may yet harbor life. By studying the differences from, and similarities to, species that evolved on our own planet, we could begin to discern what is incidental to earth life and what is fundamental to the very nature of life itself. The results could lead to breakthroughs in medicine, genetic engineering, and all the biological and biochemical sciences.

To search for resources and signs of life, the first explorers on Mars will have to range across the Martian landscape. The pressurized ground rover will have enough fuel to rack up more than 24,000 kilometers on the odometer in the course of numerous week-long sorties. As the rover crew travels, it will leave behind small remote-controlled robots that will allow the base crew, and those of us on earth, to continue exploring a multitude of sites via television.

September 2009. After a year and a half on the Martian surface, the astronauts clamber aboard the ERV and blast off, receiving a hero's welcome on earth some six months later. They leave behind Mars Base 1, consisting of the *Beagle* hab, a rover, a greenhouse, power and chemical plants, a stockpile of methane/oxygen fuel for use by follow-up missions, and nearly all of their scientific instruments.

May 2010. Shortly after the first crew returns to earth, a second crew reaches Mars in Hab 2 and lands at what will become Mars Base 2. The crew of the second mission will spend most of its time exploring the territory around its own site but will probably drive over to the old *Beagle* at Mars Base 1 to continue scientific investigations begun there.

Every two years, a pair of Ares boosters will blast off, one delivering a hab to a prepared site, the other an earth return vehicle to open up a new region for the next mission. Over time, a network of exploratory bases will be established, turning large areas of Mars into human territory. But eventually one base region will be chosen as the site for building an actual Mars settlement. Ideally, this will be situated above a geothermally heated reservoir to supply hot water and electric power. In time, a set of structures resembling a small town will slowly take form.

The high cost of transportation between earth and Mars will create a strong financial incentive to find astronauts willing to extend their tours to four years, six years, and more. But over time, the transportation costs to Mars will steadily fall, driven down by new technologies and competitive bids from contractors offering to deliver cargo to support the base. Photovoltaic panels and windmills manufactured on site from Martian materials will add to the power supply, and locally produced inflatable plastic structures will multiply the town's pressurized living space. With more arrivals and longer stays, the population of the town will grow. In the course of things, children will be born, and families raised, on Mars —the first true colonists of a new branch of human civilization.

It is possible that someday millions of people will call Mars their home. Ultimately, we may even be able to devise techniques to alter the frigid, arid climate of Mars and return the planet to the warm, wet climate of its distant past. If a way were found to warm the polar caps, for example, enough carbon dioxide might be released into the atmosphere to warm the rest of the planet by means of the greenhouse effect. In so doing, we could transform Mars from a lifeless or nearly lifeless planet to a living, breathing world.

That is for the distant future. Yet we today have a chance to pioneer the way. All it takes is present-day technology mixed with some nineteenth-century chemical engineering, a dose of common sense, and a little bit of moxie.

GETTING DOWN TO BUSINESS

Now it's time to come back to earth. The greatest obstacle to gaining a foothold on Mars won't be found in the engineering details of a human Mars mission. It won't be found in the rigors of the journey or the long days exploring a new world. It won't be found on Mars. The greatest obstacle to sending humans to Mars resides here on our home planet in the guise of earthly politics. How can we raise the money needed to get the program off the ground?

President Bush's Space Exploration Initiative failed politically, but it was never seriously pushed. In actuality, there is plenty of latent political support in this country for a humans-to-Mars program. I have experienced this firsthand when speaking before numerous groups, from Rotary Clubs to plumbers' conventions—groups with no vested interest in a Mars program. The burning question is always, "How come we're not doing this?" In 1994, more than half of those queried in polls by *Newsweek* and CBS supported a piloted mission to Mars. And an ABC News/*Washington Post* poll conducted in early 1996 found that a majority of Americans believed that the space program had brought enough benefits to the country to justify its costs. In fact, the main public complaint I hear about the space program is not that it costs too much but that it's not going anywhere.

Any serious call for a low-cost near-term human mission to Mars—namely, Mars Direct—would engender a growing juggernaut of public support. Of course, translating enthusiasm into funding is another thing entirely. But I see three principal models—each very different—for accomplishing this. I call them the JFK model, the Carl Sagan model, and the Newt Gingrich model.

The JFK model is the most widely understood because it's the only one that has been done—it's how we reached the moon. In this model, the president of the United States calls upon the nation to meet the challenge of the future. In his call for the Apollo program, JFK said it would create new technologies, new jobs, and new knowledge, but that going to the moon was fundamentally "an act of faith and vision, for we do not know what benefits await us." Kennedy's Apollo speeches rang with destiny. Few people hearing those speeches could fail to recognize they were witnessing history being made. . . .

Although the JFK model succeeded once, the question may well be asked whether the nationalistic foundations that supported Apollo exist today. Instead of demonstrating American superiority, wouldn't it perhaps be better for a humans-to-Mars program to promote international cooperation? This brings us to an alternative approach, the one I call the Sagan model, after its most consistent, eloquent, and vocal advocate.

Carl Sagan has been promoting an international approach to Mars exploration for more than a decade. His original call focused on U.S.-Soviet collaboration, which he saw as a way to bind two adversarial nations together in a historic undertaking. The scientific talents of both nations would work toward ends other than enlarging their nuclear stockpiles.

Today, nations that decided to collaborate on such a project would still enjoy an obvious economic benefit: more partners means more pockets. Technologies as well as costs can be shared. The United States, for example—having long since abandoned the trusty Saturn V rocket—lacks a launch vehicle with enough power to launch a Mars Direct–style initiative. But Russia has the Energia, the most powerful rocket on the planet, capable of lofting 100 tonnes to low earth orbit. Energia has flown just twice, in part because it has lacked a mission; a humans-to-Mars program would fit the bill nicely. The design of the international space station calls for using several Russian modules.

. . . Sagan's proposal for a joint venture may make the Mars program a hostage to stability in Russia or elsewhere. But on the other hand, the prestige and economic benefits that can accrue from participation in such a project may help preserve the very peace and stability on which the project would depend.

EYES ON THE PRIZE

I call the third approach to getting humans to Mars the Gingrich approach because I devised it under the prodding of the speaker of the House. In the summer of 1994 I was invited to dine with Rep. Newt Gingrich (R-Ga.) and some of his staff to explain my ideas about Mars exploration. Gingrich was enthusiastic. "I want to support this with legislation," he told me, but he proposed doing it "in a more free-enterprise kind of way than just gearing up the NASA budget to go to Mars." He invited me to talk more about it on his TV show, which I did.

I then met several times with Jeff Eisenach, President of the Progress and Freedom Foundation, Gingrich's D.C. think-tank. What we came up with was the idea of a Mars Prize bill: the U.S. government would post a $20 billion reward for the first private organization to successfully land a crew on Mars and return it to earth, as well as several prizes of a few billion dollars each for attaining various technical milestones along the way.

This is, to say the least, a novel approach to human space exploration, which until now has been entirely government run. But it has a number of remarkable advantages. In the first place, this approach renders cost overruns impossible. The government will not spend a penny unless the desired results are achieved, nor spend a penny more than the award sum agreed upon at the start. Success or failure will depend solely upon the ingenuity of the American people and the workings of the free enterprise system, not upon political wrangling. After all, when Charles Lindbergh flew the Atlantic, he did it not as part of a government-funded program but in pursuit of a privately posted prize. Many such prizes were offered for breakthrough technical accomplishments in aviation's early years, and collectively they played a major role in raising the art of flight from its infancy to a globe-spanning transportation network.

There are other advantages as well. Because it would attract billions of dollars in private investment and technical development, the prize would spur economic growth even before any government expenditure. Moreover, this approach would call into being not only a private space race but a new kind of aerospace industry, one based on minimum-cost production. Under today's "cost plus" method, where the government pays aerospace companies 10 to 15 percent above whatever it costs them to do a job, contractors are overstaffed and have little incentive to control overhead. Under the prize system, the profit would be the value of the prize minus the company's costs, period. Firms would have a big incentive to drive costs down and would bear a much lighter accounting and documentation burden....

No doubt many people would be skeptical that a piloted Mars mission could be flown for $6 billion—but that wouldn't matter. If the Mars Prize bill were passed, the only thing that would matter was whether a few investors think it could. We wouldn't have to convince a majority of Congress, only a Bill Gates. And if nobody took up the challenge, the whole exercise would have cost the taxpayers absolutely nothing....

Establishing the first human outpost on Mars would be the most significant act of our age. People everywhere today remember King Ferdinand and Queen Isabella of Spain only because they are associated with the voyage of Christopher Columbus. In contrast, few can name the predecessors and successors of these monarchs, and all the wars, atrocities, palace coups, scandals, booms, and bankruptcies that must have seemed so important to the people of that time are today nearly forgotten. Similarly, almost no one 500 years from now will know what Operation Desert Storm was, let alone the Whitewater scandal. But they will remember who first settled Mars.

NO

John Merchant

A NEW DIRECTION IN SPACE

From the very beginning of the Space Age, a fundamental issue has been whether space exploration should be conducted by manned missions or by unmanned missions. In a so-called unmanned mission, only inanimate systems are transported into space. In a manned mission, human beings are on board the spacecraft.

Manned missions are very much more expensive because of the difficulty of maintaining human life in the implacably hostile environment of space. Many in the scientific community complain that the very expensive manned missions drain resources from unmanned scientific missions. For a variety of reasons, manned missions have attracted political support in spite of their very high cost. This support is eroding, however, and may not now be sufficient to achieve goals such as the human exploration of Mars. As a result, "America's manned space program is at a crossroads."

A third option may now be available: an unmanned mission providing virtually the same subjective and objective effect as a manned mission, but with the much lower cost of an unmanned mission. The ongoing revolution in digital processing technology may now make it possible to develop the means for a human to be effectively there, without going there, over interplanetary distances. Unlike conventional robotic rover missions, this third option would be, both subjectively and objectively, virtually equivalent to a manned mission.

Manned space exploration is currently based upon transportation technology that physically moves the human body to the space environment to be explored. This is a technology that has little or no terrestrial application. The progress of manned space exploration has been slow, and is likely to continue that way, because of the prohibitive cost and the danger to human life of the current technology. It is conceivable that manned space exploration, based upon the current technology, might even be abandoned at some point. The new (third) option for manned-equivalent space exploration is based upon semi-autonomous control technology. Investment in this technology is likely to pay off with critically important industrial applications on Earth in the coming decades, no matter what may happen in space. Manned-equivalent

From John Merchant, "A New Direction in Space," *IEEE Technology and Society Magazine*, vol. 13, no. 4 (Winter 1994), pp. 22–29. References and some notes omitted.

space exploration by this new approach will be much less expensive, and much more likely to prosper and to produce knowledge and commercial opportunities in space....

THE REMOTE PROJECTION OF HUMAN ACTION

In normal human action, an operator interacts directly with the local environment in which the operator is located. The operator senses that environment with sense organs and then effects changes in it by the direct neuromuscular effort of arms, hands, fingers, legs, etc,. The operator is 100% coupled to the local environment. Our entire species experience has been that the only way of effecting human action is by the direct location of the body. To allow humans to explore space, the technological response has therefore been to provide the means, however expensive, of transporting the human body into space. However, human action is an information process. Information can be transmitted, or projected, over great distances. Thus, at least in principle, as in a thought experiment, human action can be effected at a remote location, without locating the body there, by the remote projection of human action.

Remote projection allows an operator to effect human action on a remote environment exactly as if the operator's body were there. In our thought experiment, the operator interfaces directly with a Local Projection Unit (LPU), which is located, with the operator, in the local environment. The LPU is connected via a data link to a Remote Projection Unit (RPU) located in the remote environment. Physical sensors on the RPU derive exactly the same sensory information that the operator would sense if the operator's body were actually located in the remote environment. This information is relayed back to the LPU and presented to the operator's sense organs so that the operator experiences exactly the same subjective sensation as if the operator's body were in the remote environment. Likewise, the neuromuscular effort exerted by the operator in response to this sensory information is measured by transducers in the LPU, relayed to the RPU, and then applied to mechanical transducers in the RPU to cause the same mechanical effort to be exerted on the remote environment as if the operator's body were indeed located there.[1] In this idealized thought experiment the operator is 100% coupled to the remote environment.

Conventional remote control falls between normal human action and the remote projection of human action. The operator remains strongly coupled to the local environment and is only weakly coupled to the remote environment. Only a small part of human action is projected to the remote environment. For example, in remote control of an underwater robot (Remotely Operated Vehicle or ROV) the operator's visual field is almost entirely covered by the local scene of the cabin in the mother ship where the operator is located (strong coupling to the local environment), whereas the operator can see only a small TV image of the remote scene (weak coupling to the remote environment). Likewise, the ROV operator can exert full neuromuscular effort on the local environment on the ship but has only a crude capability, via the robot's manipulator arms, to exert neuromuscular effort on the remote underwater environment.

The remote projection of human action does not violate physical law and is, therefore, a valid thought experiment.

(It is assumed, for the moment, that the distance over which human action is projected is small enough [e.g., less than 1,000 miles] so that the round trip transmission delay involved [0.005 sec over 1,000 miles] is negligible relative to the human physiological response time [about 0.1 sec]. Practical implementation, and the extension to very much longer distances, will be considered later.)

By the remote projection of human action an operator can effect human action on a remote environment, exactly as if the operator's body were there. The subjective sensation would be the same as being there. Objectively, the ability to effect change would be the same as being there. For all practical purposes, the operator is there. But the operator's body is not.

What exactly does it mean to "be there"? For inanimate objects there is no doubt about what it means. A car can unambiguously be said to be in the parking lot when it is physically located in that parking lot. However, for a person, there are two entirely different meanings, corresponding to which of the two entirely different types of human interaction apply—physical interaction or human action. For physical interaction, which supports the physical existence of the body by exchange of physical entities, to be there necessarily means to locate the body there. But this is not the case for the information process, defined as human action. Suppose a person is carrying out human action in an environment in such a way as to be indistinguishable from the way it would be carried out if that person's body were present in that environment. This would be virtually equivalent to that person being present in that environment. (The physical interaction of the person with the environment is irrelevant in the present context.)

If, indeed, this could be done over interplanetary distances, then transporting the human body into space—at enormous expense and risk—would make absolutely no sense. But this is true only if the great leap can be made from conventional remote control to a practical implementation of the remote projection of human action of such fidelity as to be virtually equivalent to going there.

It will be seen that the ongoing revolution in digital processing technology may make it possible to implement the remote projection of human action to this standard over interplanetary distances. It may be possible, for example, to operate a space station in Earth orbit, to set up a lunar base, or to explore Mars, without humans ever leaving the Earth.

This is the proposed new direction in space. It will be described here in the context of a mission to Mars, but is applicable to any mission of space exploration for which a manned mission might otherwise be considered.

SPECIFICATION

Can what was described in principle actually be done in practice? In one very limited, but extremely important, way it already has been done—by the telephone.

The sense of hearing and the neuromuscular action of speech comprise a very important, self-contained, subset of human action by which much of the communication between human beings is effected. Words spoken by another person are received by the sense of hearing. This information is processed by the brain to derive an appropriate response. The vocal cords are then exercised by neuromuscular effort to send this response back by speech.

Relative to the full set of human action, the telephone is conventional remote control. However, relative to the self-contained action subset of hearing and speech the telephone implements the remote projection of human action over terrestrial distances—with great economic advantage over going there.

To project human action into space it is not necessary to project the full range of human sensory and neuromuscular functions but only those that an astronaut would be able to, and would need to, deploy while exploring and developing space.

The visual sense is fundamental for exploration, whether on Earth or in space. Accordingly, if an operator on Earth is to explore Mars (for example) by the remote projection of human action, then that operator should be provided with remote vision of Mars virtually equivalent to the vision of Mars that would be experienced by an astronaut actually on Mars. The remote display of the Martian scene should therefore cover the operator's entire field of view no matter in which direction the operator's head is pointed. The operator should be able to instantly derive the dynamic visual effect of looking around and moving over the surface of Mars at will, going in for close-up views of detail of special interest. The subjective sensation of viewing the Martian scene remotely should in all respects be virtually the same as going there, moving around, and viewing the scene directly.

In order for the remote projection of human action to be a valid alternative to going there, the operator on Earth should also be provided with a capability for exerting remote neuromuscular effort on Mars virtually equivalent to that of an astronaut on Mars. That is, the operator on Earth should be able, in effect, to move around on Mars, pick up and examine Martian artifacts, operate scientific equipment, operate construction equipment, assemble structures, etc. The operator should also be able to repair equipment on Mars. In particular, when operating remotely via one RPU, the operator should be able to repair another RPU.

It is not necessary that these tasks be performed as fast as they could be by the direct action of an astronaut. Unlike exploration by a manned mission, exploration by the remote projection of human action could be almost continuously active over an essentially indefinite period with a number of RPUs operating simultaneously. An overall mission-effectivity at least equal to that of a manned mission could therefore be achieved even if the average rate of task performance of each RPU was significantly (e.g., up to ten times) lower than by the direct action of an astronaut.

These are the basic sensory and neuromuscular requirements of a system to project human action into space. A very challenging aspect of this specification is that it must be satisfied over very long operating ranges. For a space station in Earth orbit, the operating range would be of the order of 10,000 miles, for exploration of the moon 250,000 miles, and for exploration of Mars 100 million miles.

The transmission delay introduced by these long round-trip distances has a devastating effect on conventional remote control. Suppose that an operator on Earth were trying to aim a TV camera on Mars. If the operator commanded the camera to pan over to a new direction it would take about half an hour before the TV picture relayed back from Mars would change in response to this command. If the operator started to pick up a rock on Mars with a remotely controlled mechanical hand it would be half an hour before the operator saw the mechanical hand begin to move in the TV picture relayed back from Mars. The hand-eye coordination needed to grasp and pick up objects would be virtually impossible.

Another challenging aspect of the specification is that a remote vision capability far in excess of conventional TV is required. Even a single standard TV channel would strain the capacity of the Mars-Earth data link.

In the next section it will be shown how the new technology of semi-autonomous systems, AI, and image processing might be applied to satisfy these challenging technical requirements. The processing power needed to apply these sophisticated techniques may now be available as a result of the digital processing revolution.

IMPLEMENTATION

To project human action to Mars (for example) a Remote Projection Unit (RPU) is located on Mars and connected, via an interplanetary data link, to a Local Projection Unit (LPU) located on Earth. A human operator on Earth is interfaced to the LPU to allow the operator to effect human action on Mars.

The RPU is equipped with one or more video cameras, means of locomotion, a navigation device giving its current position in local Martian coordinates, and a digital processor. The LPU has a powerful digital processor and a spherical dome display screen consisting of approximately 100 high-resolution display panels. This screen provides the operator, at the center of the dome, with a [spherical] visual display.

New technology is applied to provide the specified functions of *remote vision* and *remote neuromuscular effort* on Mars. As a result of the ongoing digital processing revolution, the processing power needed to implement these sophisticated functions may now be available.

Remote Vision

As the RPU moves around on Mars, a TV camera mounted on the RPU will generate a continuous sequence of 30 video frames per second of the Martian scene. Single frames are selected from this camera at a very low frame rate (e.g. one frame every three seconds) and transmitted back to Earth. (At this low frame rate, only a small part [e.g. 0.1 Mbps] of the channel capacity of the data link will be used.) As these frames are received on Earth they are [used to construct a]... 3-D model of the Martian scene [that] is stored in the LPU. It is continuously extended and updated as new video frames are received. As soon as enough of the model becomes available, it is processed in real time by the LPU processor to generate whatever view of the Martian scene the operator has currently selected. Images of RPUs and other human-made objects will be automatically inserted into this view based upon current estimates of their position and orientation on Mars. This high-resolution view of the Martian scene, covering the operator's entire visual field at all times, is displayed on the operator's... spherical display screen. The operator sees nothing of the local environment. If the operator's head should be turned to look to the side, or even all the way around to look backwards, the display will always present to the operator the corresponding view of the Martian scene.

The displayed view is computed, 30 times a second, according to the viewing point and viewing direction on Mars as currently selected by manual action of the operator. The operator is able to command continuous motion of this viewing point along any path. The corresponding sequence of images generated by the LPU instantly provides the operator with the same dynamic visual effect, over the operator's entire visual field, as if the operator were actually walking, driving, or even flying around the Martian scene. The operator therefore experiences virtually the same total subjective visual sensation as an astronaut would, actually there on Mars.

The derivation of a 3-D model... and then the generation of selected views from this model, has been demonstrated in video releases from NASA to the TV networks after an encounter of a spacecraft with a planet or moon.... [It will] not only make it possible to overcome the effect of transmission delay on the operation of the remote vision system, it [will] also make much better use of the limited capacity of the data link between Mars and Earth. Instead of continuously transmitting the highly redundant sequence of video frames at 30 frames/sec from the camera, new scene structure information is transmitted only when needed to update the current 3-D model in the LPU, and then at a very low frame rate.

This remote vision system requires a powerful digital processor in the LPU to perform the stereo matching and also to generate the very-wide-angle, high-resolution output imagery at 30 frames per second. About 100 high-resolution flat-panel displays (FPDs) would be needed to present this imagery to the operator. To limit the processing load, the operator's head direction can be measured and the resolution of the computed imagery reduced in the peripheral areas of the operator's visual field. In particular, no picture information at all need be computed over the [half-sphere] immediately behind the operator.

By terrestrial standards this remote vision system is expensive, but not relative to space systems and operations. Although complex and highly sophisticated, no technological breakthrough is needed.

In actual operation a number of RPUs would be deployed at one location on Mars, together with scientific instruments and support equipment for the mission of exploration and development. A central control unit equipped with a powerful processor would control all of the RPUs. Roving RPUs would be used to generate the video frames to be transmitted to Earth. These frames would be buffered and processed in the LPU, on Earth, to derive the 3-D structure of the region. As the operator in the LPU explored the scene currently being displayed, the operator's areas of interest would be noted and relayed back to the central control unit on Mars. The control unit would then redirect the roving RPUs to generate more detailed views of these particular areas so that a correspondingly more detailed 3-D model of them could be computed by the LPU.

On moving into a new scene area it might take several hours for the roving RPUs to gather video imagery of that area and transmit it back to Earth, and for the 3-D model of that Martian scene to be computed by the LPU processor. If the operator should designate a particular region as being of special interest, there would be a similar delay before a close-up 3-D model of that region became available on Earth. These delays would be accommodated by appropriate scheduling of the operator's work. While close-up views of one region were being developed by one RPU, the operator could direct other RPUs to perform other tasks such as operation of scientific equipment, gathering of samples, etc.

Remote Neuromuscular Effort

Because of the approximately 30-minute transmission delay, the operator on Earth cannot exert neuromuscular effort on Mars by direct, micro-managing, remote control of the RPU. Instead, functioning as a supervisor, the operator will generate a sequence of high-level instructions, transmit them to the RPU, and then leave it to the RPU to carry them out. For example, the operator might designate a point on the operator's display of the Martian scene and instruct the RPU to move to that point, pick up a particular rock, and then bring that rock to a scientific test station for analysis. This packet of high-level directives would be transmitted to Mars and then executed autonomously by the RPU during the transmission-delay period. The operator, like a good supervisor, would leave it to the RPU to do the work, checking only afterwards that the instructions had been properly carried out.

The RPU uses its on-board navigational sensors and TV cameras to sense its position relative to the terrain. Range information is derived by laser ranging and/or stereo matching (in the RPU) of views of the terrain immediately ahead. All of this information is used by the RPU to help it move, autonomously, from point to point as commanded by the operator.

The RPU uses its manipulator arms to perform the manual action tasks that have been commanded by the operator. Laser ranging and/or stereo matching (by the RPU processor) of the images from TV cameras on the RPU will provide whatever 3-D information may be required for "hand-eye" coordination in the autonomous execution of these tasks.

The RPU on Mars must be able to operate autonomously for a period of the order of the transmission delay. The operator will send a detailed, time-sequenced, list of tasks to be performed during this period. These tasks will be defined according to the operator's goals and by the remote visual sensing by the operator of the Martian environment. Because the environment in which the RPU will operate is highly predictable, all of these tasks will generally be as appropriate when they are received on Mars as when they were formulated on Earth some time earlier. In other words, the RPU needs only enough intelligence to perform each individual task with a high probability of success, determine when it has been completed, or determine when the task cannot be completed. Thus, although the RPU must operate autonomously during the transmission delay period, the bulk of the intelligence applied during that time is the human intelligence of the operator on Earth who formulates the task list.

The new subsumption (robot control) architecture developed by Brooks, in which there is a tight coupling of sensing to action rather than the top-down approach of traditional artificial intelligence, may be particularly appropriate to provide the limited task-based intelligence needed by the RPU.

For navigation, the task-list from the operator will specify the route to be followed as a series of arcs along the Martian surface defined in Martian spatial coordinates. A number of systems (robots) have been developed that demonstrate autonomous movement over rough terrain.

In addition to navigation, the RPU must be able to use its manipulator arms to autonomously assemble, operate, and repair human-made systems as commanded by the operator. In many cases the operator will issue a list of task macros for these mechanical tasks. The detailed task list for each macro will be pre-defined as part of the design of these human-made systems. To facilitate task execution, the human-made components will be specially designed, marked, and color-coded for automatic recognition and ranging by image processing, and for handling by mechanical hands.

Highly sophisticated mechanical arms and hands (manipulators) have been demonstrated that, for example, can hold an egg and crush a can. Autonomous control of manipulators to perform mechanical tasks has been demonstrated.

The RPU will also collect samples from the Martian surface for detailed visual examination or analysis by scientific instruments. The operator will designate on the visual display of the Martian scene in the LPU the samples to be collected as well as the route to be taken by the RPU to get to them. Once again, only a limited intelligence in the RPU is needed to move along a designated path, grasp and pick up the designated object, and retrieve it if it should be dropped.

Just as terrestrial systems are human-engineered for easy operation and maintenance by humans, all of the scientific equipment, construction equipment, forklifts, and other support equipment deployed on Mars would be "RPU-engineered" to facilitate operation by RPUs. For example, control and operation of this equipment would be exerted by the RPUs by wireless (radio) control, not by knobs, levers, and switches. Diagnostic information for servicing would likewise be derived mainly by wireless interface, rather than by test probes. The support equipment would be marked and mechanically designed to facilitate assembly and maintenance by means of tools and techniques specially designed to exploit the positive features of a manipulator arm rather than those of the human hand.

Once Remote Projection Units (RPUs) were deployed on Mars, or other space locations, they would remain in operation there indefinitely, repairing each other as necessary. On Earth, there would be continuous access to these space locations for exploration and commercial development.

Processing Power

The processing power that can be implemented in human-made systems, virtually zero just a few decades ago, is now approaching that of sophisticated physiological systems. Raw processor speed has increased dramatically, and this increase is still continuing. New device technologies, for example based upon quantum effects or the use of photons instead of electrons, may provide additional dramatic increases in speed. Massively parallel computing is another, potentially very important, route to increased computational power.

From the powerful computers commonplace in all aspects of daily life to virtual reality and the projected information superhighway, the digital processing revolution—a great watershed event in the history of human development—is already transforming life here on Earth. This new technology may now make it possible to implement the functions of remote vision and remote neuromuscular effort—that is, to project human action over interplanetary distances.

CONCLUSION

The benefits of the development and application of the remote projection of human action for space exploration are seen as:

1. The same mission result is achieved as would be derived from a manned space mission, but at much lower cost. This is because only inanimate systems, not human bodies, are transported into the alien environment of space.
2. No danger to human life.
3. Exploration and development of a space environment could be conducted indefinitely, since there is no requirement to return human bodies to Earth.
4. Because of items 1, 2, and 3, commercial operations in space would be more likely to develop.
5. Exploration could be conducted by anybody—not just by astronauts.
6. In the next century the semi-autonomous robotic, vision, and computer technologies upon which the remote projection of human action is based will likely find very important industrial application for enhanced productivity. There are already a number of terrestrial applications for telerobots.[2]
7. The development of the remote projection of human action would preserve and utilize the talents of scientists and engineers, and the resources of industry, now becoming available as military R&D [research and development] programs wind down.

The basis of the technology required to project human action into space already exists: high-resolution display panels, manipulator arms, autonomous execution of elementary tasks, generation of 3-D models of the object scene from moving video and from stereo pairs, powerful digital processors, etc. Can all of this technology be extended, developed and assembled to provide, within an acceptable cost and schedule, virtual equivalence to going there?

Initial design studies should be undertaken to assess feasibility. Among the many issues to be considered are the algorithms for remote vision and remote neuromuscular effort, the processing power (gigaflops) and electrical power (watts) needed at the remote space location, the processing power needed in the LPU to drive the display, and the display system itself. Non-real-time, and real-time, laboratory demonstrations of the critical functions of remote vision and remote neuromuscular effort would follow the initial design study. Finally, a full terrestrial demonstration of the remote projection of human action over interplanetary distances would be undertaken. The interplanetary transmission delay would be simulated by passing the high-level, and thus very low bandwidth, operator commands through a digital delay line.

This step-by-step program would establish the feasibility, quantify the cost, and demonstrate performance of the remote projection of human action into space. Technical risk would be low since each step would be taken only when justified by prior results.

The information-processing revolution is a watershed event in human history. Looking forward from the vantage of this great technological divide, development of the remote projection of human action over interplanetary distances may already, or will likely soon, be possible. Then, the impasse of prohibitive cost that has bedeviled the manned space program from its inception would be broken.

NOTES

1. This combination of RPU, LPU, and data link is known as a teleoperator if it functions as a direct mindless extension of the operator's senses and motor effectors, or as a telerobot if it has sufficient intelligence so that the operator's role is supervisory.

2. Undersea oil operations, undersea science, nuclear power plants and radioactive "hot cells," toxic waste cleanup, construction, agriculture, mining, warehousing and mail delivery, firefighting and lifesaving, policing, military operations, assistive devices for the disabled, telediagnosis, telesurgery.

POSTSCRIPT

Should the Goals of the U.S. Space Program Include Manned Exploration of Space?

Interest in Mars expeditions is hardly new. During the Bush administration, the Synthesis Group of the White House Science Office, given the task of finding the best way to go to Mars, concluded that the benefits of such a mission would include job creation, investment in science and technology, and stimulated innovation. The Synthesis Group's report included a call for the development of a Heavy Lift Launch Vehicle (HLLV) to reduce the cost and difficulty of getting materials into orbit. Such a cargo vehicle has been discussed for years as a supplement to the space shuttle, with a great deal of emphasis on its value for constructing space stations as well as interplanetary spaceships. The dream continues—see Stanley Schmidt and Robert Zubrin, eds., *Islands in the Sky: Bold New Ideas for Colonizing Space* (John Wiley, 1996) —but an HLLV has never been approved.

In the fall of 1993 a new technology was introduced: the McDonnell-Douglas Delta Clipper DC-X, which boasted vertical-takeoff-and-landing capabilities and total reusability. Despite successful test flights and promising HLLV potential, however, NASA chose to award the contract for the next-generation space shuttle to Lockheed Martin's VentureStar, a more conventional, shuttle-like design. See G. Harry Stine, *Halfway to Anywhere* (M. Evans, 1996).

Do we need to send people to explore Mars? Won't robots do? The question is timely because in 1997—two decades after the two Viking landers extensively mapped out and characterized the Red Planet—scientists returned to Mars with the Pathfinder lander and its accompanying Sojourner rover. The Mars Global Surveyor arrived in September to photograph the planet from orbit, inventory rock types, and map future landing sites. It seems likely that as long as the robots continue to succeed in their missions, manned missions will continue to be put off because funding shortages will probably continue. Funding for space exploration remains low largely because problems on Earth (environmental and other) seem to need money more urgently than space exploration projects do. There is also a strong drive to reduce government spending and shrink the federal deficit. As a result, experts expect major new missions—manned or unmanned—to become rare events. See Christopher Anderson, "The Coming Crunch for Space Science," *Science* (February 18, 1994). Certainly the prospects for a renewal of manned space exploration, much less a trip to Mars, seem very dim.

ISSUE 10

Is It Worthwhile to Continue the Search for Extraterrestrial Life?

YES: Frank Drake and Dava Sobel, from *Is Anyone Out There? The Scientific Search for Extraterrestrial Intelligence* (Delacorte Press, 1992)

NO: A. K. Dewdney, from *Yes, We Have No Neutrons: An Eye-Opening Tour Through the Twists and Turns of Bad Science* (John Wiley, 1996)

ISSUE SUMMARY

YES: Professor of astronomy Frank Drake and science writer Dava Sobel argue that the search for radio signals from extraterrestrial civilizations has only just begun and that scientists must continue to search because contact will eventually occur.

NO: Computer scientist A. K. Dewdney argues that although there may indeed be intelligent beings elsewhere in the universe, there are so many reasons why contact and communication are unlikely that searching for them is not worth the time or the money.

In the 1960s and early 1970s the business of listening to the radio whispers of the stars and hoping to pick up signals emanating from some alien civilization was still new. Few scientists held visions equal to Frank Drake, one of the pioneers of the search for extraterrestrial intelligence (SETI) field. Drake and scientists like him utilize radio telescopes—large, dish-like radio receiver–antenna combinations—to scan radio frequencies (channels) for signal patterns that would indicate that the signal was transmitted by an intelligent being. In his early days, Drake worked with relatively small and weak telescopes out of listening posts that he had established in Green Bank, West Virginia, and Arecibo, Puerto Rico. (See Carl Sagan and Frank Drake, "The Search for Extraterrestrial Intelligence," *Scientific American*, May 1975.)

There have been more than 50 searches for extraterrestrial radio signals since 1960. The earliest ones were very limited. Later searches have been more ambitious, culminating in the 10-year program known as the High Resolution Microwave Survey (HRMS). The HRMS, which began on Columbus Day of 1992, uses several radio telescopes and massive computers to scan 15 million radio frequencies per second. This has been the most massive SETI to date and the one with the greatest hope of success.

At the outset, many people thought—and many still think—that SETI has about as much scientific relevance as searches for Loch Ness Monsters and

Abominable Snowmen. However, to Drake and his colleagues, it seems inevitable that with so many stars in the sky, there must be other worlds with life upon them, and some of that life must be intelligent and have a suitable technology and the desire to search for alien life too.

Writing about SETI in the September–October 1991 issue of *The Humanist*, physicist Shawn Carlson compares visiting the National Shrine of the Immaculate Conception in Washington, D.C., to looking up at the stars and "wondering if, in all [the] vastness [of the starry sky], there is anybody out there looking in our direction.... [A]re there planets like ours peopled with creatures like us staring into their skies and wondering about the possibilities of life on other worlds, perhaps even trying to contact it?" That is, SETI arouses in its devotees an almost religious sense of mystery and awe, a craving for contact with the *other*. Success would open up a universe of possibilities, add immensely to human knowledge, and perhaps even provide solutions to problems that our interstellar neighbors have already defeated.

SETI also arouses strong objections, partly because it challenges human uniqueness. Many scientists have objected that life-bearing worlds such as Earth must be exceedingly rare because the conditions that make them suitable for life as we know it—composition and temperature—are so narrowly defined. Others have objected that there is no reason whatsoever to expect that evolution would produce intelligence more than once or that, if it did, the species would be similar enough to humans to allow communication. Still others say that even if intelligent life is common, technology may not be so common. Richard C. Teske, for example, in "Is This the E.T. to Whom I Am Speaking?" *Discover* (May 1993), argues that the geological processes that have supplied humans with the raw materials of technology—metals—are too unlikely to have been repeated elsewhere. A similar criticism is that technology may occupy such a brief period in the life of an intelligent species that there is virtually no chance that it would coincide with Earth scientists' current search. Whatever their reasons, SETI detractors agree that listening for extraterrestrial signals is futile.

In the selections that follow, Drake and science writer Dava Sobel discuss Drake's career-long search for messages from distant stars. Today's technology, the authors note, has made it possible to duplicate the work of the very first search (Project Ozma) in a fraction of a second, making it that much more probable that Earth will soon make contact with extraterrestrials. Computer scientist A. K. Dewdney discusses several reasons why listening for extraterrestrial signals is a waste of time and money.

YES

Frank Drake and Dava Sobel

NO GREATER DISCOVERY

My scientific colleagues raise their eyebrows when I speculate on the appearance of extraterrestrials. But about 99.9 percent of them agree wholeheartedly that other intelligent life-forms do exist—and furthermore that there may be large populations of them throughout our galaxy and beyond.

Personally, I find nothing more tantalizing than the thought that radio messages from alien civilizations in space are passing through our offices and homes, right now, like a whisper we can't quite hear. In fact, we have the technology to detect such signals *today,* if only we knew where to point our radio telescopes, and the right frequency for listening.

I have been scanning the stars in search of extraterrestrial intelligence (an activity now abbreviated as SETI, and pronounced *SET-ee*) for more than thirty years. I engineered the first such effort in 1959, at the National Radio Astronomy Observatory in Green Bank, West Virginia. I named it "Project Ozma," after a land far away, difficult to reach, and populated by strange and exotic beings. I used what would now be considered crude equipment to listen for signals from two nearby, Sunlike stars. It took two months to complete the job. With the marvelous technological advances we have made in the intervening years, we could repeat the whole of Project Ozma today in a fraction of a second. We could scan for signals from a *million* stars or more at a time, at distances of at least a *thousand* light-years from Earth....

Until the late 1980s, the fact that we had not yet found another civilization, despite continued global efforts and better equipment, simply meant we had not looked long enough or hard enough. No knowledgeable person was disappointed by our inability to detect alien intelligence, as this in no way proved that extraterrestrials did not exist. Rather, our failure simply confirmed that our efforts were puny in relation to the enormity of the task—somewhat like hunting for a needle in a cosmic haystack of inconceivable size. The way we were going about it, with our small-scale attempts, was like looking for the needle by strolling past the haystack every now and then. We weren't embarked on a search that had any real chance of success.

Then many people began to grasp the nature and scope of the challenge, the consequent investment required to succeed, and the importance of success to all humanity. They pushed relentlessly for a serious search. And won. The National Aeronautics and Space Administration (NASA) committed $100 million to a formal SETI mission spanning the decade of the 1990s, making the work a priority for the space agency and guaranteeing that coveted telescope time will be devoted to the search.

Now, after all our efforts over the past three decades, I am standing with my colleagues at last on the brink of discovery.... I see a pressing need to prepare thinking adults for the outcome of the present search activity—the imminent detection of signals from an extraterrestrial civilization. This discovery, which I fully expect to witness before the year 2000, will profoundly change the world....

I want to show that we need not be afraid of interstellar contact, for unlike the primitive civilizations on Earth that were overpowered by more advanced technological societies, we cannot be exploited or enslaved. The extraterrestrials aren't going to come and eat us; they are too far away to pose a threat. Even back-and-forth conversation with them is highly unlikely, since radio signals, traveling at the speed of light, take *years* to reach the nearest stars, and many *millennia* to get to the farthest ones, where advanced civilizations may reside. But one-way communication is a different story. Just as our radio and television transmissions leak out into space, carrying the news of our existence far and wide, so similar information from the planets of other stars has no doubt been quietly arriving at Earth for perhaps billions of years. Even more exciting is the likelihood of *intentional* messages beamed to Earth for our particular benefit. As we know from our own efforts at composing for a pangalactic audience, reams of information about a planet's culture, history, and technology —the entire thirty-seven-volume set, if you will, of the "Encyclopedia Galactica" —could be transmitted (and received) easily and cheaply.

As a scientist, I'm driven by curiosity, of course. I want to know what's out there. But as a human being, I persevere in this pursuit because SETI promises answers to our most profound questions about who we are and where we stand in the universe. SETI is at once the most technical of scientific subjects, and also the most human. Every tactical problem in the search endeavor rests on some age-old philosophical conundrum: *Where did we come from? Are we unique? What does it mean to he a human being?...*

* * *

[W]e have only just begun to search.

So many individuals I meet seem to think that we have already searched the sky completely and continuously over the past thirty years. The deed is done, they assume. And since we found nothing out there, to search further is to beat a dead horse. But in fact, the combinations of frequencies and places to look have hardly been touched.

In my historical analysis; the search for extraterrestrial intelligence divides itself into four eras. The first dates back at least three thousand years, to the time when people started contemplating the universe....

I trace the start of the second era to the coming of the Copernican Revolution in the sixteenth century. That was when astronomers such as Kepler and Galileo, who used a real telescope, recognized that some of the other objects in the Solar System were planets similar to the

Earth. Scientific observations could now support the philosophical argument in favor of other life in the cosmos—and perhaps even within the Solar System....

The third era began in 1959–60, when scientists first employed quantitative measures to compute the strength of possible signs of life crossing interstellar space. In other words, we made precise calculations of the detectability of alien signals, and acted on them. Projects —beginning with [Cornell physics professors Philip] Morrison and [Giuseppe] Cocconi's proposal to search for radio waves and my strategy for Project Ozma —sprang from a greater knowledge of the universe and a real sense of the numbers involved. For the first time, SETI embodied philosophical, qualitative, *and* quantitative elements. Scientists conducted some sixty "third era" extraterrestrial searches in the 1960s, 1970s, and 1980s. Most of these, however, were low-budget productions, done with leftover funds in borrowed time on equipment built for other purposes.

The fourth era, which starts now, is not only quantitative, it is also, finally, *thorough*. The projects of the 1990s represent the most exhaustive probing to date of the cosmic haystack. Here I am referring especially to the NASA SETI project....

My involvement in SETI activities has actually increased over the years, because SETI itself has grown so much. It occupies more people than ever before, and demands more of their time. Jill Tarter, for example, is the first astronomer to work full-time as a SETI scientist. When she isn't fully engaged in her role as project scientist, the senior scientific position in the NASA SETI project, she is in Washington, explaining the project to congressional representatives. Paul Horowitz runs a close second in activity. Despite his teaching duties at Harvard, Paul has had one search or another in progress since 1977. In some years he devotes nearly 100 percent of his time to these efforts—masterminding a new project and then personally soldering the thousands of joints that hold the equipment together....

I finally got my turn to meet Paul in 1977, when he was already a full professor of physics at Harvard....

A short time later,... Paul accepted a 1981–82 NASA Ames fellowship, which enabled him to work on SETI at the Ames Research Center and at Stanford University. He joined the Ames-Stanford group trying to create a SETI machine that could analyze a huge number of separate channels—128,000 of them, more than anyone had ever been able to monitor simultaneously....

The sheer number of channels in this multichannel analyzer was a big advance in itself, but Paul also made the components portable, so they could be packed up in three small boxes and hand-carried to any observatory, anywhere in the world. The system, which he dubbed "Suitcase SETI," traveled first to Arecibo [Ionospheric Observatory in northern Puerto Rico, home of the largest radio telescope ever built]. After examining 250 stars with it, Paul took it back to Harvard in 1983. He hooked it up to the same telescope I had partially built and calibrated in my student days —the one I had used to observe the Pleiades for my doctoral thesis. Suitcase SETI's rambling days were over at that point. Portable though it was, it never ventured out of Harvard's Oak Ridge Observatory again, A new name, Project Sentinel, recognized the fact that Paul's multichannel analyzer was now connected to a dedicated telescope, with funding from The Planetary Society to run a permanent SETI facility.

In time, Sentinel begat "META-SETI" —the Megachannel ExtraTerrestrial Assay—which boosted the number of channels from 128,000 to more than 8 million.... Paul needed the extra channels, he said, to respond to a new concept put forward by Phil Morrison, who had reminded him in a letter that everything in the universe is in motion....

Intelligent radio signals from distant civilizations could [therefore] be expected to arrive shifted in frequency, just as the starlight from distant suns is shifted toward the red or the blue end of the optical spectrum by stellar motions. There was no way to predict which way a signal's frequency would shift without knowing how its home star was moving. Thus a message transmitted on the hydrogen frequency could wind up far above or far below that frequency by the time it reached a radio telescope on Earth.

With META, Paul could scrutinize myriad frequencies in the vicinity of the hydrogen line and sift through them, narrow bandwidth by narrow bandwidth, on millions of channels at once to detect the displaced signals.

In 1991 Paul set up a second META, also financed by The Planetary Society, called META II, in the Southern Hemisphere, at the Instituto Argentino de Radioastronomia in Villa Elisa, Argentina. This allowed Argentinian astronomers led by Raul Colombo to observe the portion of the southern sky that's not visible from Cambridge. META II opened up very important new regions of the Milky Way as well as a clear line of sight to the two galaxies that are the Milky Way's nearest neighbors: the Magellanic Clouds. Now, with META and META II thriving, Paul is already dreaming of BETA. This would be a new system ("It'll be *betta* than META," he promises), with one hundred million channels.

Paul has obviously done more searching, with more sensitivity, than anybody who preceded him, so it shouldn't be too surprising to learn that he's actually heard things through his systems. Indeed, Paul has records of about sixty signals that are all excellent candidates for being the real thing. But Paul's searches run themselves, automatically. By the time he recognizes the candidates in the recorded data, hours or days later, it's too late to check them. Looking for them later proves fruitless, as they are no longer where they were. No doubt the civilizations are still there—if that's what made the signal—but they've stopped talking, at least for the moment.... If only Paul's strategy included a human operator who could double-check the signals on the spot! However, Paul has severe budget constraints, and I know that he can't afford to pay someone to sit there through the long nights and wait.

The new NASA SETI Microwave Observing Project will change all that, because I'll be sitting there myself. Or Jill will, or some other radio astronomer who will be able to react immediately to chase down a candidate signal the moment it appears. This project, which has been in various stages of planning and development since 1978, is just now beginning its methodical hunt. Because of its great power and sensitivity, it outstrips all previous search activities combined. Three days' operation can accomplish more than was done in the preceding three decades. Indeed, it gives me a strange chill to acknowledge that it takes this new setup only one one-hundredth of a second to duplicate what

Project Ozma did in its full two hundred hours....

What does NASA SETI have that no other search had? The short answer is "everything." It has everything that early searches had, and everything we could think of that had never been done before.

Like Ozma, NASA SETI scrutinizes a group of relatively close, Sun-like stars for signs of intelligent life. But where Ozma had only two targets, NASA SETI has one thousand. This much more extensive "targeted search," however, is still only half the mission. The other half is an "all-sky survey" that repeatedly scans the whole grand volume of outer space for alien signals from any star, anywhere. Our dual search strategy deals with two alternate possibilities for our cosmic neighbors: Either the easiest aliens to detect are right nearby (targeted search), or they are very far away but very bright (all-sky survey and targeted search).

Like the Ohio State project, NASA SETI is an ongoing endeavor that will run for years. But unlike the low-cost efforts that preceded it, this project fought for and won a total of more than $100 million in federal funding. While other searches started up and faded out without so much as a nod from NASA, this one enjoys the same position as a mission to send a small spacecraft to another planet. Mission status means that SETI is supported all through NASA management, right up to the topmost level.

Like META and META II, NASA SETI spans the globe and the heavens. It utilizes at least five telescopes—at Arecibo, Green Bank, the Observatoire de Nançy in France, the Goldstone Tracking Station in California, and an identical NASA tracking station at Tidbinbilla, Australia. It is the first truly global cooperative effort to search for interstellar signals.

Unlike... Suitcase SETI, NASA SETI is no backseat or part-time visitor. It constitutes the largest single program running at Arecibo and will soon dominate a fully dedicated telescope at Green Bank. It employs more than one hundred people, including a rotating team of radio astronomers who stand ready to respond to candidate signals in real time.

Most American searches until now have sought narrow-band signals on magic frequencies, such as the hydrogen line. We call them "magic" because they seem to have some real rationale for being logical channels of communication. Part of their magic is that they occupy quiet regions of the electromagnetic spectrum. What's more, the hydrogen line, considered the most magical frequency of all, is such a fertile field for making general discoveries in radio astronomy that scientists of all civilizations probably keep close tabs on it. Thus, a signal on that particular frequency should have the greatest chance of being detected. The hydrogen line is the frequency Morrison and Cocconi suggested in their original paper, and the actual frequency searched in Project Ozma....

Magic frequencies have special appeal, but even human beings disagree as to which ones are best.... The point is, any search based on a magic frequency assumes first of all that extraterrestrials are broadcasting on a chosen frequency, and furthermore that we can know what that frequency is.

The NASA SETI project makes no such assumptions. It scans most of the frequencies in the waterhole that penetrate the Earth's atmosphere. This means we'll have a much greater chance than ever before to detect a message, whether the aliens choose a frequency for convenience' sake or some numerology of their own. Our new equipment frees us from the need to select just one or two frequencies from among the vast field of possibilities....

META set a world's record with 8 million channels, but NASA SETI has 28 million. At the core of its hardware is a device called a multichannel spectrum analyzer (MCSA in NASA's beloved alphabet soup), which divides the incoming radio noise into 14 million narrowband channels. The MCSA also combines the signals from several adjacent channels to create another 14 million broader bandwidths, just in case the extraterrestrials use them.

The MCSA relies on ultra-advanced software to make sense out of the millions of data points pouring in every second. Software analyzes the data, looking for patterns that reveal intelligence—and that could not possibly be intercepted as fast or as well by human intelligence. The human operator, whose presence is so important to me, steps in *after* computers sound the alarm that a candidate signal has just been detected....

In the course of gushing about the great power of NASA SETI compared to any and all of its predecessors, I've dropped several huge numbers, referring to everything from frequencies and sensitivities to dollars and cents. That said, do I really need one more quantitative comparison to make my point? Would it really clarify things further to say that NASA SETI is a ten-millionfold improvement over past efforts? Maybe not. Maybe the more important thing to say now is that the magnitude of our current efforts creates so much promise that we find ourselves contemplating what we should do when we actually receive signal evidence of extraterrestrial life. When and how do we inform the people of Earth?

John Billingham [a former aerospace physician with England's Royal Air Force] has probably given more thought to this delicious dilemma than anyone else. Working with other members of the SETI committee of the International Academy of Astronautics (IAA), he has drawn up a "Declaration of Principles Concerning Activities Following the Detection of Extraterrestrial Intelligence." It lists all the steps to be taken to verify the authenticity of a signal and inform the proper authorities that extraterrestrial word has been heard.

This document has been approved or endorsed by every major, international, professional space society, including the IAA, the International Institute of Space Law, the Committee on Space Research, Commission 51 of the International Astronomical Union, and Commission J of the Union Radio Scientifique Internationale. In essence, Billingham's protocol says, *Make sure you've got something; then tell EVERYBODY.*

I've spoken at some length about how one goes about checking a candidate signal for authenticity—how to establish extraterrestrial origin, and how the special hallmarks of artificiality can distinguish a signal as being of intelligent design. But to announce to the world at large that you've made the greatest discovery in the history of astronomy—perhaps in history, period—takes an even wider margin of certainty.

On the NASA SETI project, you probably can't ask another observatory to help you verify your findings. If the long-awaited signal is intercepted at Arecibo, and it is weak, which is the most likely possibility, then no other observatory in the world could make the desired verification. This is because Arecibo has the greatest collecting area of any telescope, as well as the Gregorian feed and other specialized equipment. Even the other participants in NASA SETI, in France and Australia, will not match Arecibo's wide range of frequency coverage. And if the signal did fall within their frequency range, they might lack the sensitivity to hear it. Arecibo is so much more sensitive than the others—ever so much more capable of picking a faint, fragile *"We are here!"* out of a welter of cosmic noise.

In lieu of interobservatory checks and balances, the people at Arecibo (I hope I'm one of them when this happens) will have to spend several days checking and rechecking their data, locating the signal, if possible, a second, third, and fourth time rather than risk setting off a false alarm. After several days, however, repeated observations would build up a chink-free wall of evidence that would justify going public....

Hard upon detection of an intelligent signal, there follows the delicate matter of a reply to the civilization that sent it. I've thought a lot, of course, about what to say in that happy situation. I have waited a lifetime for the opportunity, and the waiting has not diminished my confidence or my enthusiasm. I can't be specific about it, though, because when you really think about it, the only answer to the question "What do you say?" is "It depends."

It depends on the nature of the signal and what it's telling us. It depends on the world's reaction. It depends on the distance the message traveled, because we couldn't establish true dialogue with civilizations far removed from us—only lengthy monologues, crossing each other eternally in the interstellar mail. It depends on whether we can understand it. Certainly no stock reply, prepared in advance and stashed in someone's file cabinet, could match more than one of the infinite possibilities for the message's content. Certainly any reply should be crafted on a worldwide basis, and only after lengthy deliberation by knowledgeable individuals.

I have a recurring dream in which we receive our much-anticipated intelligent signal from across the Galaxy. The signal is unambiguous. It repeats over and over, allowing us to get a fix on its source, some twenty thousand light-years away. The signal is... apparently dense with information content. It is so full of noise, however, that we can't extract any information from it. And so we know only that another civilization exists. We cannot decipher the message itself.

If this dream becomes real, such documented detection of alien signals will, of course, be big news in itself. It will be a call to action, too, beckoning us to do whatever is required—build a much larger radio-telescope system, for example—to obtain information about that civilization, to learn whatever secrets the extraterrestrials will share with us.

Indeed, our response to a message from an alien civilization may thus be a response to the *situation* instead of an actual reply to the senders. We will tell the world at large what has happened, and that we're taking the next step by building better equipment to understand the message we've received. How I would love to have to go to Congress with a budget request for that project. I don't imagine I'd encounter much opposition....

I do not wonder *whether* this will happen. My only question is *When?*

The silence we have heard so far is not in any way significant. We still have not looked long enough or hard enough. We've not explored a large enough chunk of the cosmic haystack. I could speculate that "they" are watching us to see if we are worth talking to. Or perhaps the ethic exists among them that rules, "There is no free lunch in the Galaxy." If we want to join the community of advanced civilizations, we must work as hard as they must. Perhaps they will send a signal that can be detected only if we put as much effort into receiving it as they put into transmitting it. NASA SETI is the beginning of the first truly meaningful effort to demonstrate the sincerity of our intentions.

Thus, the lesson we have learned from all our previous searching is that the greatest discovery is not a simple one to make. If there were once cockeyed optimists in the SETI endeavor, there aren't any now. In a way, I am glad. The priceless benefits of knowledge and experience that will accrue from interstellar contact should not come too easily. To appreciate them, we should expect to devote a substantial portion of our resources, our assets, our intellectual vigor, and our patience. We should be willing to sweat and crawl and wait.

The goal is not beyond us. It is within our grasp.

NO

A. K. Dewdney

SURFING THE COSMOS: THE SEARCH FOR EXTRATERRESTRIAL INTELLIGENCE

It was noon, April 8, 1960. The recently completed 85-foot radio telescope dish at Green Bank, West Virginia, had just lost the star Tau Ceti below the horizon. Steering motors hummed and the great dish swung grandly to the south along the horizon until, like a great ear, it listened to another star, Epsilon Eridani. Up in the control room, radio astronomer Frank Drake and his colleagues listened eagerly to sounds coming from a loudspeaker. The sounds enabled the astronomers to hear the signals being intercepted by the dish.

Gathered in the 85-foot parabolic surface, electromagnetic waves, some of them from Epsilon Eridani and some from much further away, reflected to the focus of the dish where a large cylindrical housing sheltered a precisely tuned amplifier. The signals from the amplifier were fed to a chart recorder in the control room and, of course, to the speaker. It was a propitious day, the dawn of Drake's dream of intercepting messages from an alien civilization.

Called Project Ozma, the dream reflected Drake's conviction that somewhere out there, alien intelligences were transmitting helpful messages to less developed civilizations or, failing that, were at least inadvertently broadcasting their radio and television programs. Given the air of anticipation that surrounded the inaugural evening of Project Ozma, Drake and his colleagues can perhaps be forgiven for what happened next:

> [S]carcely five minutes had passed before the whole system erupted. WHAM! A burst of noise shot out of the loudspeaker, the chart recorder started banging off the scale, and we were all jumping at once, wild with excitement. Now we had a signal—a strong, unique pulsed signal. Precisely what you'd expect from an extraterrestrial intelligence trying to attract attention.

To check that the source of the signal really was Epsilon Eridani, Drake had the telescope taken off the target. The sound disappeared, meaning that this star (or a planet near it) may actually have been the source. Unfortunately, when they returned the telescope to track the star, the noise had disappeared.

An even more significant incident followed on the heels of the first one: One of the telescope operators told a friend about the signal and the friend contacted a newspaper. Before Drake knew it, he was deluged by calls from the media demanding to know what had happened.

"Have you really detected an alien civilization?"

"We're not sure. There's no way to know."

This answer could not have been better calculated to raise curiosity about the incident still further, guaranteeing a great deal of publicity for Project Ozma. A better answer would have been, "As far as we know, the anomalous signal originated right here on Earth." Both responses are true, of course, but the second would have a more chilling effect on the media. Drake, after all, was no stranger to anomalous signals.

At the tender age of twenty-six, he had been observing the Pleiades star group when a new signal suddenly appeared on his chart recorder. Drake recalls:

> It was a strikingly regular signal—too regular, in fact, to be of natural origin. I had never seen it before, though I had repeated the spectrum measurement countless times. Now, all of a sudden, the spectrum had sprouted this strong added signal that looked unusual and surely of intelligent design.... I still can't adequately describe my emotions at that moment. I could barely breathe from excitement, and soon after my hair started to turn white.

Drake never succeeded in recapturing the signal and today suspects that it may have been a military aircraft.

Since the exciting early days of extraterrestrial probing, Project Ozma has been succeeded by SETI, the search for extraterrestrial intelligence. Sponsored by NASA, the National Aeronautics and Space Administration, the SETI project, along with similar schemes, has absorbed over a billion dollars in congressional appropriations. Is the money well spent? The project has had and continues to have many critics, but few have gone to the heart of the matter.

As I will show in a later section, the problem with SETI lies at the very beginning of the scientific method—the hypothesis. Not only is it unavoidably geocentric, it is essentially nonfalsifiable. There is a troublesome formula, moreover, that is supposed to make the hypothesis seem more reasonable. As I will also show, the formula is a two-edged sword that actually argues against the hypothesis.

At this writing, none of the SETI projects have revealed so much as a whisper of alien intelligence. This has not stopped Drake from going out on a limb. He seems eager to "prepare thinking adults for the outcome of the present search activity—the imminent detection of signals from an extraterrestrial civilization. This discovery, which I fully expect to witness before the year 2000, will profoundly change the world."

When confronted with the failure of SETI programs up to this point, Drake wisely opines, "Absence of evidence is not evidence of absence."

SCANNING THE SKIES

Radio telescopes supplement ordinary optical telescopes by giving us a picture of the cosmos by the light of radio waves. I say "light" because radio waves are just another part of the great electromagnetic [EM] spectrum, which also includes light waves....

Many of the stars that appear as sharp points in the light telescope map also show up as blobs in the radio telescope map. This means that such stars not only emit light, they emit radio waves. By the same token, some of the radio sources that show up in the radio maps have no visual counterparts, or if they do, turn out to be clouds of gas, regions of violent galactic activity. Radio telescopy has been an invaluable tool in learning more about the structure of our own galaxy and that of other galaxies, not to mention a string of amazing discoveries such as pulsars and quasi-stellar objects, or quasars.

Nevertheless, the relatively long length of radio waves makes it very difficult to get much resolution out of a single telescope. Indeed, the signal from a radio telescope is essentially one-dimensional, like a sound track. The signal from an optical telescope is, of course, two-dimensional—a picture. These days, radio astronomers squeeze more resolution from their instruments by using several receivers at widely separated locations, as if to construct an effective dish that has the same dimensions relative to radio waves that optical mirrors do in relation to light waves.

Consider a typical dish as it tracks a distant star. Radio waves pour onto the dish from all directions. Some of the radiation comes from the Earth itself, stray radio or television broadcasts, ham radio operators, taxi dispatchers, truckers on CB radios, cellular phone callers, direct-to-home-satellite broadcasts, and so on. These signals, the only evidence we have of intelligent life so far, all come from the planet Earth. They sometimes bedevil the life of normal radio astronomers.

Electromagnetic waves of more natural origin arrive from the ionosphere, where energetic particles from the Sun collide with molecules of air at the very top of the Earth's atmosphere. Electromagnetic waves also come from radio sources in our own solar system, such as Jupiter and the Sun. From beyond the solar system, faint waves arrive from other stars in our own galaxy, from pulsars, and from stellar clouds. Other ripples, ancient and feeble, arrive at the great dish from other galaxies, not to mention those primordial sources, the unimaginably remote quasars.

Life is nevertheless relatively easy for the normal radio astronomer who has techniques for eliminating many kinds of interference from earthly sources. He or she may listen largely undisturbed for the random hiss of ancient stars and galaxies or the repetitive clicking of a pulsar. But the radio astronomer who searches for intelligent life must stand this rationale on its head, listening for the whispers of intelligent transmission amid a welter of natural electromagnetic hisses, clicks, chirps, and buzzes. The SETI astronomer is even more bedeviled by earthly signals. They tend to sound just like the thing he or she is searching for.

Is it just possible that somewhere, among all the radiation flooding the dish from a myriad of sources, one or two indescribably faint signals amount to whispers from distant and ancient civilizations? Perhaps the signals patiently repeat the recipes for astounding scientific and technical breakthroughs, as some SETI enthusiasts have dreamed.

In the meantime, even as radiation from a multitude of sources pours onto the dish, a kind of reverse process goes on. All of those television and radio signals that interfere with the radiation from outer space are themselves zooming away from Earth in all directions at the speed of light. In their entirety, the signals form a vast, expanding ball of radiation. Since radio broadcasts began about ninety years ago, the radius of that ball of radiation is now about ninety light years. It is large enough to contain a few hundred stars in our near galactic neighborhood, albeit still tiny compared to our galaxy as a whole. It is nevertheless possible, of course, that a technological civilization on Alpha Centauri or Ophiuchus has picked up our broadcasts, including enough episodes of *The Three Stooges* to place the Earth in a state of permanent galactic quarantine.

The point of that rapidly expanding sphere of programming has certainly not escaped the SETI theorists. Other civilizations with the ability to monitor electromagnetic radiation should be able, sooner or later, to hear our signals, however faint. Should we not, by the same token, be able to pick up the signals of other civilizations? The prospect has an unquestioned fascination about it. Imagine what an alien signal might be like, the very stuff of science fiction! But is it science? Or is it fiction? In particular, is Drake behaving like an apprentice?

In this case, it all depends on the hypothesis and your opinion of it. As the well-known astronomer Carl Sagan once speculated, the sheer numbers of stars in our galaxy lends enormous weight to even the most slender estimates of probability for the evolution of technological civilizations elsewhere: What is the chance of Earth-like planets? What is the chance of life spontaneously emerging on such a planet? Even small probabilities, when multiplied by the enormous number of stars out there, turn into something almost definite. The hypothesis is this: Given the near-ubiquity of life in our galaxy, some life forms have surely developed intelligence, including the ability to communicate by radio waves. Such waves should be detectable by suitable receivers right here on Earth.

There are a few minor flaws in this hypothesis and one major one. The minor flaws involve unstated assumptions that enter the hypothesis....

The hypotheses of the theoretical physicist or cosmologist "work" because the model is precise and one can tell, almost as soon as new observational evidence arrives, whether it confirms the model or not. If the speculation was far-fetched to begin with, the physicist should not be too surprised if further observations fail to confirm the model. ...[A] hypothesis must be falsifiable.

Consider now the scientist who looks up at the night sky and asks the age-old question, Is there anyone out there? The question seems perfectly reasonable. It means, Is there another race of beings, living somewhere else, whom we would call intelligent? Apart from the fact that we as yet have no formal scientific definition of intelligence..., most people think they know intelligence when they see it, at least among fellow human beings.

Perhaps the best laboratory in which to consider alien cultures is right here on Earth. Consider a country that is dominated by Zen Buddhism, for example. Many people would say that the Zen monk represents a very high level of human development (without being exactly sure what that means). If the world were full of Zen monks, however, we would be very unlikely to have radio. The technology would contribute very little to the insights necessary on the fivefold way, and one could argue that the technology and its development would constitute a completely unnecessary distraction from the real work of the monk, which is to rid himself of attachments to things of the world. As for advice from beyond, Zen monks have all they can handle in advice from the teacher.

With the peculiar myopia that characterizes Western culture, we have come to regard our own development as more or less inevitable, an extension of the Darwinian imperative into the technocultural realm.

The real question is What is the chance of a Western-style scientific-technological civilization developing out there? The "Western" qualifier is crucial, for we in the Western world may be living in a spell, trapped in yet another aberrant vision of our place in the universe, one no less misleading than the pre-Copernican idea of a central Earth. If the sorcerer is under a spell, he will hardly do better than the apprentice!

LIVE BY THE FORMULA, DIE BY THE FORMULA

The unquestioned pioneer of the SETI project is the well-respected radio astronomer Frank Drake. Early in his career as a radio astronomer, Drake developed an interest in the possibility of life on other planets, particularly intelligent life. He became intrigued, some might say obsessed, with the prospect of intelligent beings broadcasting radio signals into space, signals that we on Earth might intercept—to our infinite advantage.

Intuitively, Drake understood that with 200 billion stars in our galaxy, there might be a very good chance that someone out there was already sending the very signals he dreamed of receiving. To put the project on a quantitative footing, Drake devised the equation below. To some people it may appear complicated, but mathematically speaking, it could hardly be simpler. The right-hand side of the equation consists merely of a bunch of variables all multiplied together:

$$N = R^* \times Fp \times Ne \times Fl \times Fi \times Fc \times L$$

The equation attempts to estimate the number N of "radio civilizations" in our galaxy. A radio civilization is simply a race of intelligent beings that have developed the ability to broadcast and receive messages via electromagnetic radiation, and do so on a regular basis. The equation estimates the number N by taking into account a variety of factors in the product:

R^* number of new stars that form in our galaxy each year

Fp fraction of stars having planetary systems

Ne average number of life-supporting planets per star

Fl fraction of those planets on which life develops

Fi fraction of life forms that become intelligent

Fc fraction of intelligent beings that develop radio

L average lifetime of a communicating civilization

At first glance, the equation seems perfectly definite. If you happen to know the value of each variable, you can come up with a pretty good estimate for N. If the estimate you arrive at is reasonably large, you may use the equation to squeeze endless amounts of money out of Congress to support a search for intelligent life. The equation, after all, is mathematical, and that means real science.

An estimate of the number R^* is based on an assumed rate of star formation of about ten a year. This is an extremely crude estimate based on current observations of regions where stars appear to be forming in our galaxy. The actual number has undoubtedly varied enormously over time, particularly in the remote past. For the Drake formula, it's all uphill from this point on.

The fraction *Fp* of stars having planetary systems is completely unknown. Although a few relatively nearby stars are suspected of having very large Jupiter-like companions or planets that are nearly stars in their own right, we have yet to observe a single star with a planetary system even remotely like our own. Period. It follows that we haven't the slightest idea what the real value of *Fp* might be, and any "estimates" would better be called wild guesses.

If we haven't a clue how many stars have planetary systems, then we're even more in the dark about the average number *Ne* of life-supporting planets per star. Some of them may well have such planets. Perhaps they all do. Perhaps our sun is the only such star. We simply have no idea.

Will a "life-supporting planet" ever develop life? I'm not sure how a planet could support life if it didn't already have it. The Earth has oxygen, for example, only because photosynthesizing organisms evolved here a long time ago and eventually filled the atmosphere with this (for us) vital gas. Perhaps the rather silly variable *Fl* should be set equal to 1 and simply dropped from the equation.

As you will see from a glance at the remaining variables, it gets worse.

The fraction *Fi* of life forms that become intelligent is even less well known, if that is possible, than the previous variables. What do we mean by "intelligent," anyway? As you may have already discovered . . . , we're not even sure what we mean by our own "intelligence"! Once again, my guess for this variable is as good as Frank Drake's.

The fraction *Fc* of intelligent life forms that develop radio is likewise completely unknown and pointless to estimate. Finally, the lifetime *L* of the average radio civilization is the only variable about which we have any information, and that information may be about to improve. We know, for example, that our own radio civilization has existed for about ninety years. There is a real possibility that it may reach one hundred. In any case, this sample of one is our only basis for an estimation of *L*.

How do Drake and his disciples use the formula? Here are two examples that have appeared in popular magazine articles on the subject. I have no doubt that the guesses come directly from the SETI school.

$$N = 10 \times 0.3 \times 1 \times 0.1 \times 0.5 \times 0.5 \times 10^6 = 125{,}000$$

$$N = 10 \times 1 \times 1 \times 1 \times 0.01 \times 0.1 \times L = 0.01 \times L$$

In the first guess, *L* was given a value of 10^6, or one million years. The second guess refused to assign a definite value to *L*, which is strange, considering that we already know more about *L* than the other variables. Nevertheless, using the value for *L* from the first equation in the second, we get a more conservative estimate:

$$N = 10{,}000$$

That's still quite a few radio civilizations. Why haven't we heard from any of them yet? We might find the answer by taking a closer look at ourselves, in particular, and our probable destiny as a radio civilization. It is not nuclear holocaust that will seal our fate as a spherical broadcaster of invaluable cultural and scientific information to the cosmos, but the incredible inefficiency of antenna broadcasting!

As every radio engineer knows, broadcasting electromagnetic waves in all directions at once is an enormously wasteful way to transmit information. Although the emissions from standard mast antennas can be directed somewhat in the form of lobes, only the tiniest fraction of broadcast energy ever reaches receiving antennas. The evidence is now very clear that the Earth is rapidly fading as a source of electromagnetic energy. Increasingly, we transmit radio and television signals by cable, not to mention the exponentially increasing Internet traffic on phone lines and fiber-optic cables. An even more powerful trend involves the broadcast of television signals toward Earth from satellites, signals that are completely absorbed by the ground. The Earth may be about to vanish as a radio source.

If this is true, then 100 might be taken as a perfectly reasonable estimate for the crucial variable L. In this case, plugging the new value for L into the last equation..., we get

$$N = 0.01 \times 100 = 1$$

That must be us.

Another implication of current developments in the dissemination of information points up another minor flaw in the Drake hypothesis. Increasingly, radio signals between points in deep space will be beamed ever more precisely at the target receivers, somewhat like a laser beam. This would make their reception by nontargeted civilizations increasingly less likely. Can anyone believe that these vastly "superior" alien civilizations would themselves employ any method so incredibly wasteful as spherical broadcasting to communicate with each other? The implications for SETI enthusiasts are clear: Don't hold your breath waiting for that magic signal.

Finally, there may well be radio signals that SETI will eventually intercept but the signals will present us with an enormous headache. Seemingly intelligent, they will only be meaningful to beings of a similar mindset, whatever that might mean. Neither I nor anyone in the SETI team can imagine what a distinctly inhuman mentality might be like.

POSTSCRIPT

Is It Worthwhile to Continue the Search for Extraterrestrial Life?

The modern, high-tech version of SETI, the High Resolution Microwave Survey (HRMS), almost never came to pass. As Donald Tarter of the International Space University, in "Treading on the Edge: Practicing Safe Science With SETI," *Skeptical Inquirer* (Spring 1993), writes, "SETI's recent history has been one of fighting for scientific respect and then fighting for funding.... SETI has been so frequently ridiculed and singled out as [a program that could be eliminated by budget-cutting congressional members] that officially SETI no longer exists." He then notes that, shortly before NASA began its current search for extraterrestrial intelligence, the name was changed to HRMS.

However, the name change did not solve the problem. A year after HRMS was born, the budget was cut. By October 12, 1993, the $1 million a month needed to sustain it had been eliminated from the budget by a House-Senate conference committee. It was not the sort of arguments raised by critics such as Dewdney that defeated HRMS; it was image. SETI smacked too much of science fiction and Hollywood. It might not be terribly expensive—the cost of a single space shuttle flight could pay SETI's bills for several years—but whatever it cost seemed to the budget cutters pure waste when compared to the many other programs and problems requiring funds.

Yet SETI was not dead. Many scientists—including nonastronomers such as David M. Raup of the University of Chicago's Department of Geophysical Sciences and Committee on Evolutionary Biology—disagree with those who believe that humans are probably alone in the universe and who say that the search for intelligent extraterrestrials is not worth the effort. So do many nonscientists. The private SETI Institute has been able to garner $7.5 million in donations so far, chiefly from prominent business leaders in California's computer industry. "They're in it for the adventure. They have vision," says Seth Shostak, a SETI Institute scientist (*Science*, January 28, 1994).

In 1995 that vision permitted the observation of 209 sunlike stars (see *Science*, February 23, 1996), and backers have promised money for another five years of searching. Frank Drake, the SETI Institute's president, says that the institute hopes to raise an endowment sufficient to maintain the search for many more years.

Another effort worth noting is that of the SETI League, which is recruiting owners of obsolete satellite TV dishes (3–5 meters in diameter; new dishes are much smaller) to connect the dishes to home computers and let them listen to the sky as part of Project Argus. See Marcus Chown, "The Alien Spotters," *New Scientist* (April 19, 1997).

ISSUE 11

Should Genetic Engineering Be Banned?

YES: Andrew Kimbrell, from *The Human Body Shop: The Engineering and Marketing of Life* (HarperSanFrancisco, 1993)

NO: James Hughes, from "Embracing Change With All Four Arms: A Post-Humanist Defense of Genetic Engineering," *Paper Presented at the University of Chicago Health and Society Workshop* (May 6, 1994)

ISSUE SUMMARY

YES: Andrew Kimbrell, policy director of the Foundation on Economic Trends in Washington, D.C., argues that the development of genetic engineering is so marked by scandal, ambition, and moral blindness that society should be deeply suspicious of its purported benefits.

NO: James Hughes, assistant director of research at the MacLean Center for Clinical Medical Ethics in the Department of Medicine at the University of Chicago, argues that the potential benefits of genetic engineering greatly outweigh the potential risks.

In the early 1970s scientists first discovered that it was technically possible to move genes—biological material that determines a living organism's physical makeup—from one organism to another and thus (in principle) to give bacteria, plants, and animals new features and to correct genetic defects of the sort that cause many diseases, such as cystic fibrosis. Most researchers in molecular genetics were excited by the potentialities that suddenly seemed within their grasp. However, a few researchers—as well as many people outside the field—were disturbed by the idea; they thought that genetic mix-and-match games might spawn new diseases, weeds, and pests. Some people even argued that genetic engineering should be banned at the outset, before unforeseeable horrors were unleashed.

Researchers in support of genetic experimentation responded by declaring a moratorium on their own work until suitable safeguards could be devised. Once those safeguards were in place in the form of government regulations, work resumed. Before long, it became clear that the hazards were less than had been feared, although the benefits were going to take years of hard work to achieve. James D. Watson and John Tooze document the early years of this research in *The DNA Story: A Documentary History of Gene Cloning* (W. H. Freeman, 1981). For a shorter, more recent review of the story, see Bernard D. Davis, "Genetic Engineering: The Making of Monsters?" *The Public Interest* (Winter 1993).

By 1989 the technology had developed tremendously: researchers could obtain patents for mice with artificially added genes ("transgenic" mice); firefly genes had been added to tobacco plants to make them glow (faintly) in the dark; and growth hormone produced by genetically engineered bacteria was being used to grow low-fat pork and increase milk production by cows. While these developments were being made, a storm of protest was gathering strength. Critics argued that genetic engineering was unnatural and violated the rights of both plants and animals to their "species integrity"; that expensive, high-tech, tinkered animals gave the competitive advantage to big agricultural corporations and drove small farmers out of business; and that putting human genes into animals, plants, or bacteria was downright offensive. See Betsy Hanson and Dorothy Nelkin, "Public Responses to Genetic Engineering," *Society* (November/December 1989).

By 1990 the first proposals to add genes to *human* cells in order to restore normal function were being made (Inder M. Verma, "Gene Therapy," *Scientific American*, November 1990). Not long after that, the first gene therapy attempts were approved by the National Institutes of Health (NIH), despite objections that altering a human being's genes meant violating that person's nature and identity at the deepest possible level. To avoid producing genetic changes that would be passed on to future generations, which cannot consent to the changes, researchers have been restricted to modifying only somatic (body) cells, not germ cells (sperm and eggs). Still, in 1994 Mark A. Findeis, a group leader at OsteoArthritis Sciences in Cambridge, Massachusetts, described numerous genetic therapies under development in "Genes to the Rescue," *Technology Review* (April 1994).

Anti-genetic-engineering activist Jeremy Rifkin, president of the Foundation on Economic Trends in Washington, D.C., has stressed that because we do not know the future undesirable side effects of genetic engineering, and because those side effects may be horrible, we should reject the technology. In the following selection, Andrew Kimbrell echoes this view of genetic engineering and argues that its history is so marked by scandal, ambition, and moral blindness that it poses an exceedingly disturbing precedent for the future.

James Hughes, representing the voice of optimism, argues in the second selection that genetic engineering offers "such good that the risks are dwarfed" and finds "faith in the potential unlimited improvability of human nature and expansion of human powers far more satisfying than a resignation to our current limits."

YES

Andrew Kimbrell

ENGINEERING OURSELVES

In an age of protests, this was the first of its kind. It was early March 1977, and hundreds of demonstrators had flocked to the futuristic, domed auditorium of the National Academy of Sciences (NAS). The protesters chanted slogans such as "We will not be cloned," and they carried signs bearing warnings, including "Don't Tread on My Genes."

The object of the protest was a three-day symposium being held under the auspices of the NAS. The forum was intended to bring together scientists, government officials, and business leaders to discuss the future prospects of genetically altering life-forms, including humans. The chairman of the meeting, Dr. David Hamburg, president of the NAS Institute for Medicine, undoubtedly had anticipated that this would be the usual scientific conference, a collegial discussion of current scientific and legislative issues that had been cropping up as a result of advances in genetic manipulation. It was not to be.

The demonstrators, led by activist Jeremy Rifkin, crowded the auditorium with their signs and dominated the session with their chants and shouted questions to the symposium's panels. They relentlessly prodded the scientists and bureaucrats, urging them to confront the moral and ethical implications of engineering the genetic code of life. They also repeatedly demanded that speakers disclose who was financing their research. (The forum was supported in part by funds from a variety of drug manufacturers.) Finally, under a barrage of questions about the eugenic [breed- or race-improving] and discriminatory potential of biotechnology, the chairman had no choice but to offer the podium to Rifkin and others to air their concerns.

Speaking up with the protesters were many prominent scientists. At a press conference prior to the demonstration, Nobel Prize winner George Wald called the use of genetic engineering "the biggest break in nature that has occurred in human history." Renowned biochemist Dr. Erwin Chargoff warned against the use of genetic research to attempt to control the evolution of humans and other life-forms.

The activists and scientists who voiced their concerns that day were part of a growing chorus of those who feared the engineering of life. As early as

From Andrew Kimbrell, *The Human Body Shop: The Engineering and Marketing of Life* (HarperSanFrancisco, 1993). Notes omitted.

1967, Marshall Nirenberg, the Nobelist who first described the "language" of the genetic code, had delivered a stern lecture about engineering human beings, along with a remarkably prescient prophecy:

> My guess is that cells will be programmed with synthetic messages within 25 years.... The point that deserves special emphasis is that man may be able to program his own cells long before he will be able to assess adequately the long-term consequences of such alterations, long before he will be able to formulate goals, and long before he can resolve the ethical and moral problems which will be raised.

The fears of the early gene engineering critics focused on proposals to engineer the human germline—to permanently alter the genetic makeup of an individual that is passed on to succeeding generations. Many scientists were predicting that, by manipulating the genes in sperm, eggs, or embryos, future physicians would be able to excise "bad" genes from the human gene pool. Critics envisioned a future human body shop industry in eliminating the genes responsible for sickle-cell anemia or cystic fibrosis by mass engineering of these "problem" genes from the sex cells (the sperm and ova) of individuals. Future genetic engineers could also add foreign genes to a patient's genome, genes from other humans or even different species. These genes might protect an individual from various diseases, or confer desired qualities like better looks or brains. Ultimately, they believed that as scientists learned more about the relationship of genes to disease and other human traits, there would be an inevitable push to treat life-forms as so many machines whose working parts, genes, could be engineered or replaced if they were "defective."

Moreover, it was clear that if the genetic engineering of human beings should come, and most believed it would, there would be a quantum leap in both negative and positive eugenics. No longer would it be necessary to attempt to carefully control generations of breeding to create "good" characteristics, or to resort to sterilization, abortion, or genocide in order to remove abnormal or undesirable traits. Individuals could be altered through genetic surgery that would repair or replace bad genes and add good ones. Nobel Prize winner Jean Rostand's early visions of the eugenic potential of gene engineering went even further: "It would be no more than a game for the 'man farming biologist' to change the subject's sex, the colour of his eyes, the general proportions of body and limbs and perhaps the facial features." Many agreed with scientists such as Wald and Chargoff that the genetic alteration of people could eventually change the course of evolution. In 1972, ethicist Dr. Leon Kass wrote, "The new technologies for human engineering may well be 'the transition to a wholly new path of evolution.' They may therefore mark the end of *human* life as we and all other humans know it."

For over two decades, scientists, activists, ethicists, and the media have engaged in the debate over the medical and moral questions surrounding the germline genetic engineering of human beings. Editorials have appeared with headlines questioning "Whether to Make Perfect Humans" and how to arrive at "The Rules for Reshaping Life." Many critics have continued to argue against the entire enterprise of "the remaking of man." They question the wisdom of having scientists decide which part of the human genome should be eliminated and which enhanced. And if not scientists, who, they ask, will determine which human genes are bad and which good? They warn that even supposedly "bad" genes may bring extraordinary benefits to humanity. Recently, it was discovered that cystic fibrosis genes appear to provide individuals with protection from melanoma, an increasingly common form of skin cancer. Research conducted in the 1980s determined that sickle-cell anemia genes appear to help provide individuals with immunity to malaria. Excising such genes from the human gene pool in the effort to eliminate human disease could backfire with potentially catastrophic results.

There is also the question of how and when society will ensure that the powerful technology of germline gene engineering will be limited to the treatment of serious human diseases.... [G]enetic screening of embryos is already being used for eugenic purposes, including sex selection; and genetically engineered drugs are being used for cosmetic purposes in a way that helps foster certain forms of discrimination. Who will ensure that germline therapy is not abused in the same discriminatory and eugenic way? Will those with under normal height or I.Q. become key targets of the future entrepreneurs of germline therapy? Other novel legal questions arise from the prospect of germline therapy, issues similar to those being asked in reference to advances in prenatal genetic screening. Do children have the right to an unmanipulated germline? Or, conversely, do they have a right to the best germline that genetic surgery can offer and money can buy?

As the debate around germline gene therapy continues, another form of human genetic engineering has already begun. This form of genetic manipulation does not involve sex cells, but rather those cells that do not partake in reproduction. These cells are called *somatic cells*. Engineering these cells is both easier and far less controversial than attempting to manipulate germ cells. Altering somatic cells triggers far less concern about eugenics, in that the cells being repaired or added affect only the single individual being engineered. They do not affect the inheritance of genetic traits. Early uses of somatic cell engineering include providing individuals with healthy or repaired genes that might replace those that are faulty and causing disease.

Though somatic cell gene therapy does not affect the genetic inheritance of future generations, there are still fears. Will individuals with "poor" genetic readouts —those predisposed to a variety of disorders or abnormal traits—be under pressure by parents, education providers, insurance companies, and employers to undergo gene therapy to remove their "bad" genes? Will the therapy be used "cosmetically" to add or eliminate nondisease traits, such as growth, skin color, or intelligence? Will victims of discrimination be pressured by societal prejudice to alter in themselves those traits society views as negative?

The early concerns about germline and somatic cell genetic engineering relied primarily on future projections of the potential abuse of the technology. However, two early cases involving misuse of gene therapy contributed significantly to the controversy that marked the early years of experimentation on the genetic manipulation of humans. The first scandal involving the nascent technology happened over two decades ago.

[In the next section of the original source, which is not reprinted here, Kimbrell discusses two instances of early gene therapy experimentation, one in 1970 and one in 1980, that were considered unethical because of the scientists' seeming lack of regard for the treated patients. The first case, which involved Dr. Stanfield Rogers, led to the first proposed legislation on genetic engineering and eventually provoked the National Institutes of Health (NIH) to produce guidelines regulating the use of human gene engineering. The later case, which involved Dr. Martin Cline of the University of California, also contributed to a promulgation of legislative and regulatory action on human gene engineering, including the establishment of a White House commission led by ethicist Alexander Morgan Capron and of the Biomedical Ethics Review Board to explore the ethical implications of human gene technology.—Ed.]

Throughout the 1980s, the criticisms of gene therapy continued. In 1983, Jeremy Rifkin organized a religious and scientific coalition against the use of genetic engineering on humans. The coalition and its signed statement opposing germline engineering were front-page stories around the United States. Unlike Capron's commission, the coalition's resolution on germline therapy was unambiguous: "Resolved, the efforts to engineer specific genetic traits into the germline of the human species should not be attempted." Its logic on prohibiting heritable gene alterations was also straightforward: "No individual, group of individuals, or institutions can legitimately claim the right or authority to make such decision on behalf of the rest of the species alive today or for future generations." The resolution, which was presented to Congress, was signed by a remarkable variety of religious leaders, including mainstream Jewish, Catholic, and Protestant religious organizations, as well as by many prominent scientists.

Six years later, an important and detailed religious statement on biotechnology was issued by the World Council of Churches (WCC). It contained a strong policy statement calling on all churches to support a "ban on experiments involving the genetic engineering of the human germline." The WCC was also deeply concerned about somatic cell gene experiments. The report called upon member churches to urge "strict control on experiments involving genetically engineered somatic cells, drawing attention to the potential misuse of... [this technique] against those held to be 'defective.'" The timing of the WCC statement could not have been more pertinent, for 1989 was to be the year that the age of human genetic engineering officially began.

PLAYING GOD?

On January 30, 1989, almost twelve years after the first demonstration on human genetic engineering, another such protest took place. The protesters came to a meeting of the National Institutes of Health Recombinant DNA Advisory Committee (RAC). Since publishing its guidelines in 1976, RAC had met dozens of times to discuss and approve experiments in genetic engineering. The advisory committee, composed mainly of scientists, held meetings that were usually staid affairs replete with lengthy discussion of arcane data and procedures.

This RAC meeting was like no other. There, demanding to be heard by the NIH scientists and genetic engineers, were fifteen of the nation's most prominent leaders in disability rights, many themselves suffering from disabilities. Additionally, several biotechnology activists were present to demand accountability of the scientists on the RAC. Many of the scientists appeared visibly uncomfortable at the prospect of discussing human gene engineering with people concerned about a new age of eugenics—and all under the unaccustomed glare of TV cameras. Those present knew that they were at a historic moment in the genetic engineering revolution, for this RAC meeting had as an agenda item discussion of approval for the world's first legally sanctioned genetic engineering experiment on humans.

The experiment involved genetic engineering but was not intended to be a cure. Researchers wished to insert novel genetic "markers" into certain immune cells taken from the bodies of terminally ill cancer patients, and then transfuse those cells back into the patients. With the help of the markers, they hoped to track which cells were working effectively and which were not. The procedure was to be carried out by the NIH's prime genetic engineering team of Drs. French Anderson, Steven A. Rosenberg, and Michael Blaese.

Minutes after RAC chairman Dr. Gerard J. McGarrity called the meeting to order, critics began to express deep concern that the NIH had begun the historic process of approving human gene engineering protocols while still doing nothing to put in place a review process on the ethical and legal implications of human genetic alteration. Jeremy Rifkin announced that his Foundation on Economic Trends had filed suit that morning, calling on a federal court to halt the experiment until the NIH committed itself to allowing the public a greater voice in decisions on gene therapy. Rifkin also noted that the lawsuit was based on the fact that the historic experiment was approved by a secret mail ballot, the first in RAC's history. He repeated the concerns he and other demonstrators had expressed over a decade before: "Genetic engineering raises unparalleled ethical and social questions for the human race. They cannot be ignored by the NIH. If we are not careful we will find ourselves in a world where the disabled, minorities, and workers will be genetically engineered." Another protesting voice at the meeting was Evan Kemp, then Commissioner of the Equal Opportunity Commission (EEOC), and himself disabled:

> The terror and risk that genetic engineering holds for those of us with disabilities are well grounded in recent events.... Our society seem to have an aversion to those who are physically and mentally different. Genetic engineering could lead to the elimination of the rich diversity in our peoples. It is a real and frightening threat.

Those present asked the RAC to set up an outside review board for human genetic engineering experiments that would include experts in the rights of minorities, workers, and the disabled. They insisted that the RAC scientists, though astute on advances in genetics, were no experts in the public policy implications of their work. "This group cannot play God when it comes to deciding what genes should be engineered in and out of individual patients," Rifkin said during heated arguments with members of the committee. "What will be the criteria for good or bad genes? Who will decide what genes, and which people, will be engineered?" he continued. "The people in this room are just not qualified to raise these monumental social issues. You're just not going to be able to maintain that control of power within a small group. We need to broaden this group." A few members of the RAC board became belligerent, denying, sometimes angrily, the suggestion that they lacked the expertise to oversee the larger social and political implications of their work. Others simply ignored the proposal. When the vote came, the RAC board unanimously (twenty in favor, three abstentions) turned down the proposal to set up a public policy review committee.

The RAC critics lost the NIH vote, but they won the battle in court. On May 6, the NIH settled the law case filed against the NIH, agreeing to immediately make changes in the RAC guidelines that would forbid mail or secret ballots and would also provide more review for gene therapy experiments. The legal settlement cleared the way for the first legally sanctioned gene engineering experiment on humans. The gene "marker" experiment took place a few days later, on May 22, 1989.

CLAIMING IMMUNITY

The second gene experiment on humans was performed just over a year after the first. It was the first official attempt to use somatic cell human gene engineering as a therapy for disease. On September 14, 1990, a four-year-old girl from Cleveland with the immune disorder popularly known as the "bubble boy syndrome" was injected with a billion cells into which a new gene had been inserted. The girl was born without the gene that controls successful functioning of certain immune cells called T lymphocytes. The rare condition (it affects only about twenty children worldwide), known as adenosine deaminase (ADA) deficiency, leaves victims helpless in the face of disease and infection. Many children suffering from ADA deficiency have been kept alive by isolating them in a germ-free capsule, as was "David," the famous "Boy in the Bubble" at Baylor College of Medicine in Houston, Texas.

Dr. French Anderson and a team at NIH intravenously infused the child with blood cells containing the missing ADA gene in hope that it would help her recover normal functioning of her immune system. On the surface the medical procedure looked little different from a normal blood transfusion. The procedure, which took place in the Pediatric Intensive Care Unit of the Clinical Center of NIH, in Bethesda, Maryland, lasted twenty-eight minutes. One hour later the young patient was wandering around the hospital playroom, eating M&Ms.

The young girl who had become the first human gene therapy patient to be legally engineered with human genes became something of a celebrity, as did Dr. Anderson. The media reported the historic occasion in glowing terms. Soon reporters were writing about "Dr. Anderson's Gene Machine." After some initial reports of success, it was not uncommon to hear that genetic engineering had cured the "bubble boy syndrome." A second patient began gene treatment in January 1991. It was hard to imagine a more altruistic beginning for a technological development that so many had feared as the beginning of a new eugenic movement.

The experiment had its dark side, however, including some unfortunate parallels with Rogers's scandalous experiments on children in the early 1970s. A careful examination revealed that Anderson's procedure may have been more hype than cure. The "bubble boy syndrome" cases were now a misnomer:

None of the handful of existent cases required the bubble to protect the immunologically impaired children from disease. Since the mid-1980s, these children were being adequately treated with a new drug therapy. Anderson, however, had started his research into ADA before the drug therapy was available. Many felt that he continued on with his protocol more out of stubbornness and ambition than medical necessity. Months before the experiment took place, members of the Human Genome Subcommittee had openly questioned Anderson on the rationale for subjecting children to the risks of gene therapy when they were already being treated successfully. So concerned were the RAC members about the effectiveness of Anderson's therapy that they restricted Anderson and his team to working only with patients who were already receiving the drug therapy. This in turn led to the question of how Anderson could accurately assess the results of his experiment. One scientist noted that it would be a little like attempting to assess the results of aspirin on a patient who was being treated with antibiotics.

Whether or not Anderson is using his patients as gene therapy guinea pigs, his experiments appear to violate the general bioethical rule that the expected benefits to an individual from an experimental therapy should equal or exceed the potential harm. The experiment's protocol was clear. The procedure did not offer children suffering from the genetic disorder a cure, but merely a supplemental therapy. The beneficial results of the experiment are at best marginal. A cure awaits improvements in bone marrow transplantation.

By contrast, the dangers to children from Anderson's experiment could be quite real. Anderson and others involved in inserting genes into patients use animal retroviruses to carry those genes. The retrovirus used in all early gene therapy experiments, including the ADA experiment, is one called murine leukemia virus (MuLV). It is a retrovirus obtained from mice. Anderson engineers the ADA gene into the retrovirus and then injects the gene package into a patient. Once inside the patient, the retrovirus invades cells and drops off the genes. Genetic engineers like Anderson attempt to render these carrier retroviruses harmless, but there are still concerns that these viruses could cause cancer or other serious disease in patients. Except in the case of Anderson's ADA experiments, MuLV had only been approved for use in terminally ill patients in whom the retrovirus could do little additional harm. Yet Anderson used this suspect retrovirus on children who were living relatively normal lives with potentially long life spans ahead of them.

In December 1991, less than a year after Anderson began genetically engineering his second patient, an unsettling report was made public. A researcher, Arthur Nienhaus, described his discovery that the MuLV virus had caused cancer in primate. The researcher suspects that the cancer may have been caused by a contaminant that leaked into the virus during production. Anderson and others were quick to note that they used a different system to produce their MuLV, one less prone to contamination. However, the discovery bolsters the view that much more needs to be learned before MuLV is widely used as a gene therapy tool.

In a rare demonstration of scientific breaking of ranks, several fellow genetic engineers openly expressed their displeasure with the Anderson experiment. One gene therapy expert called the Anderson procedures "absolutely crazy." Dr. Arthur Bank, professor of medicine and human genetics at Columbia University, charged that gene therapy researchers at NIH were driven by ambition and not by good science. "The main impetus [for the ADA experiment] is the need for French Anderson to be the first to do gene therapy in man.... This may turn out to be bad news for all of us," Bank told a genetics conference within a week after the experiment had started. Dr. Stuart Orkin, professor of pediatric medicine at Harvard Medical School, noted, "A large number of scientists believe the experiment is not well founded scientifically.... I'm quite surprised that there hasn't been more of an outcry against the experiment by scientists who are completely objective." Dr. Richard Mulligan, a pioneer in gene therapy work and a member of the RAC board—the only one who voted against the experiment—was more direct. "If I had a daughter, no way I'd let her get near these guys if she had that defect."

Anderson has more than his experiments to defend. Critics of the approvals of the first gene therapy experiments also point out that over a five-year period, Anderson has almost singlehandedly pioneered delivering federally funded human gene engineering research to a private company with which he is a collaborator. In 1987, Anderson did what many viewed as "scientifically unthinkable" when he joined forces with venture capitalist Wallace Steinberg to help build a human gene engineering company, Genetics Therapy, Inc. (GTI), a company one observer has called the "ultimate body shop."

Steinberg had long headed the venture capital arm of Johnson & Johnson and was looking for a new market challenge in what promised to be the cutting-edge industry of the future—human genetic engineering. Traditionally, government scientists have regarded joining forces with private investors as unseemly if not unethical. Anderson's relationship to human gene engineering entrepreneurs has cast a shadow over both the science and the procedures that led to the approval of the first of several human gene therapy experiments. Concerns about conflict of interest were heightened in late 1990 when GTI hired former NIH/RAC chairman Gerard McGarrity. McGarrity had been a leading supporter of GTI's and Anderson's gene therapy experiments, and as chairman of RAC had helped shepherd the therapy proposals through the NIH approval process. In 1991, GTI's numerous maneuvers paid off: Sandoz Pharma, Ltd., one of the world's major multinational companies, bought $10 million of GTI stock and agreed to provide $13.5 million over the subsequent three years in project funding. GTI ended 1991 with cash and marketable securities of $20.8 million.

Human gene engineering is progressing quickly. Currently, over a dozen somatic cell gene engineering experiments are ongoing on three continents. Numerous other gene engineering protocols are being developed for approval in the near future. Large-scale use of gene engineering to cure disease or cosmetically change individuals is still several years away; nevertheless, the scandal, ambition, and moral blindness that have characterized the early history of human genetic engineering set a profoundly disturbing precedent for the future.

Moreover, many of the protections against abuses in the use of gene technology put in place in the 1980s are fast disappearing. The Congressional Biomedical Ethics Board, established in 1985, was disbanded in 1990. Additionally, in 1991 Dr. Anderson and others successfully urged the disbanding of the RAC Human Gene Therapy Subcommittee. Finally, in the face of a massive influx of profit-seeking and potential conflicts of interest, the viability of RAC as a responsible regulatory agency of human gene engineering is in considerable doubt.

In the future we will be genetically engineering ourselves in numerous ways—applications of biotechnology with which our society is ill prepared to deal. As researchers successfully locate genes responsible for height, weight, and I.Q., there are still no restrictions that would prevent an industry from altering these traits through somatic gene therapy. Further, researchers are now more determined than ever to begin the first germline gene engineering experiments on humans. There is general consensus that such research will become a reality over the next decade. We have no national or international mechanisms that will prevent germline engineering from permanently altering our human genome, no restrictions on the unlimited genetic alteration of sperm and eggs, or the engineering of embryos. Despite continuing controversy, publicity, and massive public funding of gene technology research, the questions demonstrators shouted at scientists over fifteen years ago have still not been answered.

NO

James Hughes

EMBRACING CHANGE WITH ALL FOUR ARMS: A POST-HUMANIST DEFENSE OF GENETIC ENGINEERING

INTRODUCTION

Nine years ago, while I rode a bus through the small, crooked, immaculate and beautiful streets of Kyoto, Jeremy Rifkin convinced me that genetic technology would determine the shape of the future. I was reading his *Algeny,* an alarmist attack on the coming of the gene age, alongside *What Sort of People Should There Be?*, a moderate defense of genetic engineering by Jonathan Glover. In a sense, in the nine years since, I have recoiled from the radical Rifkin to embrace the reformist Glover.

While extreme, Rifkin is a bellwether of Luddite tendencies in bioethics and the political Left, two of the movements within which I construct my worldview. Among bioethicists the anti-technological agenda has focused on abuses and social dangers in medical research and practice, and our alleged need to accept death and technological limits. The post-60s, environmentalist Left focuses on the ways that technology serves patriarchy, racism, imperialism, corporate profits, structural unemployment, the authoritarian state, and domination by scientific discourse. The response of bioethicists and the Left to genetic engineering has been particularly fevered, driven by accusations of eugenics and the defilement of sacred boundaries.

Since that bus ride in Kyoto my initial horrified agreement with Rifkin has shifted to determined agreement with Glover, that we can control genetic technology and make it a boon rather than a bane. Instead of a *Brave New World,* I see genetic engineering offering a grand, albeit somewhat unpredictable, future. While many of the concerns of ethicists and the Left about this technology are well-founded, I now believe they are answerable. While I still acknowledge the need for democratic control and social limits, I am now convinced that banning genetic engineering would be a profound mistake.

From James Hughes, "Embracing Change With All Four Arms: A Post-Humanist Defense of Genetic Engineering," *Eubios Journal of Asian and International Bioethics,* vol. 6, no. 4 (July 1996).

Those who set aside angst about changing human nature, and embrace the possibility of rapid diversification of types of life, are establishing a new moral and political philosophy for the 21st century, a system some refer to as "post-humanism." Like all philosophical systems, post-humanism incorporates prior philosophic and political systems but recasts them around new definitions of personhood, citizenship, and the limits of social solidarity and human knowledge. Like Glover, post-humanists view the coming of genetic technology the way most Americans now view organ transplants or chemotherapy; there are many practical questions about how the technologies get developed and tested, who needs them, and how we pay for them, but there is no question that they should be available. In this essay I will be trying to imagine what liberal democracies could be like if we allow a post-humanist flowering of genetic technology.

DISTINCTIONS WITHOUT A DIFFERENCE

Many writers on these technologies draw distinctions between "negative" and "positive" genetic modification, and the modification of the somatic versus germ-line cells. Negative genetic modification has been defined as the correction of a genetic disease, while positive modification has been defined as the attempt to enhance human ability beyond its normal limits. The somatic/germ-line distinction has been made to address the alleged ethical difference in modifying only one's own body, versus modifying one's progeny as well.

Both distinctions have been made by those who wanted to draw a line to demarcate the ethical boundaries of genetic research. The distinctions are quite fuzzy, however. Take for instance Culver and Gert's effort to define "malady" to distinguish when a genetic therapy is or isn't "enhancement":

> A person has a malady if and only if he has a condition, other than his rational beliefs and desires, such that he is suffering, or at increased risk of suffering, an evil (death, pain, disability, loss of freedom or opportunity or loss of pleasure) in the absence of distinct sustaining cause.

Doesn't any cause of illness, suffering and death, or inadequacy in the face of one's goals, fit this criterion? Take for instance a potential future genetic therapy that turned off a hypothetical aging switch, doubling the human life span; is this therapy for the diseases which result from the activation of the aging switch, such as Alzheimers or cancer, or an unconscionable intervention into the natural span of life?

As to the modification of one's own genes versus future progeny, the argument is made that current generations would be violating the self-determination of future generations by doing so. The first response is that our choice of breeding partners already "determines" the biology of future generations. Take the case of a couple who both carry a gene for latent inheritable mental illness. The only difference between their choosing not to breed with one another, and choosing to have germ-line therapy on themselves or their child to correct the illness, is that the latter choice is a far happier one.

The second response to the somatic/germ-line distinction is that advancing genetic technology will make it possible for future generations to change their genes back if they don't like them. Only modifications which remove decision-making autonomy from future generations altogether would truly raise issues of "self-determination," and I will discuss such fascist scenarios below.

These distinctions are extremely fuzzy, and do not represent important ethical boundaries. In this essay I want to defend genetic therapy and enhancement, as well as self-modification by competent adults and modification of one's progeny. Even the most liberal, and most recent, of international bioethics consensus documents, the 1995 draft UNESCO Declaration on the Protection of the Human Genome, draws the line at germ-line enhancement.

Therefore ground-zero of the terrain that I want to defend is germ-line enhancement, the modification of the genetic code such that the parent passes on the enhancements to their progeny. The defense of this practice necessarily addresses the concerns about many other technologies, such as:

- In-Vitro Fertilization
- Surrogate Mothering
- Extra-uterine Gestation
- Genetic Screening and Diagnosis
- Genetic Selection, including Sex Selection
- Cloning of Embryos

In a more fundamental sense I am writing in defense of our control of our bodies, individually and collectively. I want to build a broad enough defense to cover any technology offering modification of human abilities, whether a specific genetic application has been imagined for that purpose or not.

ETHICAL STARTING POINTS FOR A DEFENSE

Rule Utilitarianism

In general I assume the ethical stance of Millsian rule utilitarianism: acts are ethical which lead to the greatest good or happiness for the greatest number. Rule utilitarianism means that, when confronted with a distasteful case, such as throwing a Christian to a lion for the amusement of thousands of Romans, I fall back on general rules of thumb: "In general, societies that respect individual rights and liberties will lead to greater happiness for all."

In the case of genetic engineering my broad assertion is that gene-technologies can, and probably will, give people longer, healthier lives, with more choices and greater happiness. In fact, these technologies offer the possibility that we will be able to experience utilities greater and more intense than those on our current mental pallet. Genetic technology will bring advances in pharmaceuticals and the therapeutic treatment of disease, ameliorating many illnesses and forms of suffering. Somewhat further in the future, our sense organs themselves may be re-engineered to allow us to perceive greater ranges of light and sound, our bodies re-engineered to permit us to engage in more strenuous activities, and our minds re-engineered to permit us to think more profound and intense thoughts. If utility is an ethical goal, direct control of our body and mind suggests the possibility of unlimited utility, and thus an immeasurable good.

PRIVACY, SELF-DETERMINATION AND BODILY AUTONOMY

But there are other rules to consider, rules which are the basis of other ethical systems. Most utilitarians, and many others, accept the general rule that liberal societies, which allow maximum self-determination, will maximize social utility. The rule of, or right to, self-determination also argues that society should have very good reasons before interfering with competent adults applying genetic technology to themselves and their property. Self-determining people should be allowed the privacy to do what they want to with their bodies, except when they are not competent, or their actions will cause great harm to others.

Acknowledging self-determination as an ethical starting point addresses half of the revulsion to genetic engineering: the concern that people will be forced to conform to eugenic policies. I will discuss this fear of racist and authoritarian regimes at greater length, but suffice it to say here that individuals should not be forced to have or abort children, or to modify their own or their children's genetic code. I am addressing the desirable genetics policies of liberal societies, not of authoritarian regimes.

Within liberal societies, competent adults should generally be allowed to do as they like with their bodies, including genetically modify them. The potential risks to others from such modifications, which I will try to discuss below, are all soluble, and not sufficient to warrant contravening the right of bodily autonomy.

I also view the embryo and fetus as the biological property of the parents, and exclusively of the mother when in utero. Again, the rights of the future child and of society may restrict what we allow parents to do to their prenatal property. But I would again argue that the risks to society and to the children themselves of prenatal genetic manipulation are negligible for the near future, and regulable as they become apparent.

FREEDOM FROM BIOLOGICAL NECESSITY

Genetic technology promises freedom and self-determination at an even more basic level: freedom from biological necessity. Social domination pales before our domination by the inevitability of birth, illness, aging and death, burdens that genetic technology offers to ameliorate. As for Marx, the goal of this revolution is to move from the realm of necessity to the realm of freedom.

Social domination also builds on a biological foundation. Patriarchy is, in part, based on women's physical vulnerability, and their special role in reproduction. While industrialization, contraception and the liberal democratic state may have removed the bulk of patriarchy's weight, genetic technology offers to remove the rest. Similarly, while racism, ageism, heterosexism, and so on may be only 10% biological and 90% social construction, at least the biological factors can be made a matter of choice by genetic and biological technology.

JUSTICE AND A BETTER SOCIETY

While the biological factors in most forms of inequality are probably slight, genetic technology does promise to create a more equal society in a very basic way: by eliminating congenital sources of illness and disability that create the most intractable forms of inequality in society. We can go to great lengths to give the ill and disabled full access to society, but their disabilities place basic limits on how equal their social participation and power can be. Our ability to ameliorate these sources of congenital inequality may even impose obligations on us to do so, at least for those who are cognitively impaired and incompetent. Admittedly, we will probably have surmounted most disabilities through non-genetic technological fixes long before we do so through genetic therapy. But the general principle is that genetic technology promises to make it possible to give all citizens the physical and cognitive abilities for equal participation, and perhaps even to bring about a general enhancement of the abilities essential to empowered citizenship.

A CRITICAL DEFENSE

Unlike those libertarians who hold self-determination as a cardinal principle, I adopt more of a social democratic stance, and foresee legitimate limits that we can and should place on these technologies. For instance, some characteristics of society, such as social solidarity and general equality, are so important that they warrant the regulation of these technologies in the furtherance of these goals. Collective interests should also be pursued through active means, such as government subsidies for the research, development and application of genetic technologies.

Nor am I an unquestioning advocate of technological progress. Some technologies are so inscribed with harmful ends that no amount of regulation and social direction can make them worth the risk. If I were convinced that genetic technology, like nuclear weapons technology, had no redeeming qualities and only great risks then I would embrace a complete ban.

But the potential benefits of genetic technology far outweigh the potential risks. In short, I advocate a position of critical support, a position which reflects the suspicious optimism that most Americans have toward genetic technology.

ARGUMENTS AGAINST GENETIC TECHNOLOGY

There are at least two kinds of criticisms of genetic technology, fundamentalist and non-fundamentalist. The fundamentalist or "bio-Luddite" concerns, such as those of Jeremy Rifkin, I reject fundamentally. On the other hand, I accept the validity of many of the non-fundamental concerns, but see the problems they suggest as soluble. Few of these concerns about genetic technology raise new questions for medical ethics. The same questions have been raised by previous medical research and therapy, and those challenges have been met without bans on those technologies.

Some non-fundamentalist critics believe that, cumulatively, the risks posed by new genetic technologies are great enough to warrant postponing genetic research for some indefinite period of study and preparation. With these concerns I will argue that, with adequate technology assessment and anticipatory regulation, there will be adequate time to regulate genetic technology as we proceed; none of the risks are sufficiently weighty, individually or cumulatively, to outweigh the potential benefits.

The fundamentalist or bio-Luddite concerns I will address are:

- Bio-Luddism 1: Medicine Makes People Sick

- Bio-Luddism 2: Sacred Limits of the Natural Order
- Bio-Luddism 3: Technologies Serve Ruling Interests
- Bio-Luddism 4: The Genome Is Too Complicated to Engineer

The non-fundamentalist or pragmatic concerns I will discuss are:

- Gene Angst 1: Fascist Applications
- Gene Angst 2: The Value of Genetic Diversity
- Gene Angst 3: Genetic Discrimination and Confidentiality
- Gene Angst 4: Discrimination Against the Disabled
- Gene Angst 5: Unequal Access, Priority Setting and the Market
- Gene Angst 6: The Decline of Social Solidarity

Bio-Luddism 1: Medicine Makes People Sick

One extreme bio-Luddite position was elaborated by Ivan Illich: medicine itself makes us sick and should be done away with. A variant on this argument is that genetic screening will eventually determine that all of us are "at risk," making everyone see themselves as sick. More troubling, genetic diagnosis might create a two-tier social system, divided between those with relatively clean genes and those with genetic disease. In other words, genetic diagnosis will make us all genetically diseased. This would be even more problematic if the genetic diagnosis was for a disease which was not yet curable.

Some medicine makes some people sicker, but I hold fast to the modernist promise that scientific progress generally improves our lives and that knowledge is better than ignorance. It is unlikely that we will ever force people to know their likelihood of developing disease, though perhaps we should educate parents and physicians to be cautious about informing children of their risks. In any case, we all know that we are at risk of dying, and with or without genetic diagnosis people view the medical history of their parents and relatives as harbingers of things to come. Both knowing and refusing to know one's genetic makeup are empowering choices for competent adults; denying people the option of making this choice does not improve their lives.

This argument also presumes just the first, screening phase of the new eugenics, and not the latter correction phase. Far from making everyone sick, the advance of genetic therapy promises to make everyone well.

Bio-Luddism 2: Sacred Limits of the Natural Order

Rifkin has joined forces with religious leaders to assert another fundamentalist tenet, that genetic engineering transgresses sacred limits beyond which we should not "play God." I don't believe that divine limits are discernible, and I don't believe in any "natural order" except the one we've got. As Love and Rockets point out: "you can't go against nature, 'cause when you go against nature, it's part of nature too." There are no "natural limits" in our taking control of our biology or ecology. There is no "natural" way to have a baby or die. Even if there was a natural way to birth or die I don't believe we are morally compelled to adopt it.

Bio-Luddism 3: Technologies Serve Ruling Interests

Some hesitate to argue that medical technology is bad in and of itself, but argue instead that the powerful always shape and apply technologies to further their domination of the less powerful. While this is probably true, the conclusion is that all technology should be abandoned. The wealthy and powerful have more access to telephones than the poor and powerless, and telephones are used by the wealthy and powerful to collect more wealth and power. But I see the answer to be subsidized phone service and social reform, not banning the telephone.

Bio-Luddism 4: The Genome Is Too Complicated to Engineer

A fourth fundamentalist conviction is that the genome is too complicated to engineer, and therefore there are certain to be unpleasant, unintended consequences. This argument is directly parallel to the deep ecologists' conclusion that human management of the complex global eco-system is impossible, and that our only hope is to leave the planet alone to its own self-organization.

The genome and eco-system are both very complicated, and the ability to do more than correct local defects in either may be many decades away. But eventually we will have the capacity to write genetic code and re-engineer eco-systems, and to computer-model the structural consequences of our interventions on future bodies and planets. Of course, it will be difficult to decide when the consequences of a genetic blueprint are sufficiently well-understood that it is safe for use, and our current regulatory scheme is probably not yet adequate to the task.

Our understanding of the genome and ability to predict consequences must be very robust before we allow human applications or the release of animal applications. While Elias and Annas object to "positive" germ-line therapy, which I would defend, they propose three sensible preconditions on the application of gene-engineering:

1. that there should be considerable prior experience with human somatic cell gene therapy, which has clearly established its safety and efficacy; and
2. that there should be reasonable scientific evidence using appropriate animal models that germ-line gene therapy will cure or prevent the disease in question and not cause any harm, and
3. all applications should be approved by the NIH's [National Institutes of Health] Working Group on Gene Therapy and local Institutional Review Boards, with prior public discussion.

Those of us who believe in the possibility of effective public regulation may differ widely as to the appropriate standards the public and these regulatory bodies may use. But liberals and conservatives differ fundamentally from those bio-Luddites who believe that the natural world is so complicated, and governments so unwise, that all intervention must be forbidden.

Gene Angst 1: Fascist Applications

Another concern expressed by many critics of genetic technology is the dire consequences of the re-emergence of fascist, racist and authoritarian regimes, and their use of engineering to produce compliant, genetically uniform subjects. The first point to make about fascist uses of eugenic ideology or technology is that nothing a democratic society does to forbid itself genetic technology will have any impact on future or contemporary fascist regimes. Indeed, if there is any "national security" to be gained from genetic technology then it would behoove liberal democracies to gain them as well. For instance, public health campaigns to detect and correct the genetic predisposition to alcoholism, or to enhance the intelligence of children, could make nations much more powerful and productive than their more conservative neighbors; would it not be in the interest of democracy itself for democracies to pursue these measures?

Yet, what if the fascist regimes found strength in breeding different castes a la *Brave New World,* and democracies could only meet the challenge by becoming equally repugnant? This is a possibility, and it raises the important point: the way to stop fascist uses of genetics is to prevent the rise of fascism, not to restrict the emergence of genetic technology. As we see today with Iraq and North Korea, firm agreements by right-thinking nations that only the United States is sufficiently moral to be allowed the ownership of nuclear and chemical weapons has little impact on recalcitrant regimes. If we cannot effectively prevent the proliferation of nuclear technology, with its large radioactive facilities visible to satellites, we will have even less success with genetic laboratories. I support the strengthening of the legal, judicial and military might of U.N. so that it might begin to enforce global law, but I think the proper task for such a New World Order is the suppression of fascist regimes likely to use genetics for nefarious ends, not the policing and suppression of outlawed genetic technologies.

Genetic science does not itself encourage racism or authoritarianism. In fact, the advance of scientific knowledge may even erode the pseudo-scientific basis on which most eugenics has rested. Presumably the advance of genetic science will tell us whether there is a genetic basis for gender and racial differences in abilities, or not, and how important these are. If there are genetic factors in gender or racial difference, they will most likely be revealed as minor beside the social factors, and the genetic factors will become ameliorable through a technical fix. Some insist that knowledge itself, or knowledge about forbidden topics, will lead to fascism; I prefer the modernist optimism that knowledge is at least neutral towards, and sometimes a scourge of, obscurantism.

Gene Angst 2: The Value of Genetic Diversity

Another concern that is often expressed vis-a-vis genetic engineering is the alleged aesthetic or biological virtues of genetic diversity. Many refer to the evidence from ecology that ecosystems are more stable when they contain a greater diversity of gene-lines. Some suggest, for instance, that our very survival as a species might hinge on genetic diversity if we faced some blight that only a few were resistant to.

The first objection to this argument is that diversity is not a sufficiently compelling ethical or aesthetic virtue that it can trump the prevention of disease, or the improvement of the quality of our lives. We "reduced diversity" when we eradicated smallpox and polio, with no regrets. We "reduce diversity" when we insist on compulsory education because we don't value the diversity of extreme class inequality.

The second objection to the diversity argument is that any loss of adaptiveness through biological diversity will be compensated for by an increase in biological knowledge and control. It is unlikely that a future society would have the ability to create "superior genes" and yet be unable to meet the challenge of infectious disease.

Third, the regime of genetics I have outlined is a liberal one, which should produce as much diversity as it reduces. While I support public provision of genetic screening for disease, I oppose any eugenic coercion. People desire different attributes and abilities, for themselves and their children; for every Aryan parent that chooses a blond, blue-eyed Barbie phenotype, I expect there would be a Chinese parent choosing a classic Chinese ideal of beauty. True, this might lead to the convergence toward a few physical and mental ideals, though I suspect that phenotypic fashions will change quickly. But I see no ethical difference between permitting people to change their genes in conformity with social fashions, and permitting them to change their clothes, makeup and beliefs to do so.

Yet, perhaps there is some aesthetic and or even civic virtue in diversity. If it is valued by the public, let us establish incentives for diversity. If the number of parents choosing to raise blond boys is offensive to public opinion, we can give tax incentives for parents who bear dark-haired girls. In any case, we will quickly know if there are broad trends that we find offensive, and I trust our ability to craft non-coercive policy responses to re-establish any valued diversity we feel may be eroding.

Gene Angst 3: Genetic Discrimination and Confidentiality

Many opponents of genetic investigation are concerned that growing genetic knowledge will lead to discrimination against the "genetically diseased and disabled." Some assert that genetic therapy itself will increase this discrimination by bringing intense pressure to bear on those with genetic diseases to have the disease corrected, and not burden society and future generations with their diseases.

It is certainly true that employers are already attempting to discover the genetic risks of their employees, and deny employment or health insurance on the basis of this risk profile. A bill guaranteeing the confidentiality of genetic information has been introduced in Congress, and while it has not yet passed, some form of confidentiality is certain to be guaranteed by the turn of the millennium. In addition, the Americans with Disabilities Act and similar legislation will clearly be mustered to defend workers from genetic discrimination.

Keeping genetic information confidential from health insurers is trickier, since they would be reimbursing for any special screening or treatment that genetic risks called for. Unregulated, the use of genetic risk information could greatly strengthen the ability of insurers to exclude the illness-prone from their risk pools, or charge them premiums equivalent to the costs of their potential treatments. Again, however, popular insurance reform legislation before Congress will ban "risk-rating" and excluding clients with "pre-existing conditions." These two reforms will likely reduce the number of insurance companies in the country by half or more, and make genetic discrimination in health insurance a more or less moot point. Some have suggested further that the pervasiveness of genetic information will make private health insurance impossible; to which I say, good riddance.

There are undoubtedly many other nefarious uses to which knowledge of someone's genetic make-up can be put. But genetic information is only one small category of the information about our lives which is potentially in the public domain, and potentially injurious. The regulation of genetic technology really has very little to do with whether we establish data privacy in the 21st century.

Gene Angst 4: Discrimination Against the Disabled

Opponents of sex selection and of eugenic efforts against genetic disease argue that these decisions are acts of prejudice against women and the disabled, and perpetuate the second-class status of women and the disabled by focusing on genetic rather than social amelioration. In the first place, embryos and fetuses are not persons, and therefore their rights cannot be violated as persons or as members of oppressed social groups. While parents may make reproductive decisions for many reasons we disapprove of, such as aborting a fetus because the father was accidentally of the "wrong" race, this is not a reason to intervene.

The alleged link between choosing to abort a disabled child, or correcting their disability through genetic therapy, and the perpetuation of oppression of the disabled seems tenuous at best. Perhaps by reducing the population of disabled we reduce their power at the ballot box. But a parent's moral obligation to give their children the greatest quality of life, and the fullest range of abilities, includes not only the obligation to treat a disabled child with respect and love, but also the obligation to keep them from having disabilities in the first place. It also seems likely that a society with fewer disabled would increase rather than decrease their per capita expenditures on the disabled.

Gene Angst 5: Unequal Access, Priority Setting and the Market

As a social democrat, one of my gravest concerns is how social inequality will constrain access to genetic technology, and how genetic technology may reinforce social inequality. Establishing the appropriate balance of state and market in genetics starts with the creation of a national health budget, most likely through the creation of a national health system, such as the [President Bill] Clinton plan or some other form of national health insurance. Such a system allows the ethical determination of utility trade-offs, from what the level of health care expenditures should be, to what should be included in the basic package of guaranteed medical services and what should be consigned to the private medical market.

If we had such a system, I don't think most fertility treatments would make the cut, nor would future positive genetic "enhancements." On the other hand, genetic screening and corrective genetic therapy would clearly be socially acceptable, cost-effective, and therefore a plausible positive right. This leaves me in a quandary; I want fertility treatments and positive genetic enhancement to be legal and available, but I'm not prepared to argue that they are a positive right worthy of public subsidy. Yet, if gene products are left in the market, only the wealthy will have access to them, with the upper-classes having more life opportunities and potentially becoming genetically healthier and more intelligent than the poor, which is unethical in an equal opportunity society.

These problems are really a subcategory of the larger task of determining which medical tests and procedures should be:

- required by law, e.g. vaccinations
- publicly funded, but not obligatory, e.g. abortion in progressive states
- encouraged, but unsubsidized, e.g. exercise
- discouraged, but not banned, e.g. smoking
- banned., e.g. heroin

Any assignment of genetic technologies to the categories between obligatory and forbidden allows for potential inequality. Most opponents of genetic technology, when pressed, would stop short of banning these technologies out-right, and thus leave them to be inequitably distributed by the market. At the other extreme, there are no audible voices calling for a program of mandatory, universal genetic redesign. This leaves me with Glover in the usual social democratic, mixed-market middle: try a little public, and a little private, and we will tinker with it as we proceed.

Gene Angst 6: The Decline of Social Solidarity

Finally some critics suggest that parents would become alienated from their genetically engineered children. Dator and other post-humanists suggest that genetic engineering and other technologies may create conflict between humans and post-humans, and threaten social solidarity. I think this is a serious concern, and one goal of the social regulation of genetic technology would be to moderate the rapidity with which society genetically advances and diversifies. The gaps between the bodies and abilities of parents and children should not be so great as to make parenting impossible. Also the unenhanced public's concerns will inevitably be a factor in regulating the enhancement of the modified minorities. While some of these conservative concerns may be warranted, if the enhanced feel they have no responsibility to the unenhanced and seek to dominate or exploit them, we must also avoid allowing simple chauvinism and fear of the unknown to stop genetic enhancement.

While tremendous social conflicts can be imagined, they are not that different from the conflicts between ethnic minorities and majorities, or between the First World and the Third, or between social classes. Like other sources of social division, the relations between new genetic communities will hopefully be mediated by the same institutions, courts and legislatures, minority rights and majority rule. The real challenge faced by a post-human ethic is to define new parameters for which forms of life should be considered property, social wards (neither property nor competent persons, such as children), and persons with full citizenship.

CONCLUSION

While humanists and economists urge us to embrace financial and existential limits, and give up the quixotic quest for immortality, the post-humanists say "Some alive today may never die." The potential problems created by new medical technology are numerous, and we must work hard to ensure that our societies are free and equal enough that these tools create more good than harm. But I believe this to be an achievable goal, and that genetic technology offers, if not immortality, such good that the risks are dwarfed. Like all speculation (and all utilitarian judgments are based on social speculation) this optimism is founded on numerous points of faith. But I find faith in the potential unlimited improvability of human nature and expansion of human powers far more satisfying than a resignation to our current limits.

POSTSCRIPT

Should Genetic Engineering Be Banned?

Genetic engineering has had and will have applications in agriculture (improved crop plants and animals) and in the pharmaceutical industry (drug production). In the last few years, much of the excitement and the alarm have centered on its use to treat diseases by modifying human genes. Gene therapy has not yet become a multimillion-dollar industry, but there have been some successes. In October 1993 researchers reported that giving hemophiliac dogs—dogs with a hereditary defect that delays blood clotting—a copy of the gene for the blood-clotting agent they lacked improved the ability of their blood to clot. In March 1994 a similar approach repaired mice that had an autoimmune condition similar to the human disease lupus erythematosus. In April 1994 researchers announced that giving a woman with familial hypercholesterolemia—a rare genetic disorder that is marked by very high levels of blood cholesterol and early death from heart attack—the proper version of her defective gene reduced her cholesterol levels by 20 percent. For more detailed discussions of how researchers are transferring genes into human cells, see Mark A. Findeis, "Genes to the Rescue," *Technology Review* (April 1994) and Mario R. Capecchi, "Targeted Gene Replacement," *Scientific American* (March 1994).

By 1995 over 100 genetic therapies were being tested in humans. However, technical difficulties remain, and the successes have not been decisive (see Eliot Marshall, "Gene Therapy's Growing Pains," *Science*, August 25, 1995).

The technology is still young, its growth is still largely ahead, and its promise is yet to be fulfilled. *Will* that promise be fulfilled? Many people remain worried about the negative possibilities: In Germany tough regulations have made genetic engineering research of any kind almost impossible to carry out. In England a cancer research project was shut down in February 1994 because of fears that common cold viruses engineered to carry cancer genes might escape and cause a cancer plague (the actual risk was almost zero because the virus had been made unable to reproduce in cells, but government regulators judged the lab's containment measures to be inadequate). The availability of genetic information has raised fears of discrimination by insurers and employers (see Kathy L. Hudson et al., "Genetic Discrimination and Health Insurance: An Urgent Need for Reform," *Science*, October 20, 1995). And religious groups have objected to the patenting of genes (see Ronald Cole-Turner, "Religion and Gene Patenting," *Science*, October 6, 1995).

On the Internet . . .

CDT Privacy Issues Page
At this site, the Center for Democracy and Technology (CDT) demonstrates how easy it is for people to find out about you by using the Internet as a research tool.
http://www.cdt.org/privacy/

Kasparov vs. Deep Blue: The Rematch
This site is all about the chess match of the century.
http://www.us.chess.ibm.com/

Project Gutenberg
Project Gutenberg is an ongoing project to convert the classics of literature into digital format.
http://gutenberg.etext.org/

Hans Moravec
From here, take a peek at Hans Moravec's new book *Universal Robots: Object to Person to Spirit* (Oxford University Press, 1998).
http://www.frc.ri.cmu.edu/~hpm/

The Electronic Frontier Foundation
The Electronic Frontier Foundation is concerned with protecting individual freedoms and rights such as privacy as new communications technologies emerge.
http://www.eff.org/

PART 4

The Computer Revolution

Fans of computers are sure that the electronic wonders offer untold benefits to society. When the first personal computers appeared in the early 1970s, they immediately brought unheard-of capabilities to their users. Ever since, those capabilities have been increasing. Today children command more sheer computing power than major corporations did in the 1950s and 1960s. Computer users are in direct contact with their fellow users around the world. Information is instantly available and infinitely malleable.

Some observers wonder about the purported untold benefits of computers. Specifically, will such benefits be outweighed by threats to traditional institutions (will books, for example, become an endangered species?) or to human pride (a computer has already outplayed the human world chess champion)? Also, scientists debate quite seriously whether or not there will ever be a computer that genuinely thinks like a person.

- **Will the Information Revolution Benefit Society?**
- **Are Computers Hazardous to Literacy?**
- **Will It Be Possible to Build a Computer That Can Think?**

ISSUE 12

Will the Information Revolution Benefit Society?

YES: John S. Mayo, from "Information Technology for Development: The National and Global Information Superhighway," *Vital Speeches of the Day* (February 1, 1995)

NO: Andrew L. Shapiro, from "Privacy for Sale: Peddling Data on the Internet," *The Nation* (June 23, 1997)

ISSUE SUMMARY

YES: John S. Mayo, president emeritus of Lucent Technologies Bell Laboratories, formerly AT&T Bell Laboratories, argues that the information revolution will benefit society by slowing migrations from rural to urban areas, aiding economic development, and improving access to education, health care, and other social services.

NO: Andrew L. Shapiro, a fellow of the Twentieth Century Fund, argues that information technology turns personal information into a commodity and threatens the basic human right to privacy.

Not all the effects of technological development are foreseeable. New technologies may promise—and even deliver—wondrous capabilities. They may make their inventors and developers rich beyond their wildest dreams. They may also create social problems and harm or even destroy some subgroups of society. The mechanization of the textile industry in the nineteenth century, for example, led directly to oppressive child labor and wage slavery. Railroads were a boon to farmers and ranchers, but they hastened the destruction of the American Indian cultures. Agricultural mechanization drove blacks from small farms to the industrialized cities, where other forces destroyed community and family cohesiveness. And when the automobile moved many high-paying jobs from cities to suburbs, inner-city residents were denied the hope of escaping from poverty and ghettos were created.

Technology also has environmental effects. Something as ordinary as agriculture leads to wastelands, deserts, dust bowls, erosion, pesticide-poisoned water, and more. The printing press created a demand for paper that has helped to drive the clearing of forests. Paper-making machinery boosted that demand; added demands for toilet paper, tissues, and paper towels; and contributed enormously to water pollution. The automobile created both air and water pollution, as did the discovery and application of electricity, which added the problem of nuclear waste. Refrigerators, air conditioners, and aerosol cans helped to create the hole in Earth's protective ozone layer.

The list of the side effects of technology is endless. It is also unavoidable to a huge extent, for it is impossible to do anything without affecting the world around us. What we can do is be aware of the potential for damaging side effects of our technologies and strive to keep them to a minimum.

The latest technology to spread rapidly through society is "information technology," best known to many of us in the form of computers and the "information superhighway" (the National Information Infrastructure, or NII), or the Internet. Currently using telephone lines (and soon cable TV lines or direct links to orbiting satellites), the superhighway moves electronic data—words, numbers, voices, music, images—very rapidly in and out of homes and offices. It increases access to information, education, and entertainment for everyone who is hooked up to the system. And that access is spreading rapidly, for not only does the technology become more capable every year, it also becomes cheaper.

Are there environmental effects resulting from the information revolution? Many people argue that transmitting information uses much less energy than transporting commuters, so it should be less polluting. Also, improved access to information should improve decision making in general and thereby help to minimize the environmental effects of human actions. "Virtual reality" even hints at a day when national parks and wilderness areas will be less congested because people will be "visiting" these areas via computer simulations. On the other side are people like political scientist James H. Snider. In "The Information Superhighway as Environmental Menace," *The Futurist* (March/April 1995), Snider argues that some of these advantages have ominous implications because if information-processing workers can work at home, they will be free to spread out across the countryside, carrying with them a myriad of environmental impacts. We may face, he says, "an environmental disaster of the first magnitude."

John S. Mayo has been intimately involved in the development of the information revolution. He sees its future in the rosiest of terms, saying that the information revolution will stimulate development throughout the world, reduce the environmental problems posed by urbanization, and increase human welfare. Andrew L. Shapiro argues that because information technology turns personal information into a commodity, it threatens the basic human right to privacy. To the extent that respect for human rights is essential to a just society, the information revolution therefore threatens society's underpinnings.

YES

John S. Mayo

INFORMATION TECHNOLOGY FOR DEVELOPMENT: THE NATIONAL AND GLOBAL INFORMATION SUPERHIGHWAY

Delivered at the National Research Council/World Bank Symposium, Marshalling Technology For Development, Technology Trends And Applications Session, Irvine, California (by video), November 28, 1994

[In late 1995 and early 1996, AT&T underwent restructuring that resulted in three separate companies. The new systems and technology company is called Lucent Technologies. All references to AT&T in this selection should be considered synonymous with Lucent Technologies.—Ed.]

I plan to discuss major trends in information technology by first examining the driving forces propelling the emerging multimedia communications revolution and the evolution of the so-called information superhighways—to use the popular term. Then I will glance at this multimedia revolution and at AT&T's vision of information superhighways. And I will conclude by touching on the impact of all this information technology on developing countries.

Now, it's clear that the key underlying information technologies are the prime drivers and the key enablers behind the emerging multimedia communications revolution and the evolution of information superhighways—as well as a host of other advances that together are changing the way we live, work, play, travel and communicate. Because these key information technologies are changing the work and home environments, these same technologies are helping to address customer needs. The more they can do, the more new products and services the customer wants. It has been an upward spiral that has lasted over three decades, and will surely last at least one or two decades more.

What are these key underlying information technologies? They are silicon chips, computing, photonics or lightwaves, and software. And we've seen technology capabilities doubling every year in a number of such domains—

From John S. Mayo, "Information Technology for Development: The National and Global Information Superhighway," *Vital Speeches of the Day* (February 1, 1995). Originally delivered at the National Research Council/World Bank Symposium, Marshalling Technology for Development, Technology Trends and Applications Session, Irvine, California (by video), November 28, 1994.

for example, in computing and photonics—and doubling every 18 months in silicon chips. Even software—once a "bottleneck" technology because of quality and programmer-productivity problems—is beginning to advance rapidly in major areas like telecommunications, thanks to advanced programming languages and reuse of previously developed software modules.

To cite perhaps the most widely known example, we've witnessed explosive growth in the power of silicon chips—one measure of which is the number of transistors we can cram onto a chip the size of a fingernail. And this number, now in the millions, is moving steadily toward known physical limits. In the early part of the next century, today's familiar solid state devices may mature with transistors measuring about 400 atoms by 400 atoms each—the smallest such transistors likely to operate reliably at room temperature. The new frontier then will not be in making the devices smaller, but in creatively and economically using the vast increase in complexity and power made possible by this remarkable technology.

The amazing progress of silicon chips forms a microcosm of the broad thrust of information technology and all the associated forces that are leading to the multimedia communications revolution and the evolution of information superhighways. Let's look at the progress and impacts of these related driving forces.

After the invention of the integrated circuit, every time the number of transistors on a silicon chip increased by a factor of a thousand, something had to be reengineered—that is, something had to be radically changed or improved, because it was a new ball game. So the first reengineering that we did—as we headed toward that first thousand-fold increase—was to change all of our design processes, which had been based on discrete components.

When we reached a thousand transistors per chip, we used the new digital circuitry to reengineer our products from analog to digital, as did many other industries. Let me stress that this early progress toward digital products, enabled by silicon chips and software, brought about the digitalization of most systems and services—domestically and, more and more, globally. This digitalization created a powerful force that is driving us toward multimedia communications and information superhighways.

Then, about a decade ago, we reached toward a million transistors per chip—and powerful microcomputers became possible, along with all the periphery related to microcomputers and the needed software systems. All this led to an explosion of advanced communications services that forced the judicial process that led to the reengineering of our company: from a company that provided largely voice and data-on-voice telecommunications services to a company focused on universal information services. The theme of universal information services is voice, data and images anywhere, anytime with convenience and economy. Providing advanced services on an increasingly intelligent global network was the beginning of multimedia communications, now emerging as the revolution of the 1990s and beyond.

We are currently in the era of yet another thousand-fold increase in transistors per chip. And reengineering has now extended beyond our company and is leading to the merging of communications, computers, consumer electronics and entertainment. The bringing together of these four industries has started out in obvious ways—that is, through joint projects, joint ventures, mergers, acquisitions and some new start-up companies. This reengineering of our industry appears to be the next-to-the-last step of the information revolution brought on by the invention of the transistor.

The last step, and one that may go on forever, is the reengineering of society—of how we live, work, play, travel and communicate. It will create a whole new way of life. For example, it will change education through distance learning and school at home; it will change work life through virtual offices and work at home; and it will diminish the need to transport our bodies for work or routine tasks such as visiting and shopping. Let me quickly add, however, that it will take social change as well as technology to make many of these changes happen.

Another driving force toward multimedia communications and information superhighway evolution is the worldwide push toward common standards and open, user-friendly interfaces that will encourage global networking, and maximum interoperability and connectivity.... [S]ervice providers and customers will be able to use equipment from many different vendors without worry about compatibility. This will facilitate the upgrading of existing networks and the construction of new networks on a worldwide basis.... Similar standards in domestic networks will enable digital communications to the workplace and home, and will make possible high data-rate services.

But let me be clear on this point: although we have a lot of good work on standards, universal connectivity and interoperability will remain a big challenge as the communications and computing industries merge....

Now, the pacing force behind the multimedia and information superhighway revolution is not so much the technology as it is marketplace demands. For the greater part of this century, the user willingly accepted whatever technological capabilities we were able to achieve. Thus, the telecommunications industry was supplier-driven, and the suppliers managed the evolution of the industry and the information highways. But, as you may know, the technology became so rich that it made many more capabilities possible than the user could accept. To put it differently, we could design a lot more products and services than the customer was willing to pay for. That marked the transition from a supplier-driven industry to today's customer-driven industry—from supplier push to marketplace pull.

And, importantly, the global transfer and assimilation of information technology are combining with political and regulatory forces such as the move to privatization of telecommunications around the world—in both developed and developing countries. The result is the growth of ever-stronger global competition in the provision of communications products and services. Such emerging competition is another force driving the evolution of both multimedia communications and information superhighways. And there is an on-going challenge to public policy—not just in the U.S., but globally—to provide a framework for that evolution to occur, a framework that ensures full and fair competition for all players.

These, then, are some of the important forces driving us into the multimedia communications revolution and the associated evolution of information superhighways.

Let's look a bit further into these subjects and start with the multimedia revolution. After all, the pursuit of multimedia is creating social pressures on the evolution of information superhighways—both here and around the world. So what is "multimedia?" A reasonable working definition is that the term "multimedia" refers to information that combines more than one medium, where the media can include speech, music, text, data, graphics, fax, image, video and animation. And we at AT&T tend to focus on multimedia products and services that are networked; that is, connected over a communications and information network.

Examples of such networked multimedia communications range from videotelephony and videoconferencing; to real-time video on demand, interactive video and multimedia messaging; to remote collaborative work, interactive information services such as electronic shopping, and multimedia education and training. Eventually, we will have advanced virtual reality services, which will enable people to indirectly and remotely experience a place or an event in all dimensions.

Now, we are excited about multimedia because public switched networks—or information highways, if you will—can presently accommodate a wide array of networked multimedia communications, and the evolutionary directions of those networks will enable them to handle an increasingly vast range of such communications. Moreover, there is also a potentially vast market for multimedia hardware and supporting software. Although actual projections differ widely, the most commonly quoted projection for the total worldwide market for multimedia products and services is roughly $100 billion by the year 2000.

We at AT&T are playing a major role in facilitating the emerging multimedia revolution—as a service provider, as a provider of network products to local service providers, and as a provider of products to end users. These are familiar roles for AT&T, so let me briefly describe another, perhaps less familiar, major role we are studying in relation to the multimedia revolution. That is the role of what we call "the missing industry"—and that role is a "host" for a wide variety of digital content and multimedia applications developed by others. Hosting is a function that connects end users to the content they seek. Customers will gain easy, timely and convenient access to personal communications, transactions, information services and entertainment via wired and wireless connections to telephones, handheld devices, computers and eventually television sets. Independent sources for this digital content eventually will range from publishers to large movie studios to small cottage-industry software houses.

This role is also of interest here because of the key information superhighway challenge it illustrates—specifically, because openness of critical interfaces and global standards are vital to this complex hosting function. The entertainment industry, for example, must have software systems that are compatible with those of the hosting industry, and these software systems must, in turn, be compatible with those of the communications and information-networking industry, which then must be compatible with the customer-premises equipment industry.

In addition, the tremendous growth in available information and databases will stimulate the need for personal intelligent agents. These "smart agents" are software programs that are activated by electronic messages in the network, and that find, access, process and deliver desired information to the customer. They can perform many of the time-consuming tasks that have discouraged a number of users from taking advantage of on-line services and the emerging electronic marketplace. "Smart agents" are one feature of AT&T's recently announced enhanced network service called AT&T PersonaLink℠ Services.

Let me say I'm looking forward to these "smart agents"—software that can take the hassle out of life. Shopping for the best mortgage, or finding the best new car deal, or finding out which store has the item I want is a hassle, and has people at the interface who add negative value. Just last week I needed a replacement part. I called the store twice and got no satisfactory response to my calls. So I went to the store, waited in line, and then the salesperson queried the database and said, "We don't have it in stock." My "smart agent" could have queried their database and saved them and me a big investment in a zero-revenue operation. There was never a problem with the database; the problem was that people were inadvertently in the way of my ability to access it—adding negative value, but diligently trying to do their jobs. A "smart agent" could simply have done it better.

Now, it's important to note that in the age of multimedia communications, people who are geographically separated from each other will not, for example, just play games together over networks—they will visit and find what is emotionally nourishing, and build their relationships. According to AT&T's vision, in this evolving age, consumers and business associates will seek new relationships based on telepresence, a new type of community, and a social experience independent of geography. This potential for interactive networks is quite unlike that found in the proposed availability in the U.S. of 500 pre-programmed TV channels on the CATV cable. The beauty is that people will have the freedom to choose any subject or service from the intelligent terminals in their homes and offices. A key point is that they will be able to network clusters of friends or associates to enjoy such services as a group.

I must stress that networked multimedia communications will dramatically change the nature of work, and will therefore have a broad impact on business—first in developed nations and eventually in developing nations. Videoconferencing, for example, is first coming into businesses to enhance productivity, save time, and reduce travel. And current developments in multimedia telephony are making the possibility of remote collaborative work more and more realistic. In a few years, for example, a person could be working with colleagues or suppliers in branch offices in New York, Irvine, Hong Kong, Paris, and Sydney. Working in real time, they could accomplish the combined task of producing printed materials, presentation slides, and a videotape introducing a new product line.

As I noted, the pursuit of multimedia communications is driving social issues relating to the evolution of the information superhighway. Now, what is AT&T's vision of the information superhighway?

Our vision is to bring people together, giving them easy access to each other and to the information and services they want and need—anytime, anywhere. In our view, the information superhighway is a seamless web of communications and information networks —together with other elements of our national information infrastructure, such as computers, databases, and consumer electronics—which will put vast amounts of information at the fingertips of a variety of users. And we see the information superhighway, quite simply, as a vast interoperable network of networks —embracing local, long distance and global networks, wireless, broadcast and cable, and satellites. In addition, the information superhighway also embraces the Internet.... Importantly, the information superhighway is *not* a uniform end-to-end network developed and operated by government or any one company. It is the totality of networks in our nation, interconnected domestically and globally. And it is an important part of evolving global information superhighways.

Now, let's turn to the impacts of these technology trends on developing nations. These advanced information technology trends, multimedia communications and information superhighways will have a variety of broad, beneficial social impacts on developing nations, including the following:

Item 1. Advanced communications, growing in ubiquity, could slow the migration of rural people to urban areas—a traditional problem in countries such as The People's Republic of China.

Item 2. Access to jobs and services. People living in rural areas would be less inclined to move to the cities if advanced communications systems gave them access to jobs and sophisticated social services where they already live. (In the U.S., for example, our pervasive communications infrastructure has enabled information-intensive businesses to flourish anywhere in the country.)

Item 3. Information superhighways could alleviate congestion and commuter-traffic pollution in cities by making telecommuting possible—by bringing good jobs to people, wherever they are. (In the U.S., as you know, the work-at-home movement is gaining momentum, and trials with certain types of jobs show that employees can be even more highly productive without leaving their homes. One side benefit here is reduced costs for urban office space.)

Item 4. Information superhighways could also revolutionize education and eliminate differences in quality between rural and urban education systems—by enabling a limited number of the very best teachers and professors to reach huge numbers of students. Both students and teachers could be located practically anywhere, in "virtual" classrooms—and they could enhance learning by accessing multimedia network databases on a great variety of content areas.

Item 5. Information superhighways could also revolutionize medical care by helping to deliver high-quality medical care far from large population centers. Advanced communications would permit frequent meetings between rural health workers and physicians located in more populated areas. The same capability would also permit direct doctor-to-patient consultation and follow-up.

Item 6. Advances in information technology are stitching together a truly global society and a global economy —which developing nations would be able to participate in fully.

Item 7. Peoples and countries would be able to retain their ethnic and cultural identities, but at the same time they would be able to communicate, transact and interact seamlessly across geographic and political boundaries.

Item 8. In addition to these capabilities, a modern information infrastructure would help strengthen the ties that hold a nation's people together. In a large country such as China, for example, the huge distances between cities and regions, and the enormous complexity of regional dialects, have made communication among the Chinese people exceptionally difficult. So an information superhighway would have the potential to help lessen both the obstacle of distance and the barrier of language. And information technology will also eventually make possible real-time translation of languages.

These, then, are some of the social impacts of information technology on developing nations. In addition to social impacts, the key information technology trends, multimedia communications and information superhighways will have some broad public-policy impacts on developing nations, including the following:

Item 1. In general, investment in communications infrastructure would contribute greatly to a nation's overall economic development. Moreover, the new technologies that developing countries would be investing in are becoming more and more cost-effective. So there is a strong need to ensure sufficient investment in construction and management of a country's communications infrastructure.

Item 2. There is an opportunity to choose a technology path that would move a developing nation into the information age most directly. The opportunity is to "leapfrog" many of the older technologies that preceded today's advanced network systems—for example, to install glass fiber in local distribution networks. A country thus has the opportunity to economize on scarce capital resources by investing in a national information superhighway in the initial stages.

Item 3. There is technology to "jump-start" a developing nation. For example, cellular radio can provide telephony almost overnight and serve large markets while the fiberoptic infrastructure is put in place.

Item 4. In addition, there is a need for heavy investment in the development of the *human* infrastructure, not just the *physical* infrastructure. The global leaders of the 21st century will be those countries that have not only invested in the right technologies, but also in the intellectual growth of their people.

Item 5. Information technology is vital to economic reform and development—to improving the economic and social life of a nation's people, and to attracting and meeting the needs of foreign investors.

Item 6. Information technology would also enhance financial management—for example, by enabling a country to move away from a cash economy to one in which electronic transactions are not only faster, but also provide for much greater visibility into economic activity.

Item 7. Information technology would both *facilitate and complicate* the job of governing; *facilitate* by making available to decision-makers vastly expanded resources of timely information; *complicate* by vastly expanding the numbers of people who would be informed about important issues and who would inevitably want to play a role in deciding them.

These are some of the broad public-policy impacts of information technology on developing nations. Although the government of the U.S. clearly does not have all the answers, some of our steps, as well as missteps, might be helpful for such nations to consider.

As you know, the U.S. government has played a crucial role in nurturing rapid technological progress, as well as rapid application of new technologies in the marketplace. In the communications sector, for example, the government has established a clear set of national objectives—such as universal service, technological leadership, and broadband capability into all population centers. The government has also created a strong, independent regulatory structure designed to ensure that private companies serve the public interest in a fair and competitive marketplace—although we still have a long way to go toward genuine and effective competition in the local exchange. Many, if not most, developing nations are still evolving their policies, laws and regulations governing the communications industry. And I cannot overemphasize the importance of this task.

In summary, rich information technology, the worldwide push toward global standards, ever-increasing customer demands, and growing global competition are key driving forces behind the emerging multimedia communications revolution and the evolution of national information superhighways. The growth of multimedia communications and the further competitive evolution of these information superhighways, as well as of global information superhighways promise a broad range of Information Age benefits to virtually every citizen of our nation. And they also promise to extend these Information Age benefits to virtually every citizen of the world, including the developing nations.

NO

Andrew L. Shapiro

PRIVACY FOR SALE: PEDDLING DATA ON THE INTERNET

I've got Ted Turner's Social Security number here, along with Rush Limbaugh's home address and a couple of phone numbers for Bob Dole in Kansas. I found this information for free on the Internet in about ten minutes. With a little money and some wily sleuthing, I could probably use this data to get their credit histories, financial records and maybe some confidential medical facts. I might even be able to screw around with their bank accounts.

It is this naked vulnerability that has, quite justifiably, made Americans increasingly anxious about privacy over the past two decades. A 1995 Louis Harris poll found that 82 percent of respondents were concerned about their personal privacy, up from 64 percent in 1978. Over the same period, the proportion of those who were "very concerned" about privacy increased almost 50 percent. This new fear reflects the emergence of a sophisticated system of private surveillance—or dataveillance, as David Shenk calls it in his new book, *Data Smog*—that is rapidly overshadowing threats from the state.

It was once too expensive for anyone but the government to collect, store and coordinate data, creating profiles on hundreds of millions of citizens. But the creeping ubiquity of digital computer technology has ushered in a major industry of high-tech data pushers who are dedicated to gathering and selling personal information about practically everyone, mostly for marketing purposes. (Privacy experts estimate that the average American is profiled in at least twenty-five, and perhaps as many as 100, databases.) "Marketers can follow every aspect of our lives, from the first phone call we make in the morning to the time our security system says we have left the house, to the video camera at the toll booth and the charge slip we have for lunch," said President Clinton recently—a somewhat unexpected remark given his poor record on privacy. With the rise of online commerce and communication, this collection increasingly happens imperceptibly and without the consent of the observed. The result is a broad and lucrative market for personal information that allows anyone with a buck to find out a whole lot about anyone else—often just by trolling around the Net. It's Orwell meets Adam Smith, introduced by Bill Gates.

From Andrew L. Shapiro, "Privacy for Sale: Peddling Data on the Internet," *The Nation* (June 23, 1997).

* * *

Traditionally, privacy advocates have responded to this plight with calls for broad federal legislation to replace the current patchwork of state and federal law that leaves personal data woefully unprotected. Proposals usually require conspicuous notice of what information is being collected and for what purpose; meaningful and informed consent by consumers (for example, allowing them to "opt in" to data collection rather than having to "opt out"); the ability to access files about oneself and to correct inaccuracies; a scheme of redress for violations; and creation of an independent federal privacy protection agency to enforce compliance.

But in the current deregulatory climate, the Clinton Administration and some privacy defenders are taking a different approach. They're calling for the creation of a market for privacy to compete with or complement the growing market for personal information. (A report released in April by a presidential advisory panel, for example, mentioned "the intriguing possibility that privacy could emerge as a market commodity in the Information Age.") Just as there is demand for consumer data among profiteers, so there is a counterdemand on the part of individuals to keep that information private. The answer, say these advocates, is to have consumers bargain with vendors over acceptable rules for data collection and use.

For example, if I'm a real stickler for privacy, I may want to pay more to use an Internet service provider or a Web site that will guarantee me Level 5 privacy (on a hypothetical 1 to 5 scale where 5 represents a commitment not to gather any data). Someone else who doesn't care at all about privacy can pay less to use a Level 1 provider, the kind that sucks up data like a Dustbuster. From the company's standpoint, this makes sense because there is monetary value in that data. If they get it, they charge you less (or give you more); if they don't, they charge more (or provide less).

This, in some sense, is how the World Wide Web works today. Web sites generally offer their material for free; in return, users give them personal information. This may mean typing your name, phone number or whatever in some blank registration field. But using something called "cookies," Web sites also surreptitiously collect data such as what Internet service provider you use, what site you most recently visited, what computer and browser you're using (for a demonstration, see www.cdt.org/privacy). Since the Net is already so geared toward information exchange, some privacy advocates figure they might as well formalize that process in an open market. That market would extend beyond cyberspace to every exchange of data—with the stores you shop at, your doctors, maybe even your friends.

* * *

Now, before you go postal about how your privacy rights are being sold down the river, consider some appealing features of the market for privacy: Recognizing the value of information as an asset, it seeks to give consumers property rights in that information. Your data and sanctuary are your own; you sell them only if you choose—and you can, at least in theory, choose exactly who knows what about you. This would seem to be better than today's free-for-all, where the few rules that exist are vague and, even worse, our data are routinely stolen from us by invisible thieves.

Consider also that this market approach has received support not just from the Netscape-led business consortium looking into it but from many of the leading digital civil liberties organizations, including the Center for Democracy and Technology and the Electronic Frontier Foundation. C.D.T. is working with the World Wide Web Consortium, the Direct Marketing Association and others on the Platform for Privacy Preferences, a technical standard that will allow users to negotiate privacy practices with data collectors in a way similar to my Level 1 to 5 example. E.F.F. has teamed up with other industry players to create eTrust, a coalition that rewards privacy-friendly Web sites with a sort of Good Housekeeping seal of approval. Even stalwarts like Marc Rotenberg of the Electronic Privacy Information Center believe that consumers should start bargaining over the flow of their facts and figures. "There are already now markets for personal data," says Rotenberg. "The goal is to make them more fair, to give individuals more control."

The problem is, the data pushers will be fighting tooth and nail to see that this doesn't happen. And even if it does, the privatization of privacy will create as many dilemmas as it solves, if not more.

First, it may make privacy even more elusive than it is today, particularly online. For example, while dataveillance is the norm on the Net, cybersavvy privacy hawks have their ways of evading it. One trick is to use technologies that allow for anonymous Web surfing. Another is the ever popular low-tech option of providing false information when queried, which 34 percent of Net users admit they do, according to a Georgia Tech survey (you can bet the real number is higher). These renegade tactics will likely be unavailable in the world of formally established privacy markets. Users will have to contract with vendors in an above-board fashion and, as Pat Faley of the Direct Marketing Association sees it, "You're not going to be able to go to too many places if you want to be anonymous." That's the way the market works. You have to play by the rules, which may be different at every Web site you visit—not to mention noncyber data interactions (in a store, on the phone, etc.). This points to a bigger problem: All that time and effort spent dickering over various privacy arrangements adds up to what economists call high—even inefficiently high—transaction costs. In plain terms, it means more hassle for what may be less privacy.

Second, the privacy market will hit the poor particularly hard. As companies are able to charge increasingly higher rates for finer shades of privacy, poorer customers who can't afford these premiums will be left more exposed simply by dint of economic disadvantage. Even if the markups are small, a little added privacy may not seem worth it for those with little disposable income, especially since they are already likely to be monitored by the state if they receive welfare or live in high-crime neighborhoods. (In fact, only 39 percent of Internet users expressed a willingness to pay a markup of more than half a cent on the dollar to assure their privacy, according to eTrust.) Do we really want to perpetuate such a system of first- and second-class privacy rights?

Third, the privacy market may create a false sense of comfort, blinding us to certain unforeseeable consequences of dealing in data. For example, though a company may faithfully notify me that it collects personal information for direct marketing, I may be exposed to more than just junk-mail annoyance. Inaccurate or incomplete information in databases is routinely used to determine whether someone should be hired, insured, rented to or given credit. The readily available nature of data can lead to discrimination, harassment and even physical danger—as a Los Angeles reporter demonstrated when he bought detailed information about 5,000 children from information broker Metromail using the name of Richard Allen Davis, who was convicted of murdering 12-year-old Polly Klaas. In the arm's length transactions of the market, vendors have "no incentive to have you think about these dangers," says Oscar Gandy Jr. of the University of Pennsylvania's Annenberg School for Communication. "We're not going to be fully informed."

* * *

Fourth, there is the problem of unequal bargaining power. While most companies are less interested in your data than in having you as a customer, certain powerful firms, such as the three major credit reporting agencies, are interested exclusively in your numbers. And these companies tend to be monopolistic, presenting consumers with little real choice in the market. If you don't like the terms of the deal they offer, there's really nowhere else you can go to establish a reputable credit report that will allow you to obtain, say, a checking account or a mortgage. And then there's the person who arrives at the hospital after a care accident: Is he supposed to haggle over use of his medical data before he's treated? What about kids browsing the Web who stumble upon, say, the *Batman Forever* site, which asks them to "help Commissioner Gordon with the Gotham census" by answering questions about what products they buy? (The site was recently changed after complaints from the Center for Media Education.) As Gandy argues in a recent article, "The fundamental asymmetry between individuals and bureaucratic organizations all but guarantees the failure of the market for personal information." Even a free-marketeer like E.F.F. chairwoman Esther Dyson says, "Where there is an element of coercion, you want some regulation."

Finally, looming over all of this is a commodification critique, which warns that privacy and personal information become debased when subjected to market pressures. "This is like asking people to pay to practice freedom of religion or free speech," says University of Washington professor Philip Bereano. "We do not buy and sell civil liberties. This is commodity fetishism. It is capitalism run amok." So it would seem. Yet Bereano is actually referring to well-established privacy rights, like the Fourth Amendment right to be free from unreasonable search and seizure and the due process right to make decisions about intimate matters such as contraception and abortion. These rights, one hopes, cannot be peddled to the highest bidder. The situation is less clear, however, when it comes to personal information. In part, that's because privacy is not well defined or protected in our legal system (video rental records, for example, are protected but medical information is not). Privacy is not even mentioned in the Constitution, and our courts and legislatures have made it the somewhat insecure stepchild of legal rights.

* * *

Look, for example, at the compromised status of privacy in the ongoing debate over encryption, the text-scrambling technology that keeps electronic communications secure. Civil libertarians have argued strenuously and persuasively against law enforcement's attempts to tap encrypted messages—first with the Clipper Chip, now with an equally problematic "key recovery" scheme. But their arguments based on pure privacy principles have fallen on deaf ears. Instead, modest progress now seems likely because of complaints from high-tech giants in a tizzy over their inability to compete with the foreign software companies that are dominating the growing global market for encryption tools. Crypto supporters have every reason to cheer industry's arm-twisting, since it may help secure passage of two pending bills in Congress lifting restrictions on this crucial technology. But a cynic would conclude that privacy is getting a boost just because profits—and perhaps some campaign donations—are at stake. What about the idea that privacy should be protected for its own sake?

Lately, it's an idea that has had more currency abroad than it has here. The Organization for Economic Cooperation and Development, for example, has rejected the Clinton Administration's attempts to hamstring encryption, and the European Union has enacted a strict directive limiting personal data transfer. Indeed, the E.U. will prod the United States during upcoming trade negotiations to beef up its lax standards. The Europeans can draw on any number of resources to chasten our leaders. Many instruments of international law recognize that privacy is a fundamental human right. It is also, according to scholars from various disciplines, a core value that protects dignity, autonomy, solitude and the way we present ourselves to the world.

While privacy can conceal scourges from scrutiny, it is more often a fulcrum of democracy preserving other basic freedoms, including rights of association and free speech, voting and the pursuit of liberty and happiness. As Justice William O. Douglas wrote in a 1952 dissent, echoing an idea expressed earlier by Justice Louis Brandeis, "The right to be let alone is indeed the beginning of all freedom." In this view, privacy attains special status: Just as we don't allow people to sell their vote, their body parts or themselves into slavery, we shouldn't allow them to sell their privacy.

But does this mean that I shouldn't be able to trade my own data for money or services? The market-failure problems noted above are certainly red flags. Stanford law professor Margaret Jane Radin, the author recently of *Contested Commodities,* points out that such concerns have led society to prevent other kinds of bargaining. A landlord, for example, is legally required to keep a rented apartment habitable; he can't ask the renter to waive that requirement in exchange for reduced rent. Similarly, a company can't sell a toaster at a $5 discount to a buyer who agrees not to sue in the event that a product defect causes her to be injured.

Perhaps, then, what this market needs is a safety net, a minimal level of personal information privacy that cannot be bartered away. This baseline should certainly prevent bargaining with kids. It might also include inalienable control over our most sensitive material, such as medical and financial information. Whatever its specific features, a safety net for privacy would help create an environment where personal information privacy is the norm, not the exception. The burden would be on data pushers to justify their practices rather than on hapless individuals trying to protect themselves in a perplexing new marketplace.

* * *

One more point: Some fans of the market for privacy, particularly those in industry, seem to think they've found a way to protect privacy that is an alternative to lawmaking and regulation—as if the choice was either the market *or* the government. This is just wrong. As Phil Agre of the University of California, San Diego, notes, "Governments create markets—and the more intangible the commodity, the more that is true." To work efficiently and equitably, a privacy market will require a concrete legal regime to protect what's being traded and the integrity of that trading. Some sort of federal privacy agency will likely be necessary for enforcement—to protect against data theft, and to insure fair dealing and compliance. Whether or not we trust the self-regulatory efforts of groups like the Direct Marketing Association, they surely can't control the actions of fly-by-night companies that swoop down on the Net to pick up the trail of an unsuspecting mouse. "We'll see where the market will work and where it won't, and supplement that with government action," says Christine Varney, a Federal Trade Commissioner who will be leading an F.T.C. conference on these issues....

What's clear is that the market for privacy won't do away with the need for new statutory protections and government oversight. It certainly won't give consumers the upper hand against the masterminds of dataveillance. If anything, it will further reduce privacy from an assumed right to the unceremonious status of a commodity. Folks like Ted Turner, Rush Limbaugh and Bob Dole will pay to keep meddlers from getting access to their confidential information. But what about the rest of us? If privacy is for sale, will we peddle our digits or save our data souls?

POSTSCRIPT

Will the Information Revolution Benefit Society?

Richard H. Nethe, in "Mixed Blessings: Second Thoughts on the Information Explosion," *The Humanist* (September/October 1996), writes, "Any time we make revolutionary changes, we are bound to see a great number of disrupting perturbations in the strangest and often unlikeliest of places." The information revolution, he says, is moving so rapidly that people cannot adapt, and the result may be social upheaval. Mayo clearly agrees. "Upheaval" means "drastic change," and to Mayo the most significant changes will be positive. Others are less sanguine. As James H. Snider notes, in "The Information Superhighway as Environmental Menace," *The Futurist* (March/April 1995), past improvements in transportation (especially the automobile) have led to increased dispersal of the human population. Communication has improved along with transportation, but until very recently it did not raise the possibility of dispersing the populace even further, for work was still done in central locations. Today, Snider argues, much work can be done at home, far from the office. People can therefore flee the cities and suburbs in favor of more rural areas, taking their environmental impacts wherever the roads may reach. For him, the debate is in terms of development versus the environment.

The information revolution is still too young for the flight from the cities that worries Snider to be evident, but it certainly makes it possible. Whether this is good or bad depends on what one thinks is important. Developing countries and impoverished regions of developed countries tend to be more concerned about jobs than the environment. They see anything that brings in money as good, even if it does do a bit of harm in the process.

Wealthier regions can afford to be concerned about the environment, as well as more abstract questions such as whether or not the information revolution serves the "public interest." The Telecommunications Policy Roundtable, for instance, has said that the National Information Infrastructure (NII) should support public applications such as education, libraries, public health, and the delivery of government information and services; universal access (meaning minorities and the poor should not be excluded); and protection of basic rights such as privacy, freedom of speech, and intellectual property. See "Renewing the Commitment to a Public Interest Telecommunications Policy," *Communications of the ACM* (January 1994) and Fred W. Weingarten, "Public Interest and the NII," *Communications of the ACM* (March 1994). Those who wish to read more about how the growing use of computers permits both government and private interests to intrude on individual privacy should see Erik Ness, "BigBrother@Cyberspace," *The Progressive* (December 1994). Still another way in which the information revolution threatens society has to do with security. Richard O. Hundley and Robert H. Anderson, in "Emerging Challenge: Security and Safety in Cyberspace," *IEEE Technology and Society Magazine* (Winter 1995–1996), discuss how "bad actors" may steal, spy, and sabotage by invading computers belonging to the government, businesses, and individuals; how bugs in software and hardware can cause dangerous situations; and how problems can be prevented.

Those who wish to know more about the Clipper Chip mentioned by Shapiro should see Dorothy E. Denning, "The Case for 'Clipper,' " *Technology Review* (July 1995), in which the author defends the government's right to eavesdrop on private communications. Mike Godwin, in "A Chip Over My Shoulder: The Problems With Clipper," *Internet World* (July 1994), argues that the individual's right to privacy must take precedence over the government's power to intrude, even at the risk of diminished national security. Still another article on government control of encryption is John Perry Barlow, "A Plain Text on Crypto Policy," *Communications of the ACM* (November 1993). A number of related issues are discussed in Bruce Sterling's book *The Hacker Crackdown: Law and Disorder on the Electronic Frontier* (Bantam Books, 1992).

ISSUE 13

Are Computers Hazardous to Literacy?

YES: Sven Birkerts, from *The Gutenberg Elegies: The Fate of Reading in an Electronic Age* (Faber & Faber, 1994)

NO: Wen Stephenson, from "The Message Is the Medium: A Reply to Sven Birkerts and *The Gutenberg Elegies*," *Chicago Review* (Winter 1995–1996)

ISSUE SUMMARY

YES: Author Sven Birkerts argues that electronically presented information (i.e., via computer screens) threatens traditional conceptions of literacy, the literary culture, and the sense of ourselves as individuals.

NO: Wen Stephenson, editor of the *Atlantic Monthly*'s online edition, argues that the essence of literature, literacy, and the literary culture will survive the impact of computers.

When personal computers first came on the market in the 1970s, they were considered useful tools, but their memory was limited to only a few thousand bytes, hard drives were expensive add-ons, and the Internet was a distant dream (though something similar was already connecting university and government mainframe computers). By the 1980s some people had begun to realize that computer disks could hold as much information as a book, were smaller, took less postage to mail, and could be recycled. By this time Project Gutenberg, founded in 1971 by Michael Hart at the University of Illinois, had been converting works of classic literature into digital form and making them available on disk for a number of years already. By 1990 several small companies—including Soft Press, Serendipity Systems, and High Mesa Publishing—were trying to turn this insight into profitable businesses. Today only Serendipity remains, and it exists as an Internet site (`http://www.thegrid.net/bookware/`).

The approach failed partly because people seemed to find paper books more congenial for reading than computer screens. But then the Internet came along, and the early 1990s saw an explosion of activity exploiting this new ability to put information of all kinds—including poetry, fiction, and nonfiction of precisely the sort one used to find only on paper—onto "Web pages" that Internet users could access for free. Digital publishing boomed. This is not the first time that the nature of publishing has changed. Five centuries ago the printing press greatly increased the availability of books, including Bibles, and contributed to the Protestant Reformation and the American and

French Revolutions (see Rudi Volti, *Society and Technological Change*, 3rd ed., St. Martin's Press, 1995, chapter 11). The printing press also utterly destroyed the primacy of the scroll and delivered the first hard blow to literacy as a distinguishing characteristic of the social elite. There were, of course, protests aimed at the way this new technology was threatening handwritten text and undermining the social order.

In the twentieth century, the invention of cheap paper ("pulp") made possible a profusion of magazines (the "pulps") and cheap novels ("dime novels") that made reading for entertainment and information accessible to vast numbers of people and did a great deal to create the modern publishing industry. It also led to loud cries of protest because the greatly increased demand for material meant that "literary quality" suffered; today, *pulp literature* is still a term of derision. Many of the protests came from those who felt that the pulps were dealing a final blow to literacy as a distinguishing characteristic of a social elite.

More recently, radio and television seemed to be restoring a kind of balance by drawing many people away from reading. Among the first to decry that change were Marshall McLuhan and Quentin Fiore, in *The Medium Is the Message* (Random House, 1967). And now we have the computer and the Internet, which promise to replace the traditional linear absorption of information—as in scrolls, books, radio and TV programs, and even storytellers—with an ever-branching web of interconnections often referred to as "hypertext" (indeed, Web pages are written in a special computer programming language known as "hypertext mark-up language," or "html").

A great many people think that the Internet and its vastly increased access to information of all sorts is an astonishing gift. Some, however, object to the way it seems to threaten the old order of things. In the following selections, Sven Birkerts argues that computer-presented information threatens traditional conceptions of literacy, the literary culture, and the sense of ourselves as individuals, among other things. Wen Stephenson, on the other hand, argues that the essence of literature, literacy, and the literary culture will survive the impact of the computer, just as it did the advent of the pulps, radio, and television.

YES Sven Birkerts

INTO THE ELECTRONIC MILLENIUM

Think of it. Fifty to a hundred million people (maybe a conservative estimate) form their ideas about what is going on in America and in the world from the same basic package of edited images—to the extent that the image itself has lost much of its once-fearsome power. Daily newspapers, with their long columns of print, struggle against declining sales. Fewer and fewer people under the age of fifty read them; computers will soon make packaged information a custom product. But if the printed sheet is heading for obsolescence, people are tuning in to the signals. The screen is where the information and entertainment wars will be fought. The communications conglomerates are waging bitter takeover battles in their zeal to establish global empires. As Jonathan Crary has written in "The Eclipse of the Spectacle," "Telecommunications is the new arterial network, analogous in part to what railroads were for capitalism in the nineteenth century. And it is this electronic substitute for geography that corporate and national entities are now carving up." Maybe one reason why the news of change is not part of the common currency is that such news can only sensibly be communicated through the more analytic sequences of print.

To underscore my point, I have been making it sound as if we were all abruptly walking out of one room and into another, leaving our books to the moths while we settle ourselves in front of our state-of-the-art terminals. The truth is that we are living through a period of overlap; one way of being is pushed athwart another. Antonio Gramsci's often-cited sentence comes inevitably to mind: "The crisis consists precisely in the fact that the old is dying and the new cannot be born; in this interregnum a great variety of morbid symptoms appears." The old surely is dying, but I'm not so sure that the new is having any great difficulty being born. As for the morbid symptoms, these we have in abundance.

The overlap in communications modes, and the ways of living that they are associated with, invites comparison with the transitional epoch in ancient Greek society, certainly in terms of the relative degree of disturbance. Historian Eric Havelock designated that period as one of "proto-literacy." . . .

But our historical moment, which we might call "proto-electronic," will not require a transition period of two centuries. The very essence of electronic transmissions is to surmount impedances and to hasten transitions. Fifty years, I'm sure, will suffice. As for what the conversion will bring—and mean—to us, we might glean a few clues by looking to some of the "morbid symptoms" of the change. But to understand what these portend, we need to remark a few of the more obvious ways in which our various technologies condition our senses and sensibilities.

I won't tire my reader with an extended rehash of the differences between the print orientation and that of electronic systems. Media theorists from Marshall McLuhan to Walter Ong to Neil Postman have discoursed upon these at length. What's more, they are reasonably commonsensical. I therefore will abbreviate.

The order of print is linear, and is bound to logic by the imperatives of syntax. Syntax is the substructure of discourse, a mapping of the ways that the mind makes sense through language. Print communication requires the active engagement of the reader's attention, for reading is fundamentally an act of translation. Symbols are turned into their verbal referents and these are in turn interpreted. The print engagement is essentially private. While it does represent an act of communication, the contents pass from the privacy of the sender to the privacy of the receiver. Print also posits a time axis; the turning of pages, not to mention the vertical descent down the page, is a forward-moving succession, with earlier contents at every point serving as a ground for what follows. Moreover, the printed material is static—it is the reader, not the book, that moves forward. The physical arrangements of print are in accord with our traditional sense of history. Materials are layered; they lend themselves to rereading and to sustained attention. The pace of reading is variable, with progress determined by the reader's focus and comprehension.

The electronic order is in most ways opposite. Information and contents do not simply move from one private space to another, but they travel along a network. Engagement is intrinsically public, taking place within a circuit of larger connectedness. The vast resources of the network are always there, potential, even if they do not impinge on the immediate communication. Electronic communication can be passive, as with television watching, or interactive, as with computers. Contents, unless they are printed out (at which point they become part of the static order of print) are felt to be evanescent. They can be changed or deleted with the stroke of a key. With visual media (television, projected graphs, highlighted "bullets") impression and image take precedence over logic and concept, and detail and linear sequentiality are sacrificed. The pace is rapid, driven by jump-cut increments, and the basic movement is laterally associative rather than vertically cumulative. The presentation structures the reception and, in time, the expectation about how information is organized.

Further, the visual and nonvisual technology in every way encourages in the user a heightened and ever-changing awareness of the present. It works against historical perception, which must depend on the inimical notions of logic and sequential succession. If the print medium exalts the word, fixing it into permanence, the electronic counterpart reduces it to a signal, a means to an end.

Transitions like the one from print to electronic media do not take place without rippling or, more likely, reweaving the entire social and cultural web. The tendencies outlined above are already at work. We don't need to look far to find their effects. We can begin with the newspaper headlines and the millennial lamentations sounded in the op-ed pages: that our educational systems are in decline, that our students are less and less able to read and comprehend their required texts, and that their aptitude scores have leveled off well below those of previous generations. Tag-line communication, called "bite-speak" by some, is destroying the last remnants of political discourse; spin doctors and media consultants are our new shamans. As communications empires fight for control of all information outlets, including publishers, the latter have succumbed to the tyranny of the bottom line; they are less and less willing to publish work, however worthy, that will not make a tidy profit. And, on every front, funding for the arts is being cut while the arts themselves appear to be suffering a deep crisis of relevance. And so on.

Every one of these developments is, of course, overdetermined, but there can be no doubt that they are connected, perhaps profoundly, to the transition that is underway.

Certain other trends bear watching. One could argue, for instance, that the entire movement of postmodernism in the arts is a consequence of this same macroscopic shift. For what is postmodernism at root but an aesthetic that rebukes the idea of an historical time line, as well as previously uncontested assumptions of cultural hierarchy. The postmodern artifact manipulates its stylistic signatures like Lego blocks and makes free with combinations from the formerly sequestered spheres of high and popular art. Its combinatory momentum and relentless referencing of the surrounding culture mirror perfectly the associative dynamics of electronic media....

Then there are the more specific sorts of developments. Consider the multibillion-dollar initiative by Whittle Communications to bring commercially sponsored education packages into the classroom. The underlying premise is staggeringly simple: If electronic media are the one thing that the young are at ease with, why not exploit the fact? Why not stop bucking television and use it instead, with corporate America picking up the tab in exchange for a few minutes of valuable airtime for commercials?...

"You have to remember that the children of today have grown up with the visual media," said Robert Calabrese [Billerica School Superintendent]. "They know no other way and we're simply capitalizing on that to enhance learning."

Calabrese's observation on the preconditioning of a whole generation of students raises troubling questions: Should we suppose that American education will begin to tailor itself to the aptitudes of its students, presenting more and more of its materials in newly packaged forms? And what will happen when educators find that not very many of the old materials will "play"—that is, capture student enthusiasm? Is the what of learning to be determined by the how? And at what point do vicious cycles begin to reveal their viciousness?

A collective change of sensibility may already be upon us. We need to take seriously the possibility that the young truly "know no other way," that they are not made of the same stuff that their elders are. In her *Harper's* magazine debate with Neil Postman, Camille Paglia observed:

> Some people have more developed sensoriums than others. I've found that most people born before World War II are turned off by the modern media. They can't understand how we who were born after the war can read and watch TV at the same time. But we can. When I wrote my book, I had earphones on, blasting rock music or Puccini and Brahms. The soap operas—with the sound turned down—flickered on my TV. I'd be talking on the phone at the same time. Baby boomers have a multilayered, multitrack ability to deal with the world.

I don't know whether to be impressed or depressed by Paglia's ability to disperse her focus in so many directions. Nor can I say, not having read her book, in what ways her multitrack sensibility has informed her prose. But I'm baffled by what she means when she talks about an ability to "deal with the world." From the context, "dealing" sounds more like a matter of incessantly repositioning the self within a barrage of onrushing stimuli.

Paglia's is hardly the only testimony in this matter. A *New York Times* article on the cult success of Mark Leyner (author of *I Smell Esther Williams* and *My Cousin, My Gastroenterologist*) reports suggestively:

> His fans say, variously, that his writing is like MTV, or rap music, or rock music, or simply like everything in the world put together: fast and furious and intense, full of illusion and allusion and fantasy and science and excrement.

Larry McCaffery, a professor of literature at San Diego State University and co-editor of *Fiction International*, a literary journal, said his students get excited about Mr. Leyner's writing, which he considers important and unique: "It speaks to them, somehow, about this weird milieu they're swimming through. It's this dissolving, discontinuous world." While older people might find Mr. Leyner's world bizarre or unreal, Professor McCaffery said, it doesn't seem so to people who grew up with Walkmen and computers and VCR's, with so many choices, so much bombardment, that they have never experienced a sensation singly.

The article continues:

> There is no traditional narrative, although the book is called a novel. And there is much use of facts, though it is called fiction. Seldom does the end of a sentence have any obvious relation to the beginning. "You don't know where you're going, but you don't mind taking the leap," said R. J. Cutler, the producer of "Heat," who invited Mr. Leyner to be on the show after he picked up the galleys of his book and found it mesmerizing. "He taps into a specific cultural perspective where thoughtful literary world view meets pop culture and the TV generation."

My final exhibit—I don't know if it qualifies as a morbid symptom as such—is drawn from a *Washington Post Magazine* essay on the future of the Library of Congress, our national shrine to the printed word. One of the individuals interviewed in the piece is Robert Zich, so-called "special projects czar" of the institution. Zich, too, has seen the future, and he is surprisingly candid with his interlocutor. Before long, Zich maintains, people will be able to get what information they want directly off their terminals. The function of the Library of Congress (and perhaps libraries in general) will change.

He envisions his library becoming more like museum: "Just as you go to the National Gallery to see its Leonardo or go to the Smithsonian to see the Spirit of St. Louis and so on, you will want to go to libraries to see the Gutenberg or the original printing of Shakespeare's plays or to see Lincoln's hand-written version of the Gettysburg Address."

Zich is outspoken, voicing what other administrators must be thinking privately. The big research libraries, he says, "and the great national libraries and their buildings will go the way of the railroad stations and the movie palaces of an earlier era which were really vital institutions in their time... Somehow folks moved away from that when the technology changed."

And books? Zich expresses excitement about Sony's hand-held electronic book, and a miniature encyclopedia coming from Franklin Electronic Publishers. "Slip it in your pocket," he says. "Little keyboard, punch in your words and it will do the full text searching and all the rest of it. Its limitation, of course, is that it's devoted just to that one book." Zich is likewise interested in the possibility of memory cards. What he likes about the Sony product is the portability: one machine, a screen that will display the contents of whatever electronic card you feed it.

I cite Zich's views at some length here because he is not some Silicon Valley research and development visionary, but a highly placed executive at what might be called, in a very literal sense, our most conservative public institution. When men like Zich embrace the electronic future, we can be sure it's well on its way.

Others might argue that the technologies cited by Zich merely represent a modification in the "form" of reading, and that reading itself will be unaffected, as there is little difference between following words on a pocket screen or a printed page. Here I have to hold my line. The context cannot but condition the process. Screen and book may exhibit the same string of words, but the assumptions that underlie their significance are entirely different depending on whether we are staring at a book or a circuit-generated text. As the nature of looking—at the natural world, at paintings—changed with the arrival of photography and mechanical reproduction, so will the collective relation to language alter as new modes of dissemination prevail.

Whether all of this sounds dire or merely "different" will depend upon the reader's own values and priorities. I find these portents of change depressing, but also exhilarating—at least to speculate about. On the one hand, I have a great feeling of loss and a fear about what habitations will exist for self and soul in the future. But there is also a quickening, a sense that important things are on the line. As Heraclitus once observed, "The mixture that is not shaken soon stagnates." Well, the mixture is being shaken, no doubt about it. And here are some of the kinds of developments we might watch for as our "proto-electronic" era yields to an all-electronic future:

1. Language Erosion. There is no question but that the transition from the culture of the book to the culture of electronic communication will radically alter the ways in which we use language on every societal level. The complexity and distinctiveness of spoken and written expression, which are deeply bound to traditions of print literacy, will gradually be replaced by a more telegraphic sort of "plainspeak." Syntactic masonry is already a dying art. Neil Postman and others have already suggested what losses have been incurred by the advent of telegraphy and television—how the complex discourse patterns of the nineteenth century were flattened by the requirements of communication over distances. That tendency runs riot as the layers of mediation thicken. Simple linguistic prefab is now the norm, while ambiguity, paradox, irony, subtlety, and wit are fast disappearing. In their place, the simple "vision thing" and myriad other "things." Verbal intelligence, which has long been viewed as suspect as the act of reading, will come to seem positively conspiratorial. The greater part of any articulate person's energy will be deployed in dumbing-down her discourse.

Language will grow increasingly impoverished through a series of vicious cycles. For, of course, the usages of literature and scholarship are connected in fundamental ways to the general speech of the tribe. We can expect that curricula will be further streamlined, and difficult texts in the humanities will be pruned and glossed. One need only compare a college textbook from twenty years ago to its contemporary version. A poem by Milton, a play by Shakespeare—one can hardly find the text among the explanatory notes nowadays. Fewer and fewer people will be able to contend with the so-called masterworks of literature or ideas. Joyce, Woolf, Soyinka, not to mention the masters who preceded them, will go unread, and the civilizing energies of their prose will circulate aimlessly between closed covers.

2. *Flattening of Historical Perspectives.* As the circuit supplants the printed page, and as more and more of our communications involve us in network processes—which of their nature plant us in a perpetual present—our perception of history will inevitably alter. Changes in information storage and access are bound to impinge on our historical memory. The depth of field that is our sense of the past is not only a linguistic construct, but is in some essential way represented by the book and the physical accumulation of books in library spaces. In the contemplation of the single volume, or mass of volumes, we form a picture of time past as a growing deposit of sediment; we capture a sense of its depth and dimensionality. Moreover, we meet the past as much in the presentation of words in books of specific vintage as we do in any isolated fact or statistic. The database, useful as it is, expunges this context, this sense of chronology, and admits us to a weightless order in which all information is equally accessible.

If we take the etymological tack, history (cognate with "story") is affiliated in complex ways with its texts. Once the materials of the past are unhoused from their pages, they will surely mean differently. The printed page is itself a link, at least along the imaginative continuum, and when that link is broken, the past can only start to recede. At the same time it will become a body of disjunct data available for retrieval and, in the hands of our canny dream merchants, a mythology. The more we grow rooted in the consciousness of the now, the more it will seem utterly extraordinary that things were ever any different. The idea of a farmer plowing a field—an historical constant for millennia—will be something for a theme park. For, naturally, the entertainment industry, which reads the collective unconscious unerringly, will seize the advantage. The past that has slipped away will be rendered ever more glorious, ever more a fantasy play with heroes, villains, and quaint settings and props. Small-town American life returns as "Andy of Mayberry"—at first enjoyed with recognition, later accepted as a faithful portrait of how things used to be.

3. The Waning of the Private Self. We may even now be in the first stages of a process of social collectivization that will over time all but vanquish the ideal of the isolated individual. For some decades now we have been edging away from the perception of private life as something opaque, closed off to the world; we increasingly accept the transparency of a life lived within a set of systems, electronic or otherwise. Our technologies are not bound by season or light—it's always the same time in the circuit. And so long as time is money and money matters, those circuits will keep humming. The doors and walls of our habitations matter less and less—the world sweeps through the wires as it needs to, or as we need it to. The monitor light is always blinking; we are always potentially on-line.

I am not suggesting that we are all about to become mindless, soulless robots, or that personality will disappear altogether into an oceanic homogeneity. But certainly the idea of what it means to be a person living a life will be much changed. The figure-ground model, which has always featured a solitary self before a background that is the society of other selves, is romantic in the extreme. It is ever less tenable in the world as it is becoming. There are no more wildernesses, no more lonely homesteads, and, outside of cinema, no more emblems of the exalted individual.

The self must change as the nature of subjective space changes. And one of the many incremental transformations of our age has been the slow but steady destruction of subjective space. The physical and psychological distance between individuals has been shrinking for at least a century. In the process,the figure-ground image has begun to blur its boundary distinctions. One day we will conduct our public and private lives within networks so dense, among so many channels of instantaneous information, that it will make almost no sense to speak of the differentiations of subjective individualism.

We are already captive in our webs. Our slight solitudes are transected by codes, wires, and pulsations. We punch a number to check in with the answering machine, another to tape a show that we are too busy to watch. The strands of the web grow finer and finer—this is obvious. What is no less obvious is the fact that they will continue to proliferate, gaining in sophistication, merging functions so that one can bank by phone, shop via television, and so on. The natural tendency is toward streamlining: The smart dollar keeps finding ways to shorten the path, double-up the function. We might think in terms of a circuit-board model, picturing ourselves as the contact points. The expansion of electronic options is always at the cost of contractions in the private sphere. We will soon be navigating with ease among cataracts of organized pulsations, putting out and taking in signals. We will bring our terminals, our modems, and menus further and further into our former privacies; we will implicate ourselves by degrees in the unitary life, and there may come a day when we no longer remember that there was any other life....

Trafficking with tendencies—extrapolating and projecting as I have been doing—must finally remain a kind of gambling. One bets high on the validity of a notion and low on the human capacity for resistance and for unpredictable initiatives. No one can really predict how we will adapt to the transformations taking place all around us. We may discover, too, that language is a hardier thing than I have allowed. It may flourish among the beep and the click and the monitor as readily as it ever did on the printed page. I hope so, for language is the soul's ozone layer and we thin it at our peril.

NO

Wen Stephenson

THE MESSAGE IS THE MEDIUM

I have before my eyes a page, and on the page, typewritten in a serif font, is a poem. It is an ode written in 1819 by John Keats. I read the first words aloud to myself, slowly, pronouncing each syllable as though it were a musical note or a percussive beat: "Thou still unravish'd bride of quietness, / Thou foster-child of silence and slow time." As I continue down the page, I linger over certain phrases and rhymes; I go back and re-read, taking the stanzas apart and putting them back together again in my mind. The words fall into their order, and I feel their rhythm somewhere in my chest, the resonance of language uttered by a human voice in solitude. I am forced back into myself by the words on the page, my mind pushed deeper and deeper into a realm of images and associations, and emotion that did not exist a moment before is conjured from some mysterious wellspring.

I repeat the last lines of the poem—an indecipherable pronouncement on the relation of art to life—and then a noise from outside draws my attention to the open window; the spell is broken. It is a sultry Sunday afternoon over the rooftops of Boston's Back Bay, and through the window of my office a humid breeze rustles the papers strewn across my desk. I notice the clock: nearly five hours have elapsed since I sat down to read, and in that time I've wandered through a collection of British poetry. It seemed like no time at all. As I stand up to stretch, there's the sensation of floating that I often experience after long immersion in literature. But the pressure of the world returns, and its gravity pulls me back. The shock of reentering the temporal zone leaves me a little dazed, disoriented. I am still inside that Keats poem. Or it is inside me—the experience proved upon my pulse, which, by the way, is beating somewhat more rapidly than normal.

Where have I been? What has happened to the sense of time and space that governed my consciousness before I came upon that text? Something has happened, something connecting me across space and time to another human being, perhaps untold others—some experience of language that is ageless, primal, and indefinable. Perhaps I have had what some would call an authentic aesthetic experience of the art of poetry. If so, then I have experienced it directly through the digital channels of the Internet, on "pages"

From Wen Stephenson, "The Message Is the Medium: A Reply to Sven Birkerts and *The Gutenberg Elegies*," *Chicago Review*, vol. 41, no. 4 (Winter 1995–1996).

of the World Wide Web, through the circuitry of an Apple computer and the cathodes of a Sony monitor, at some 28,000 bytes per second.

* * *

If imitation is the sincerest form of flattery, then I hope Sven Birkerts will take the preceding paragraphs not only as a rebuttal but as a compliment. His most recent collection of essays, *The Gutenberg Elegies: The Fate of Reading in an Electronic Age,* is one of the most engrossing, engaging, provocative, and frustrating books I've come across in a long while. Published in December, 1994, on the cusp of the millennial hype surrounding the so-called "online revolution," it has become one of the most talked about literary events of the past year. There should be nothing mysterious about its notoriety. At a time when the subject of the Internet and the new media it has spawned is everywhere you look—not just in the pages of *Wired* magazine but on the covers and in the headlines of the very print publications these new media are said to be replacing—Birkerts strikes deeply and often convincingly to the core of an anxiety felt by many in our postmodern literary culture. The strategy is simple and rather brilliant: to explore the relationship between a reader and an imaginative text at a time when serious literature is increasingly marginalized by the communications technologies that are transforming mass media and mass culture.

And yet, as one of a growing number of people with a foot in both the worlds of traditional literary publishing and the emerging online media, I can't help wondering how Birkerts could be so closed-minded to the possibilities the new media present for serious literary activity. Reading the book, especially his descriptions of the reading experience itself, I often felt as though Birkerts and I should be allies; but time and again I found myself fundamentally at odds with him, baffled by his condescension toward all forms of electronic media. Here's a typical statement in *The Gutenberg Elegies:* "[circuit and screen] are entirely inhospitable to the more subjective materials that have always been the stuff of art. That is to say, they are antithetical to inwardness."[1] Elsewhere he has elaborated on the idea that the Internet is barren of serious thought and writing. "What the wires carry is not the stuff of the soul," he said in the August, 1995, Forum in *Harper's.* Afraid of what he calls a "creeping shallowness," Birkerts complained in another context, "If I could be convinced that the Net and its users had a genuine purchase on depth, on the pursuit of things which are best pursued in stillness, in dread, and by way of patiently articulated language, then I would open wide my heart. I just don't see it." The fact that this last comment was made in the course of a live *online* conference hosted by *The Atlantic Monthly* on the America Online network (in which, I must confess, I participated as an editor), complicates things and adds perhaps more than just a touch of irony. I can only ask, as a worker in the "shallow" domain of cyberspace: should I be concerned, or merely insulted?

* * *

It isn't just that Birkerts is less than optimistic about the prospects for serious literature in cyberspace—his misgivings go well beyond a reasoned critical skepticism toward new forms of literary activity springing up on the World Wide Web and in other multimedia applications. The magnitude of his subject is spelled out in the opening pages of *The Gutenberg Elegies.* While Birkerts is hardly the first in recent years to see a "total metamorphosis" of our culture brought on by the revolution in communications technology, his stance in regard to the effects on literature and the consequences for Western culture is, I believe, radical. "Suddenly," he says,

> it feels like everything is poised for change.... The stable hierarchies of the printed page... are being superseded by the rush of impulses through freshly minted circuits. The displacement of the page by the screen is not yet total.... But, living as we do in the midst of innumerable affiliated webs, we can say that changes in the immediate sphere of print refer outward to the totality; they map on a smaller scale the riot of societal forces.(3)

At the vortex of this tranformative riot is for Birkerts the printed page. Not only is the "formerly stable system" of literary publishing being undermined and eroded, but as "the printed book, and the ways of the book—of writing and reading—are modified, as electronic communications assert dominance, the 'feel' of the literary engagement is altered. Reading and writing come to mean differently; they acquire new significations" (6). Pondering what he calls the "elegiac exercise" of reading a serious book, Birkerts concludes that "profound questions must arise about our avowedly humanistic values, about spiritual versus material concerns, and about subjectivity itself" (6). Clearly, there's more at stake here than the fate of the traditional publishing industry. For Birkerts, it is never merely a question of whether we get our reading material via the Internet or from a bookstore; what matters is how we experience what we read: "I speak as an unregenerate reader, one who still believes that language and not technology is the true evolutionary miracle.... That there is profundity in the verbal encounter itself... and that for a host of reasons the bound book is the ideal vehicle for the written word" (6). I follow along fine until his insistence on the bound book and the primacy of print stops me cold. Birkerts raises his devotion to the printed word to a nearly religious level: the book as holy relic, the page a fetish embodying everything he holds sacred.

Nowhere does Birkerts provide better insight into the underlying reasons for this devotion to print than, of all places, the inaugural issue of the electronic magazine *FEED.* As one of four participants in *FEED*'s first hypertext roundtable "Dialog," called "Page versus Pixel" (June 1995), Birkerts gets top billing and is allowed to fire off the first shot. Thus, with the kind of irony that has come to characterize so much of the self-reflexive discourse on media, Birkerts offers the crux of his argument in favor of print in the context of a hip new e-zine on the World Wide Web:

> I do not accept the argument that a word is a word is a word is a word no matter where it appears. There is no pure 'word' that does not inhabit context inextricably. I don't think the medium is absolutely the message, but I do think that the medium conditions the message considerably. A word incised in stone (to be extreme) asks to be read as a word incised; a word skywritten (to be extreme again) asks to be looked at as such. A word on the page at some level partakes of—participants in—the whole history of words on pages, plays in that arena. Reading it, we accept certain implicit notions: of fixity, of hierarchy, of opacity. By 'opacity' I mean that the physical word dead-ends on the page

and any sense of larger resonance must be established in the reader and by the reader.

Inexplicably, for the same words to be displayed on a computer screen, in Birkerts's reasoning, causes their presence to disintegrate, and with that loss so goes the entire hierarchy they represent: the whole culture disintegrates into random bits of digital information as soon as the words disappear from the screen, as if irretrievably. The image strikes me as an apt metaphor for the chaos that has been unleashed in literary studies by deconstruction and the linguistic indeterminacy associated with postmodernism. It's as though Birkerts fears for the stability and efficacy of language itself, and I begin to imagine him as Dr. Johnson's hapless lexicographer, who in the preface to his dictionary laments the vanity of the wish that language "might be less apt to decay, and that signs might be permanent, like the things which they denote."[2] Like Johnson, Birkerts appears to suffer an acute discomfort over the mutabilities of meaning and the transience of cultural values.

It is not surprising, then, that the fixity of words printed on page, and bound in a book, becomes Birkerts's last best hope for Western civilization. What's at stake for Birkerts is nothing less than the tradition of Western humanism dating back to the Renaissance and rooted in Hellenic civilization. Books are the repository of that tradition. It all rests on the stability of print. "For, in fact," he says in *The Gutenberg Elegies,* "our entire collective subjective history—the soul of our societal body—is encoded in print.... If a person turns from print... then what happens to that person's sense of culture and continuity?" (20). Birkerts observes how the narrative and syntactical structures afforded by print are for the most part linear, whereas in electronic media everything from the jump-cut in film and video to the lateral and tangential movements of hypertext works against our traditional notions of time and historical progression. That Western culture is threatened both by these technologies and by the intellectual and artistic movements associated with postmodernity is to Birkerts no mere coincidence:

> Transitions like the one from print to electronic media do not take place without rippling or, more likely, reweaving the entire social and cultural web.... One could argue, for instance, that the entire movement of postmodernism in the arts is a consequence of this same macroscopic shift. For what is postmodernism at root but an aesthetic that rebukes the idea of an historical time line, as well as previously uncontested assumptions of cultural hierarchy. The postmodern artifact manipulates its stylistic signatures like Lego blocks and makes free with combinations from the formerly sequestered spheres of high and popular art. Its combinatory momentum and relentless referencing of the surrounding culture mirror perfectly the associative dynamics of electronic media. (123)

Postmodernism's "relentless referencing" of mass culture and its rebuke of the historical time line are closely related, in Birkerts's picture, to the electronic media that are undermining the "stable hierarchies" upon which both narrative history and the novel are based. In the midst of these rioting forces, can there be any doubt that Birkerts would prefer for the spheres of high and popular art to remain sequestered? It would seem that the fate of literature in the electronic age depends upon the extent to which they do.

Accordingly, Birkerts feels the decline of print most acutely in the accompanying eclipse of the "serious" or "literary" novel as a form wielding cultural authority. Its eclipse, the direct result of the rise of electronic culture at the expense of print, represents for Birkerts no less than the death of literature itself, or at least the possibility of its death.... It is here that Birkerts's position begins to come into clearer focus, and we see a nostalgic modernist overwhelmed by the currents of postmodernism carried on the tide of electronic media. In response, Birkerts erects the physical certainty and materiality of the printed word as a levee against the flood.

Knowing as he does that the levee cannot hold—or, more likely, that it has long since broken—at least Birkerts shows a sense of humor (if a somewhat perverse one) as he takes his stand. In the final essay of *The Gutenberg Elegies,* an eerie coda titled "The Faustian Pact," we find our champion of high-modernist print culture resisting the temptations of a cyber-pop Mephistopheles decked out in the "bold colors, sans serif type fonts, [and] unexpected layouts" (210) of *Wired* magazine. Appropriately, one of Birkerts's rare bows to popular art takes the form of the blues refrain he intones from the chapter's first sentence onward: "I've been to the crossroads and I've seen the devil there" (210). He pauses. "Or is that putting it too dramatically? What I'm really saying is that I've been to the newsstand, again, to plunk down my money for *Wired.* You must have seen it—that big, squarish, beautifully produced item, that travel guide to the digital future" (210). Evidently it's not too dramatic for Birkerts; he develops the theme thoroughly before returning to it on the climactic last page. The image would be more humorous if not for the distinct impression that Birkerts is only half joking:

> Yes, I've been to the crossroads and I've met the devil, and he's sleek and confident, ever so much more 'with it' than the nearest archangel. He is causal and irreverent, wears jeans and running shoes and maybe even an earring, and the pointed prong of his tail is artfully concealed. Slippery fellow. He is the sorcerer of the binary order, jacking in and out of terminals, booting up, flaming, commanding vast systems and networks with an ease that steals my breath away. (211)

You can almost hear Birkerts pronouncing these (to him) exotic terms of the digital present, distrusting them, not quite sure what to make of them, superstitious of the mystical powers they seem to hold, as though they're poised to steal our souls—or, as Birkerts would say, "our subjective individualism" (228).

He goes on to lament the inevitable loss of this individualism in the midst of an increasing "electronic tribalism," or "hive life," and he identifies the consequences of our Faustian contract in the realization of his "core fear"—"that we are, as a culture, as a species, becoming shallower" (228). In our embrace of technology and its transformation of our culture we have sacrificed depth, "adapting ourselves to the ersatz security of a vast lateral connectedness" (228). Woven into this expanding web—pervading our academic institutions and trickling down into the mass media by way of high-brow journalism, film, and now, he fears, the Internet—is a nihilistic "postmodern culture" with its "vast fabric of competing isms," chief among them a terrorizing "absolute relativism" (228). Birkerts's answer is to resist the reflex response that all is "business as usual," and to see through the illusions that our wired Mephisto would weave seductively before our eyes. Birkerts won't be fooled: "The devil no longer moves about on cloven hooves, reeking of brimstone. He is an affable, efficient fellow. He claims to want to help us all along to a brighter, easier future, and his sales pitch is very smooth. I was, as the old song goes, almost persuaded" (229). From somewhere deep in his "subjective self," Birkerts summons his courage and with his final words heeds the inner voice that says, "Refuse it."

It's a dramatic ending, to be sure—the work of a skilled evangelist. Yet, as it turns out, in his rhetorical flourish Birkerts has inverted the meaning of the "old song," the traditional evangelical hymn in which the speaker is, as the title says, "Almost Persuaded" to respond to the call of salvation but is ultimately too late. The hymn is supposed to strike fear and contrition into the hearts of sinful listeners, and using it the way he does is a telling maneuver on Birkerts's part, turning it around so that it's the voice of an electronic-age Devil rather than the voice of God that nearly persuades him. To be honest, he almost persuades me that my soul is in danger, nearly convinces me of my shallowness—almost, but not quite.

The truth is that, like many in Birkerts's target audience, I'm susceptible to the alarm. Like the fretful soul in the revival tent I am vulnerable to the message, and the heightened passion of its delivery makes inroads where reason cannot. I recognize myself in his portrait of the "unregenerate reader." Here's my confession, my creed. Yes, I believe. I believe in aesthetic experience and in the need for literature to communicate something otherwise unknowable—and in that communication to achieve some connection with other human beings, however slight and fleeting, and however compromised by the indeterminacies of signs and the structures of meaning and power imposed by our cultural contexts. I make my own refusal: a refusal to accept that such communication is impossible. I take the very existence of... literary journal[s] as evidence that others share a need for this kind of communication, within and across personal and cultural boundaries. A postmodern pilgrim, I struggle to maintain my faith in the ability of language to transmit not just what one culture calls beauty (though that, too, is an important function) but more so, to communicate what people recognize across time and space as human experience. If I truly believed that any of this is threatened with extinction by the new electronic media, I would gladly cast my modem down the sea's throat and never look back. But, in fact, the picture I see looks considerably different, considerably less frightening.

* * *

To be fair, there are glimpses of hope even in the midst of Birkerts's apocalyptic visions, though they are rarely pursued. These are the occasions, usually at the end of a chapter, upon which he rises to affirm the possibility that language itself, even literary language, might survive despite the decline of print. "We may discover, too," he concedes, "that language is a hardier thing than I have allowed. It may flourish among the beep and the click and the monitor as readily as it ever did on the printed page" (133). Yet these moments are few and far between. The conclusion forever reached is that our literary culture and our civilization rest on print, and that the fate of our individual and collective souls depends upon the solidity of ink and paper.

It is hard to believe, despite his statements about words incised in stone versus words written in the sky, that Birkerts really accepts the McLuhanesque determinism embodied in the ubiquitous cliché about the medium being the message. What if the medium in question is language? Then what does the message become? At the end of his chapter called "Close Listening," in which he describes his experiences with books on tape, Birkerts confesses to having had an epiphanic moment, one that may suggest an answer. He recalls driving down a stretch of open road after visiting Walden Pond, near Concord, Massachusetts—a good place for a transcendental experience—and popping in a cassette of Thoreau's *Walden*, spoken by Michael O'Keefe. The effect is that Birkerts seems momentarily transformed into an all-encompassing, all-hearing ear:

> The words streamed in unmediated, shot like some kind of whiskey into my soul. I had a parenthesis of open country, then came the sentence of the highway. But the state held long enough to allow a thought: In the beginning was the Word —not the written or printed or processed word, but the spoken word. And though it changes its aspect faster than any Proteus, hiding now in letter shapes and now in magnetic emulsion, it remains. It still has the power to lay us bare. (150)

I remember first encountering that passage and putting down the book, thinking, "Precisely! I rest my case!" I can only hope that Birkerts has more experiences like this one. For he as much as admits that the essence of literature might actually survive in the valley of its saying, even outside the precincts of print.

Given such a revelation, the absence of poetry from Birkerts's discussion of literature's fate becomes all the more conspicuous. In *The Gutenberg Elegies*, he has given us a moving paean to the achievements of great prose stylists, such as Flaubert, whose *mots justes* represent the essence of what Birkerts fears we are losing. But what of the poets? How is it that they do not figure into his scheme of things? After all, sound as much as sense is the essence of poetry, and verse may bring us closer than any other use of language to that primal "Word" Birkerts communed with on the road from Walden. And that is just the point. Poetry does not fit into the design of his argument. For unlike prose, and especially the kind of prose narrative Birkerts is so keen to salvage, poetry has long been comfortable outside of print. Poetry, in many of its forms, comes much closer to the kind of aesthetic Birkerts describes as characterizing postmodern, electronic culture. And where much is made of the non-linear, cinematic techniques employed in prose narrative, so we would do well to remind ourselves that poets have been using those techniques since the first invocation of the Muse.

... [N]o matter how the words of a literary work are reproduced and transmitted, the essential qualities of the language, the sounds and meanings, survive.... Language is more than content; it is itself a medium—the medium of literature—and it transcends print, paper, silicon, electricity, even the human voice.

* * *

So, where does this leave literature—or, more to the point, literary language and the kind of communication it allows—in the computer age, and in a future of rapidly expanding online networks? One place to look is the emergence of literary publishing on the Internet. If we venture beyond the too-easy opposition of print and pixel, we might find that literature will take to the digital environment more naturally than many would expect. For one thing, computers themselves, and the experience of cyberspace, can appeal to our imaginations in ways similar to the aesthetic experience of literary language. As Robert Pinsky wrote in *The New York Times Book Review,*... in an essay titled "The Muse in the Machine: Or, the Poetics of Zork," poetry and computers may have something uniquely in common, sharing what he identifies as "a great human myth or trope, an image that could be called the secret passage: the discovery of large, manifold channels through a small, ordinary-looking or all but invisible aperture."[3] Pinsky appears to be tantalized, rather than threatened, by the possibilities implied in the comparison. "This opening up," he continues in the same essay, "the discovery of much in little, seems to be a fundamental resonance of human intelligence. Perhaps more than the interactive or text-shuffling capacity of the machine, this passage to vast complexities is at the essence of what writing through the machine might become. The computer, like everything else we make, is in part a self-portrait; it smells of our human souls." From hypertext and archival databases, to advanced language experimentation, to the increasingly sophisticated descendants of early computer text-adventure games such as Zork, the "peculiar terrain of literature-for-the-monitor" offers a vision of what the digital future may hold for the literary imagination.

Not long after that essay appeared I had the opportunity, as the moderator of another online conference for *The Atlantic Monthly,* to ask Pinsky if he would elaborate on his thoughts about poetry and its potential life in cyberspace. He confirmed a qualified optimism, emphasizing certain practical advantages of the new medium over the old, while leaving no doubt as to where, he believes, the message is to be found. "The medium of poetry—real poetry, for me—is ultimately breath," he typed (broadcasting the words to a live audience), "one person's breath shaped into meaning by our larynx and mouth. So like print, the computer is still a servant or a conduit—not the ultimate scene of poetry, which is in the ear." He then declined to predict what might be the most promising applications of computer technology to literature, opting instead to point out perhaps the most significant aspect of electronic publishing, not only for poetry but for literary activity in general: "the capacity to download what used to require a publisher, a bookstore, etc.... That compression and availability have amazing potential for freeing individuals from control, from the treatment of people as masses. In that, poetry (an ancient technology) and new technologies are potential allies in the service of individual creativity, orneriness, imagination." A few hours browsing on the World Wide Web will more than bear Pinsky out. What he tentatively projects is in fact taking shape, albeit in an infantile form, on the Internet, especially across the multimedia landscape of the Web....

It is becoming clearer that the Internet has vast potential to expand the audience for works of the literary imagination; and not only to expand access but also opportunities for interactivity, and for building communities of creative minds that could not exist otherwise. It's a lovely picture, one I'd like to believe in. But I know it is more likely that the Internet will become a vast cyberspace mall, every bit as commercialized as any other mass medium in a free-market society. And yet, if it's true, as [the poet W. H.] Auden put it, that "poetry makes nothing happen"—at least not within the realm of an expanding and virtually untapped marketplace—it is nevertheless also true that individuals do make things happen. And I will maintain—surely Birkerts would agree with me here—that literature, as a means of communication, has the power to make something happen within individuals. For this reason, it is all the more important that we do not surrender cyberspace and the new media to the purely market-driven forces of late-twentieth-century multinational capitalism. There are other values—values which cannot be measured in monetary units—that will survive only if we vigilantly carve out a space for them to breathe.

NOTES

1. Sven Birkerts, *The Gutenberg Elegies: The Fate of Reading in an Electronic Age* (Boston: Faber and Faber, 1994), p. 193. Page numbers of subsequent quotations are indicated parenthetically.

2. Samuel Johnson, "Preface to a Dictionary of the English Language," in Frank Brady and W. K. Wimsatt, eds., *Samuel Johnson: Selected Poetry and Prose* (University of California Press, 1977), p. 280.

3. Robert Pinsky, "The Muse in the Machine: Or, the Poetics of Zork," *The New York Times Book Review* (March 19, 1995): 3.

POSTSCRIPT

Are Computers Hazardous to Literacy?

In at least one sense, Birkerts is correct when he says that computers and the Internet threaten traditional conceptions of literacy and the literary culture. If we try to summon up an image to go with those conceptions, we might come up with a library reading room, vast shelves of dusty tomes, human figures slowly paging through books spread across long oak tables, silence but for a buzzing fly or two, and fingers coated with gray dust. The new order carries a very different image: modern and technological and fast, sleek casings surrounding glowing screens, and clicking mice. To a huge extent this image has already won—card catalogs in libraries have almost been completely replaced by computerized databases—and anyone with a fondness for the old image is bound to object.

Birkerts developed his objections at length and stirred up considerable controversy in some circles. But he also struck a chord. Mike Shahin, in "Internet: Boon or Bane for Literacy?" *The Ottawa Citizen* (September 8, 1996), writes, "Literacy workers, social scientists and computer nerds are beginning to realize that the Internet has the potential to radically alter the way we use and understand language, and the way we define literacy." Also, the National Literacy Secretariat and the Literacy Section of the Ontario Ministry of Education and Training are funding *CONNECT: A National Newsletter on Technology in Adult Literacy* to promote the appropriate use of technology in literacy programs.

On the other hand, Dean Blobaum, of the University of Chicago Press, reviewed Birkerts's *Gutenberg Elegies* online, saying, "Birkerts's take on electronic media follows a well-trodden path—a rehash of media theory and broad generalizations about the effects of electronic media, making it the whipping boy for the ills of western society—the decline in education, literacy, and literate culture; the financial straits of publishers; postmodernism in the arts; and the fight over the canon in literature. His thoughts on electronic media lack focus and originality." The Internet, he added, actually fulfills the 1932 call of literary giant Bertolt Brecht for a radio medium that permits the receiver of information to communicate with the provider.

In the fall of 1996 Elizabeth Murphy (at `http://www.stemnet.nf.ca/~elmurphy/emurphy/elegies.html`) wrote that Birkerts's "excessive indulgence in extreme technological determinism and reductionism make him an easy target for criticism. Indeed, Birkerts conveniently blames technology for the loss of all that he cherishes. The questions he raises about the relation between reading, technology and mind are thought-provoking [but] skirt over issues and point fingers rather than encourage understanding."

ISSUE 14

Will It Be Possible to Build a Computer That Can Think?

YES: Hans Moravec, from "The Universal Robot," in *Vision-21: Interdisciplinary Science and Engineering in the Era of Cyberspace* (National Aeronautics and Space Administration, 1993)

NO: John R. Searle, from "Is the Brain's Mind a Computer Program?" *Scientific American* (January 1990)

ISSUE SUMMARY

YES: Research scientist Hans Moravec describes the necessary steps in what he considers to be the inevitable development of computers that match and even exceed human intelligence.

NO: Professor of philosophy John R. Searle argues that a crucial difference between artificial (machine) intelligence and human intelligence—that humans attach meaning to the symbols they manipulate while computers cannot—makes it impossible to create a computer that can think.

The first primitive digital computers were instantly dubbed "thinking machines" because they were able to perform functions—initially only arithmetic—that had always been considered part of the uniquely human ability to think. Some critics of the "thinking machine" label, however, objected that arithmetic is so much simpler than, say, poetry or philosophy (after all, it is only a matter of following a few simple rules) that computers were not thinking at all. Thinking, they said, is for humans only. In fact, if a machine can do it, then it cannot possibly be real thinking.

In 1950 Alan Turing (1912–1954), an English mathematician and logician, devised a test to determine whether or not a machine was intelligent. Turing's test entailed whether or not one could converse with a person and a computer (through a teletype so that neither could be seen nor could the human be heard) and, after a suitable period, tell which was which. If the computer could pass for an intelligent conversationalist, Turing felt, then it would have to be considered intelligent.

Over the next two decades, computer scientists learned how to program their machines to play games such as chess, solve mathematical theorems, parse sentences (break them down into their grammatical components), and perform a number of other tasks that had once been thought doable by thinking humans only. In most cases the machines were not as good at these tasks as humans, but many artificial intelligence (AI) researchers believed that it was only a matter of time before the machines matched and even exceeded their creators.

The closest any machine has come to passing the Turing test may have been in the early 1970s, when Kenneth Mark Colby, then a Stanford University psychiatrist and computer scientist, programmed a computer to imitate the conversational style of paranoid humans. This was much easier than programming a computer to imitate a nonparanoid human's conversational style because paranoid individuals tend to be very rigid and predictable in their responses. When Colby had psychiatrists interview the programmed computer and a human paranoid (through a teletype, per Turing's criteria), only half could correctly distinguish between computer and human. That is, the computer did indeed come close to passing the Turing test. On the other hand, it was not trying to pass as an average human being, whose thought processes are far freer and more flexible than those of a paranoid person.

Will a computer ever be able to imitate a normal human being? And if it can, will that mean it is really "thinking" or really "intelligent"? Many computer scientists believe that it is still just a matter of time before a computer passes the Turing test with flying colors and that that machine will be truly intelligent. Indeed, many even say that the human mind is nothing more than a program that runs on a biological machine.

Others argue that machines cannot have emotions or appreciate beauty and that computers cannot be self-aware or conscious, no matter how intelligent they may seem to an interrogator. They therefore can never be intelligent in a human way.

Hans Moravec, director of the Mobile Robot Laboratory at Carnegie Mellon University's Robotics Institute, strongly believes that true artificial intelligence can be achieved. It will require computers that are much more powerful than any that exist today, he predicts, and the process of achieving intelligence will involve a series of evolutionary stages.

In contrast, John R. Searle, professor of philosophy at the University of California, Berkeley, argues that although humans *can* be regarded as biological machines, there are essential differences between natural and artificial intelligence. Furthermore, he objects to the idea that the human mind is nothing more than a computer program.

YES

Hans Moravec

THE UNIVERSAL ROBOT

Abstract. Our artifacts are getting smarter, and a loose parallel with the evolution of animal intelligence suggests one future course for them. Computerless industrial machinery exhibits the behavioral flexibility of single-celled organisms. Today's best computer-controlled robots are like the simpler invertebrates. A thousand-fold increase in computer power in the next decade should make possible machines with reptile-like sensory and motor competence. Properly configured, such robots could do in the physical world what personal computers now do in the world of data—act on our behalf as literal-minded slaves. Growing computer power over the next half-century will allow this reptile stage to be surpassed, in stages producing robots that learn like mammals, model their world like primates and eventually reason like humans. Depending on your point of view, humanity will then have produced a worthy successor, or transcended some of its inherited limitations and so transformed itself into something quite new.

INTRODUCTION: STATE OF THE ART

Instincts which predispose the nature and quantity of work we enjoy probably evolved during the 100,000 years our ancestors lived as hunter-gatherers. Less than 10,000 years ago the agricultural revolution made life more stable, and richer in goods and information. But, paradoxically, it requires more human labor to support an agricultural society than a primitive one, and the work is of a different, "unnatural" kind, out of step with the old instincts. The effort to avoid this work has resulted in domestication of animals, slavery and the industrial revolution. But many jobs must still be done by hand, engendering for hundreds of years the fantasy of an intelligent but soulless being that can tirelessly dispatch the drudgery. Only in this century have electronic sensors and computers given machines the ability to sense their world and to think about it, and so offered a way to fulfill the wish.

From National Aeronautics and Space Administration. Office of Management. Scientific and Technical Information Program. *Vision-21: Interdisciplinary Science and Engineering in the Era of Cyberspace.* (NASA Conference Publication 10129; 1993). References omitted.

As in fables, the unexpected side effects of robot slaves are likely to dominate the resulting story. Most significantly, these perfect slaves will continue to develop, and will not long remain soulless. As they increase in competence they will have occasion to make more and more autonomous decisions, and so will slowly develop a volition and purposes of their own. At the same time they will become indispensable. Our minds were evolved to store the skills and memories of a stone-age life, not the enormous complexity that has developed in the last ten thousand years. We've kept up, after a fashion, through a series of social inventions—social stratification and division of labor, memory aids like poetry and schooling, written records, stored outside the body, and recently machines that can do some of our thinking entirely without us. The portion of absolutely essential human activity that takes place outside of human bodies and minds has been steadily increasing. Hard working intelligent machines may complete the trend.

Serious attempts to build thinking machines began after the second world war. One line of research, called Cybernetics, used simple electronic circuitry to mimic small nervous systems, and produced machines that could learn to recognize simple patterns, and turtle-like robots that found their way to lighted recharging hutches. An entirely different approach, named Artificial Intelligence (AI), attempted to duplicate rational human thought in the large computers that appeared after the war. By 1965, these computers ran programs that proved theorems in logic and geometry, solved calculus problems and played good games of checkers. In the early 1970s, AI research groups at MIT (the Massachusetts Institute of Technology) and Stanford University attached television cameras and robot arms to their computers, so their "thinking" programs could begin to collect their information directly from the real world.

What a shock! While the pure reasoning programs did their jobs about as well and about as fast as college freshmen, the best robot control programs took hours to find and pick up a few blocks on a table. Often these robots failed completely, giving a performance much worse than a six month old child. This disparity between programs that reason and programs that perceive and act in the real world holds to this day. In recent years Carnegie Mellon University produced two desk-sized computers that can play chess at grandmaster level, within the top 100 players in the world, when given their moves on a keyboard. But present-day robotics could produce only a complex and unreliable machine for finding and moving normal chess pieces.

In hindsight it seems that, in an absolute sense, reasoning is much easier than perceiving and acting—a position not hard to rationalize in evolutionary terms. The survival of human beings (and their ancestors) has depended for hundreds of millions of years on seeing and moving in the physical world, and in that competition large parts of their brains have become efficiently organized for the task. But we didn't appreciate this monumental skill because it is shared by every human being and most animals—it is commonplace. On the other hand, rational thinking, as in chess, is a newly acquired skill, perhaps less than one hundred thousand years old. The parts of our brain devoted to it are not well organized, and, in an absolute sense, we're not very good at it. But until recently we had no competition to show us up.

By comparing the edge and motion detecting circuitry in the four layers of nerve cells in the retina, the best understood major circuit in the human nervous system, with similar processes developed for "computer vision" systems that allow robots in research and industry to see, I've estimated that it would take a billion computations per second (the power of a world-leading Cray 2 supercomputer) to produce the same results at the same speed as a human retina. By extrapolation, to emulate a whole brain takes ten trillion arithmetic operations per second, or ten thousand Crays worth. This is for operations our nervous systems do extremely efficiently and well.

Arithmetic provides an example at the other extreme. In 1989 a new computer was tested for a few months with a program that computed the number pi to more than one billion decimal places. By contrast, the largest unaided manual computation of pi was 707 digits by William Shanks in 1873. It took him several years, and because of a mistake every digit past the 527th was wrong! In arithmetic, today's average computers are one million times more powerful than human beings. In very narrow areas of rational thought (like playing chess or proving theorems) they are about the same. And in perception and control of movement in the complex real world, and related areas of common-sense knowledge and intuitive and visual problem solving, today's average computers are a million times less capable.

The deficit is evident even in pure problem solving AI programs. To this day, AI programs exhibit no shred of common sense—a medical diagnosis program, for instance, may prescribe an antibiotic when presented a broken bicycle because it lacks a model of people, diseases or bicycles. Yet these programs, on existing computers, would be overwhelmed were they to be bloated with the details of everyday life, since each new fact can interact with the others in an astronomical "combinatorial explosion." [A ten year project called Cyc at the Microelectronics and Computer Consortium in Austin, Texas, is attempting to build just such a common-sense data base. They estimate the final result will contain over one hundred million logic sentences about everyday objects and actions.]

Machines have a lot of catching up to do. On the other hand, for most of the century, machine calculation has been improving a thousandfold every twenty years, and there are basic developments in research labs that can sustain this for at least several decades more. In less than fifty years computer hardware should be powerful enough to match, and exceed, even the well-developed parts of human intelligence. But what about the software that would be required to give these powerful machines the ability to perceive, intuit and think as well as humans? The Cybernetic approach that attempts to directly imitate nervous systems is very slow, partly because examining a working brain in detail is a very tedious process. New instruments may change that in the future. The AI approach has successfully imitated some aspects of rational thought, but that seems to be only about one millionth of the problem. I feel that the fastest progress on the hardest problems will come from a third approach, the newer field of robotics, the construction of systems that must see and move in the physical world. Robotics research is imitating the evolution of animal minds, adding capabilities to machines a few at a time, so that the resulting sequence of machine behaviors resembles the capabilities of animals with increasingly complex nervous systems. This effort to build intelligence from the bottom up is helped by biological peeks at the "back of the book"—at the neuronal, structural, and behavioral features of animals and humans.

The best robots today are controlled by computers which are just powerful enough to simulate the nervous system of an insect, cost as much as houses, and so find only a few profitable niches in society (among them, spray painting and spot welding cars and assembling electronics). But those few applications are encouraging research that is slowly providing a base for a huge future growth. Robot evolution in the direction of full intelligence will greatly accelerate, I believe, in about a decade when the mass-produced general purpose, universal robot becomes possible. These machines will do in the physical world what personal computers do in the world of data—act on our behalf as literal-minded slaves.

THE DUMB ROBOT (ca. 2000–2010)

To be useful in many tasks, the first generation of universal robots should navigate efficiently over flat ground and reliably and safely over rough terrain and stairs, be able to manipulate most objects, and to find them in the nearby world. There are beginnings of solutions today. In the 1980s Hitachi of Japan developed a mobility system of five steerable wheels, each on its own telescoping stalk that allows it to accommodate to rises and dips in uneven terrain, and to climb stairs, by raising one wheel at a time while standing stably on the other four. My laboratory at Carnegie Mellon University in Pittsburgh has developed a navigation method that enables a robot equipped with sonar range measuring devices and television cameras to build probabilistic maps of its surroundings to determine its location and plan routes. An elegant three-fingered mechanical hand at the Massachusetts Institute of Technology can hold and orient bolts and eggs and manipulate a string in a humanlike fashion. A system called 3DPO from SRI International in Menlo Park, California, can find a desired part in a jumble seen by a special range-finding camera. The slow operation of these systems suggests one other element needed for the universal robot, namely a computer about one thousand times as powerful as those found on desks and in robots today. Such machines, able to do one billion computations per second, would provide robots approximately the brain power of a reptile, and the personality of a washing machine.

Universal robots will find their first uses in factories, where they will be cheaper and more versatile than the older generation of robots they replace. Eventually they will become cheap enough for some households, extending the reach of personal computers from a few tasks in the data world to many in the physical world. . . .

LEARNING (2010–2020)

Useful though they will be, the first generation of universal robots will be rigid slaves to simple programs. If the machine bangs its elbow while chopping beef in your kitchen making Stroganoff, you will have to find another place for the robot to do its work, or beg the software manufacturer for a fix. Second generation robots with more powerful computers will be able to host a more flexible kind of program able to adjust itself by a kind of conditioned learning. First generation programs will consist primarily of sequences of the type "Do step A, then B, then C. . . ." The programs for the second generation will read "Do step A1 or A2 or A3 . . . then B1 or B2 or B3 . . . then C1 or C2 or C3. . . ." In the Beef Stroganoff example, A1 might be to chop with the right hand of the robot, while A2 is to use the left hand. Each alternative in the program has a "weight," a number that indicates the desirability of using it rather than one of the other branches. The machine also contains a "pain" system, a series of programs that look out for problems, such as collisions, and respond by reducing the weights of recently invoked branches, and a "pleasure" system that increases the relevant weights when good conditions, such as well charged batteries or a task efficiently completed, are detected. As the robot bangs its elbow repeatedly in your kitchen, it gradually learns to use its other hand (as well as adapting to its surroundings in a thousand other ways). A program with many alternatives at each step, whose pain and pleasure systems are arranged to produce a pleasure signal on hearing the word "good" and a pain message on hearing "bad" could be slowly trained to do new tasks, like a small mammal. A particular suite of pain- and pleasure-producing programs interacting with a robot's individual environment would subtly shape its behavior and give it a distinct character.

IMAGERY (2020–2030)

Adaptive robots will find jobs everywhere, and the hardware and software industry that supports them could become the largest on earth. But teaching them new tasks, whether by writing programs or through punishment and reward, will be very tedious. This deficiency will lead to a portentous innovation, a software world-modeler (requiring another big increase in computer power), that allows the robot to simulate its immediate surroundings and its own actions within them, and thus to think about its tasks before acting. Before making Beef Stroganoff in your kitchen, the new robot would simulate the task many times. Each time its simulated elbow bangs the simulated cabinet, the software would update the learning weights just as if the collision had physically happened. After many such mental run-throughs the robot would be well trained, so that when it finally cooks for real, it does it correctly. The simulation can be used in many other ways. After a job, the robot can run though its previous actions, and try variations on them to improve future performance. A robot might even be configured to invent some of its own programs by means of a simpler program that can detect how nearly a sequence of robot actions achieves a desired task. This training program would, in repeated simulations, provide the "good" and "bad" indications needed to condition a general learning program like the one of the previous section.

It will take a large community of patient researchers to build good simulators. A robot entering a new room must include vast amounts of not directly perceived prior knowledge in its simulation, such as the expected shapes and probable contents of kitchen counters and the effect of (and force needed for) turning faucet knobs. It needs instinctive motor-perceptual knowledge about the world that took millions of years of evolution to install in us, that tells us instinctively when a height is dangerous, how hard to throw a stone, or if the animal facing us is a threat. Robots that incorporate it may be as smart as monkeys.

REASONING (2030–2040)

In the decades while the "bottom-up" evolution of robots is transferring the perceptual and motor faculties of human beings into machinery, the conventional Artificial Intelligence industry will be perfecting the mechanization of reasoning. Since today's programs already match human beings in some areas, those of 40 years from now, running on computers a million times as fast as today's, should be quite superhuman. Today's reasoning programs work from small amounts of clear and correct information prepared by human beings. Data from robot sensors such as cameras is much too voluminous and too noisy for them to use. But a good robot simulator will contain neatly organized data about the robot and its world. For instance, if a knife is on a countertop, or if the robot is holding a cup. A robot with a simulator can be married to a reasoning program to produce a machine with most of the abilities of a human being. The combination will create beings that in some ways resemble us, but in others are like nothing the world has seen before.

FIRST GENERATION TECHNICALITIES

Both industrial robot manipulators and the research effort to build "smart" robots are twenty five years old. Universal robots will require at least another decade of development, but some of their elements can be guessed from the experience so far. One consideration is weight. Mobile robots built to work in human sized spaces today weigh too many hundreds of pounds. This dangerously large mass has three major components: batteries, actuators and structure. Lead-acid batteries able to drive a mobile robot for a day contribute about one third of the weight. But nickel-cadmium aircraft batteries weigh half as much, and newer lithium batteries can be half again as light. Electric motors are efficient and precisely controllable, but standard motors are heavy and require equally heavy reducing gears. Ultrastrong permanent magnets can halve the weight and generate high torque without gears. Robot structure has been primarily aluminum. Its weight contribution can be cut by a factor of four by substituting composite materials containing superstrength fibers of graphite, aramid or the new material Spectra. These innovations could be combined to make a robot with roughly the size, weight, strength and endurance of a human.

The first generation robot will probably move on wheels. Legged robots have advantages on complicated terrain, but they consume too much power. A simple wheeled robot would be confined to areas of flat ground, but if each wheel had a controlled suspension with about a meter of travel, the robot could slowly lift its wheels as needed to negotiate rough ground and stairs. The manipulation system will consist of two or more arms ending in dexterous manipulators. There are several designs in the research labs today, but the most elegant is probably that of the so-called Stanford-JPL hand (mentioned above, now found at MIT), which has three fingers each with three controlled joints.

The robot's travels would be greatly aided if it could continuously pinpoint its location, perhaps by noting the delay from a handful of small synchronized transmitters distributed in its environment. This approach is used in some terrestrial and satellite navigation systems. The robot will also require a sense of its immediate surroundings, to find doors, detect obstacles and track objects in its workspace. Research laboratories, including my own, have experimented with techniques that do this with data from television cameras, scanning lasers, sonar transducers, infrared proximity sensors and contact sensors. A more precise sensory system will be needed to find particular work objects in clutter. The most successful methods to date start with three dimensional data from special cameras and laser arrangements that directly measure distance as well as lateral position. The robot will thus probably contain a wide angle sensor for general spatial awareness, and a precise, narrow angle, three dimensional imaging system to find particular objects it will grasp.

Research experience to date suggests that to navigate, visually locate objects, and plan and control arm motions, the first universal robots will require a billion operations per second of computer power. The 1980s have witnessed a number of well publicized fads that claim to be solutions to the artificial intelligence or robot control problem. Expert systems, the Prolog logical inference language, neural nets, fuzzy logic and massive parallelism have all had their spot in the limelight. The common element that I note in these pronouncements is the sudden enthusiasm of groups of researchers experienced in some area of computer science for applying their methods to the robotics problems of perceiving and acting in the physical world. Invariably each approach produces some simple showcase demonstrations, then bogs down on real problems. This pattern is no surprise to those with a background in the twenty five year research robotics effort.

Making a machine to see, hear or act reliably in the raw physical world is much, much more difficult than naive intuition leads us to believe....

MIND CHILDREN (2050+)

The fourth robot generation and its successors, with human perceptual and motor abilities and superior reasoning powers, could replace human beings in every essential task. In principle, our society could continue to operate increasingly well without us, with machines running the companies and doing the research as well as performing the productive work. Since machines can be designed to work well in outer space, production could move to the greater resources of the solar system, leaving behind a nature preserve subsidized from space. Meek humans would inherit the earth, but rapidly evolving machines would expand into the rest of the universe.

This development can be viewed as a very natural one. Human beings have two forms of heredity, one the traditional biological kind, passed on strands of DNA, the other cultural, passed from mind to mind by example, language, books and recently machines. At present the two are inextricably linked, but the cultural part is evolving very rapidly, and gradually assuming functions once the province of our biology. In terms of information content, our cultural side is already by far the larger part of us. The fully intelligent robot marks the point where our cultural side can exist on its own, free of biological limits. Intelligent machines, which are evolving among us, learning our skills, sharing our goals, and being shaped by our values, can be viewed as our children, the children of our minds. With them our biological heritage is not lost. It will be safely stored in libraries at least; however its importance will be greatly diminished.

What about life back on the preserve? For some of us the thought of being grandly upstaged by our artificial progeny will be disappointing, and life may seem pointless if we are fated to spend it staring stupidly at our ultra-intelligent progeny as they try to describe their ever more spectacular discoveries in baby-talk that we can understand. Is there any way individual humans might join the adventure?

You've just been wheeled into the operating room. A robot brain surgeon is in attendance, a computer waits nearby. Your skull, but not your brain, is anesthetized. You are fully conscious. The robot surgeon opens your brain case and places a hand on the brain's surface. This unusual hand bristles with microscopic machinery, and a cable connects it to the computer at your side. Instruments in the hand scan the first few millimeters of brain surface. These measurements, and a comprehensive understanding of human neural architecture, allow the surgeon to write a program that models the behavior of the uppermost layer of the scanned brain tissue. This program is installed in a small portion of the waiting computer and activated. Electrodes in the hand supply the simulation with the appropriate inputs from your brain, and can inject signals from the simulation. You and the surgeon compare the signals it produces with the original ones. They flash by very fast, but any discrepancies are highlighted on a display screen. The surgeon fine-tunes the simulation until the correspondence is nearly perfect. As soon as you are satisfied, the simulation output is activated. The brain layer is now impotent—it receives inputs and reacts as before but its output is ignored. Microscopic manipulators on the hand's surface excise this superfluous tissue and pass them to an aspirator, where they are drawn away.

The surgeon's hand sinks a fraction of a millimeter deeper into your brain, instantly compensating its measurements and signals for the changed position. The process is repeated for the next layer, and soon a second simulation resides in the computer, communicating with the first and with the remaining brain tissue. Layer after layer the brain is simulated, then excavated. Eventually your skull is empty, and the surgeon's hand rests deep in your brainstem. Though you have not lost consciousness, or even your train of thought, your mind has been removed from the brain and transferred to a machine. In a final, disorienting step the surgeon lifts its hand. Your suddenly abandoned body dies. For a moment you experience only quiet and dark. Then, once again, you can open your eyes. Your perspective has shifted. The computer simulation has been disconnected from the cable leading to the surgeon's hand and reconnected to a shiny new body of the style, color, and material of your choice. Your metamorphosis is complete.

Your new mind has a control labeled "speed." It had been set at 1, to keep the simulations synchronized with the old brain, but now you change it to 10,000, allowing you to communicate, react, and think ten thousand times faster. You now seem to have hours to respond to situations that previously seemed instantaneous. You have time, during the fall of a dropped object, to research the advantages and disadvantages of trying to catch it, perhaps to solve its differential equations of motion. When your old biological friends speak with you, their sentences take hours—you have plenty of time to think about the conversations, but they try your patience. Boredom is a mental alarm that keeps you from wasting your time in profitless activity, but if it acts too soon or too aggressively it limits your attention span, and thus your intelligence. With help from the machines, you change your mind-program to retard the onset of boredom. Having done that, you will find yourself comfortably working on long problems with sidetracks upon sidetracks. In fact, your thoughts routinely become so involved that you need an increase in your memory. These are but the first of many changes. Soon your friends complain that you have become more like the machines than the biological human you once were. That's life.

NO

John R. Searle

IS THE BRAIN'S MIND A COMPUTER PROGRAM?

Can a machine think? Can a machine have conscious thoughts in exactly the same sense that you and I have? If by "machine" one means a physical system capable of performing certain functions (and what else can one mean?), then humans are machines of a special biological kind, and humans can think, and so of course machines can think. And, for all we know, it might be possible to produce a thinking machine out of different materials altogether—say, out of silicon chips or vacuum tubes. Maybe it will turn out to be impossible, but we certainly do not know that yet.

In recent decades, however, the question of whether a machine can think has been given a different interpretation entirely. The question that has been posed in its place is, Could a machine think just by virtue of implementing a computer program? Is the program by itself constitutive of thinking? This is a completely different question because it is not about the physical, causal properties of actual or possible physical systems but rather about the abstract, computational properties of formal computer programs that can be implemented in any sort of substance at all, provided only that the substance is able to carry the program.

A fair number of researchers in artificial intelligence (AI) believe the answer to the second question is yes; that is, they believe that by designing the right programs with the right inputs and outputs, they are literally creating minds. They believe furthermore that they have a scientific test for determining success or failure: the Turing test devised by Alan M. Turing, the founding father of artificial intelligence. The Turing test, as currently understood, is simply this: if a computer can perform in such a way that an expert cannot distinguish its performance from that of a human who has a certain cognitive ability—say, the ability to do addition or to understand Chinese—then the computer also has that ability. So the goal is to design programs that will simulate human cognition in such a way as to pass the Turing test. What is more, such a program would not merely be a model of the mind; it would literally be a mind, in the same sense that a human mind is a mind.

From John R. Searle, "Is the Brain's Mind a Computer Program?" *Scientific American* (January 1990).

By no means does every worker in artificial intelligence accept so extreme a view. A more cautious approach is to think of computer models as being useful in studying the mind in the same way that they are useful in studying the weather, economics or molecular biology. To distinguish these two approaches, I call the first strong AI and the second weak AI. It is important to see just how bold an approach strong AI is. Strong AI claims that thinking is merely the manipulation of formal symbols, and that is exactly what the computer does: manipulate formal symbols. This view is often summarized by saying, "The mind is to the brain as the program is to the hardware."

* * *

Strong AI is unusual among theories of the mind in at least two respects: it can be stated clearly, and it admits of a simple and decisive refutation. The refutation is one that any person can try for himself or herself. Here is how it goes. Consider a language you don't understand. In my case, I do not understand Chinese. To me Chinese writing looks like so many meaningless squiggles. Now suppose I am placed in a room containing baskets full of Chinese symbols. Suppose also that I am given a rule book in English for matching Chinese symbols with other Chinese symbols. The rules identify the symbols entirely by their shapes and do not require that I understand any of them. The rules might say such things as, "Take a squiggle-squiggle sign from basket number one and put it next to a squoggle-squoggle sign from basket number two."

Imagine that people outside the room who understand Chinese hand in small bunches of symbols and that in response I manipulate the symbols according to the rule book and hand back more small bunches of symbols. Now, the rule book is the "computer program." The people who wrote it are "programmers," and I am the "computer." The baskets full of symbols are the "data base," the small bunches that are handed in to me are "questions" and the bunches I then hand out are "answers."

Now suppose that the rule book is written in such a way that my "answers" to the "questions" are indistinguishable from those of a native Chinese speaker. For example, the people outside might hand me some symbols that unknown to me mean, "What's your favorite color?" and I might after going through the rules give back symbols that, also unknown to me, mean, "My favorite is blue, but I also like green a lot." I satisfy the Turing test for understanding Chinese. All the same, I am totally ignorant of Chinese. And there is no way I could come to understand Chinese in the system as described, since there is no way that I can learn the meanings of any of the symbols. Like a computer, I manipulate symbols, but I attach no meaning to the symbols.

The point of the thought experiment is this: if I do not understand Chinese solely on the basis of running a computer program for understanding Chinese, then neither does any other digital computer solely on that basis. Digital computers merely manipulate formal symbols according to rules in the program.

What goes for Chinese goes for other forms of cognition as well. Just manipulating the symbols is not by itself enough to guarantee cognition, perception, understanding, thinking and so forth. And since computers, qua computers, are symbol-manipulating devices, merely running the computer program is not enough to guarantee cognition.

This simple argument is decisive against the claims of strong AI. The first premise of the argument simply states the formal character of a computer program. Programs are defined in terms of symbol manipulations, and the symbols are purely formal, or "syntactic." The formal character of the program, by the way, is what makes computers so powerful. The same program can be run on an indefinite variety of hardwares, and one hardware system can run an indefinite range of computer programs. Let me abbreviate this "axiom" as

> Axiom 1. Computer programs are formal (syntactic).

This point is so crucial that it is worth explaining in more detail. A digital computer processes information by first encoding it in the symbolism that the computer uses and then manipulating the symbols through a set of precisely stated rules. These rules constitute the program. For example, in Turing's early theory of computers, the symbols were simply 0's and 1's, and the rules of the program said such things as, "Print a 0 on the tape, move one square to the left and erase a 1." The astonishing thing about computers is that any information that can be stated in a language can be encoded in such a system, and any information-processing task that can be solved by explicit rules can be programmed.

* * *

Two further points are important. First, symbols and programs are purely abstract notions: they have no essential physical properties to define them and can be implemented in any physical medium whatsoever. The 0's and 1's, qua symbols, have no essential physical properties and a fortiori have no physical, causal properties. I emphasize this point because it is tempting to identify computers with some specific technology—say, silicon chips—and to think that the issues are about the physics of silicon chips or to think that syntax identifies some physical phenomenon that might have as yet unknown causal powers, in the way that actual physical phenomena such as electromagnetic radiation or hydrogen atoms have physical, causal properties. The second point is that symbols are manipulated without reference to any meanings. The symbols of the program can stand for anything the programmer or user wants. In this sense the program has syntax but no semantics.

The next axiom is just a reminder of the obvious fact that thoughts, perceptions, understandings and so forth have a mental content. By virtue of their content they can be about objects and states of affairs in the world. If the content involves language, there will be syntax in addition to semantics, but linguistic understanding requires at least a semantic framework. If, for example, I am thinking about the last presidential election, certain words will go through my mind, but the words are about the election only because I attach specific meanings to these words, in accordance with my knowledge of English. In this respect they are unlike Chinese symbols for me. Let me abbreviate this axiom as

> Axiom 2. Human minds have mental contents (semantics).

Now let me add the point that the Chinese room demonstrated. Having the symbols by themselves—just having the syntax—is not sufficient for having the semantics.

Merely manipulating symbols is not enough to guarantee knowledge of what they mean. I shall abbreviate this as

> Axiom 3. Syntax by itself is neither constitutive of nor sufficient for semantics.

At one level this principle is true by definition. One might, of course, define the terms syntax and semantics differently. The point is that there is a distinction between formal elements, which have no intrinsic meaning or content, and those phenomena that have intrinsic content. From these premises it follows that

> Conclusion 1. Programs are neither constitutive of nor sufficient for minds.

And that is just another way of saying that strong AI is false.

It is important to see what is proved and not proved by this argument.

First, I have not tried to prove that "a computer cannot think." Since anything that can be simulated computationally can be described as a computer, and since our brains can at some levels be simulated, it follows trivially that our brains are computers and they can certainly think. But from the fact that a system can be simulated by symbol manipulation and the fact that it is thinking, it does not follow that thinking is equivalent to formal symbol manipulation.

Second, I have not tried to show that only biologically based systems like our brains can think. Right now those are the only systems we know for a fact can think, but we might find other systems in the universe that can produce conscious thoughts, and we might even come to be able to create thinking systems artificially. I regard this issue as up for grabs.

Third, strong AI's thesis is not that, for all we know, computers with the right programs might be thinking, that they might have some as yet undetected psychological properties; rather it is that they must be thinking because that is all there is to thinking.

Fourth, I have tried to refute strong AI so defined. I have tried to demonstrate that the program by itself is not constitutive of thinking because the program is purely a matter of formal symbol manipulation—and we know independently that symbol manipulations by themselves are not sufficient to guarantee the presence of meanings. That is the principle on which the Chinese room argument works.

I emphasize these points here partly because it seems to me the Churchlands [see "Could a Machine Think?" by Paul M. Churchland and Patricia Smith Churchland, *Scientific American* (January 1990), page 321] have not quite understood the issues. They think that strong AI is claiming that computers might turn out to think and that I am denying this possibility on commonsense grounds. But that is not the claim of strong AI, and my argument against it has nothing to do with common sense.

I will have more to say about their objections later. Meanwhile I should point out that, contrary to what the Churchlands suggest, the Chinese room argument also refutes any strong-AI claims made for the new parallel technologies that are inspired by and modeled on neural networks. Unlike the traditional von Neumann computer, which proceeds in a step-by-step fashion, these systems have many computational elements that operate in parallel and interact with one another according to rules inspired by neurobiology. Although the results are still modest, these "parallel distributed processing, or connectionist," models raise useful questions about how complex, parallel network systems like those in brains might actually function in the production of intelligent behavior.

The parallel, "brainlike" character of the processing, however, is irrelevant to the purely computational aspects of the process. Any function that can be computed on a parallel machine can also be computed on a serial machine. Indeed, because parallel machines are still rare, connectionist programs are usually run on traditional serial machines. Parallel processing, then, does not afford a way around the Chinese room argument.

What is more, the connectionist system is subject even on its own terms to a variant of the objection presented by the original Chinese room argument. Imagine that instead of a Chinese room, I have a Chinese gym: a hall containing many monolingual, English-speaking men. These men would carry out the same operations as the nodes and synapses in a connectionist architecture as described by the Churchlands, and the outcome would be the same as having one man manipulate symbols according to a rule book. No one in the gym speaks a word of Chinese, and there is no way for the system as a whole to learn the meanings of any Chinese words. Yet with appropriate adjustments, the system could give the correct answers to Chinese questions.

There are, as I suggested earlier, interesting properties of connectionist nets that enable them to simulate brain processes more accurately than traditional serial architecture does. But the advantages of parallel architecture for weak AI are quite irrelevant to the issues between the Chinese room argument and strong AI.

The Churchlands miss this point when they say that a big enough Chinese gym might have higher-level mental features that emerge from the size and complexity of the system, just as whole brains have mental features that are not had by individual neurons. That is, of course, a possibility, but it has nothing to do with computation. Computationally, serial and parallel systems are equivalent: any computation that can be done in parallel can be done in serial. If the man in the Chinese room is computationally equivalent to both, then if he does not understand Chinese solely by virtue of doing the computations, neither do they. The Churchlands are correct in saying that the original Chinese room argument was designed with traditional AI in mind but wrong in thinking that connectionism is immune to the argument. It applies to any computational system. You can't get semantically loaded thought contents from formal computations alone, whether they are done in serial or in parallel; that is why the Chinese room argument refutes strong AI in any form.

* * *

Many people who are impressed by this argument are nonetheless puzzled about the differences between people and computers. If humans are, at least in a trivial sense, computers, and if humans have a semantics, then why couldn't we give semantics to other computers? Why couldn't we program a Vax or a Cray so that it too would have thoughts and feelings? Or why couldn't some new computer technology overcome the gulf between form and content, between syntax and semantics? What, in fact, are the differences between animal brains and computer systems that enable the Chinese room argument to work against computers but not against brains?

The most obvious difference is that the processes that define something as a computer—computational processes—are completely independent of any reference to a specific type of hardware implementation. One could in principle make a computer out of old beer cans strung together with wires and powered by windmills.

But when it comes to brains, although science is largely ignorant of how brains function to produce mental states, one is struck by the extreme specificity of the anatomy and the physiology. Where some understanding exists of how brain processes produce mental phenomena—for example, pain, thirst, vision, smell—it is clear that specific neurobiological processes are involved. Thirst, at least of certain kinds, is caused by certain types of neuron firings in the hypothalamus, which in turn are caused by the action of a specific peptide, angiotensin II. The causation is from the "bottom up" in the sense that lower-level neuronal processes cause higher-level mental phenomena. Indeed, as far as we know, every "mental" event, ranging from feelings of thirst to thoughts of mathematical theorems and memories of childhood, is caused by specific neurons firing in specific neural architectures.

But why should this specificity matter? After all, neuron firings could be simulated on computers that had a completely different physics and chemistry from that of the brain. The answer is that the brain does not merely instantiate a formal pattern or program (it does that, too), but it also *causes* mental events by virtue of specific neurobiological processes. Brains are specific biological organs, and their specific biochemical properties enable them to cause consciousness and other sorts of mental phenomena. Computer simulations of brain processes provide models of the formal aspects of these processes. But the simulation should not be confused with duplication. The computational model of mental processes is no more real than the computational model of any other natural phenomenon.

One can imagine a computer simulation of the action of peptides in the hypothalamus that is accurate down to the last synapse. But equally one can imagine a computer simulation of the oxidation of hydrocarbons in a car engine or the action of digestive processes in a stomach when it is digesting pizza. And the simulation is no more the real thing in the case of the brain than it is in the case of the car or the stomach. Barring miracles, you could not run your car by doing a computer simulation of the oxidation of gasoline, and you could not digest pizza by running the program that simulates such digestion. It seems obvious that a simulation of cognition will similarly not produce the effects of the neurobiology of cognition.

All mental phenomena, then, are caused by neurophysiological processes in the brain. Hence,

> Axiom 4. Brains cause minds.

In conjunction with my earlier derivation, I immediately derive, trivially,

> Conclusion 2. Any other system capable of causing minds would have to have causal powers (at least) equivalent to those of brains.

This is like saying that if an electrical engine is to be able to run a car as fast as a gas engine, it must have (at least) an equivalent power output. This conclusion says nothing about the mechanisms. As a matter of fact, cognition is a biological phenomenon: mental states and processes are caused by brain processes. This does not imply that only a biological system could think, but it does imply that any alternative system, whether made of silicon, beer cans or whatever, would have to have the relevant causal capacities equivalent to those of brains. So now I can derive

> Conclusion 3. Any artifact that produced mental phenomena, any artificial brain, would have to be able to duplicate the specific causal powers of brains, and it could not do that just by running a formal program.

Furthermore, I can derive an important conclusion about human brains:

> Conclusion 4. The way that human brains actually produce mental phenomena cannot be solely by virtue of running a computer program.

* * *

I first presented the Chinese room parable in the pages of *Behavioral and Brain Sciences* in 1980, where it appeared, as is the practice of the journal, along with peer commentary, in this case, 26 commentaries. Frankly, I think the point it makes is rather obvious, but to my surprise the publication was followed by a further flood of objections that—more surprisingly—continues to the present day. The Chinese room argument clearly touched some sensitive nerve.

The thesis of strong AI is that any system whatsoever—whether it is made of beer cans, silicon chips or toilet paper—not only might have thoughts and feelings but *must* have thoughts and feelings, provided only that it implements the right program, with the right inputs and outputs. Now, that is a profoundly antibiological view, and one would think that people in AI would be glad to abandon it. Many of them, especially the younger generation, agree with me, but I am amazed at the number and vehemence of the defenders. Here are some of the common objections.

a. In the Chinese room you really do understand Chinese, even though you don't know it. It is, after all, possible to understand something without knowing that one understands it.
b. You don't understand Chinese, but there is an (unconscious) subsystem in you that does. It is, after all, possible to have unconscious mental states, and there is no reason why your understanding of Chinese should not be wholly unconscious.
c. You don't understand Chinese, but the whole room does. You are like a single neuron in the brain, and just as such a single neuron by itself cannot understand but only contributes to the understanding of the whole system, you don't understand, but the whole system does.
d. Semantics doesn't exist anyway; there is only syntax. It is a kind of prescientific illusion to suppose that there exist in the brain some mysterious "mental contents," "thought processes" or "semantics." All that exists in the brain is the same sort of syntactic symbol manipulation that goes on in computers. Nothing more.
e. You are not really running the computer program—you only think you are. Once you have a conscious agent going through the steps of the program. It ceases to be a case of implementing a program at all.

f. Computers would have semantics and not just syntax if their inputs and outputs were put in appropriate causal relation to the rest of the world. Imagine that we put the computer into a robot, attached television cameras to the robot's head, installed transducers connecting the television messages to the computer and had the computer output operate the robot's arms and legs. Then the whole system would have a semantics.
g. If the program simulated the operation of the brain of a Chinese speaker, then it would understand Chinese. Suppose that we simulated the brain of a Chinese person at the level of neurons. Then surely such a system would understand Chinese as well as any Chinese person's brain.

And so on.

All of these arguments share a common feature: they are all inadequate because they fail to come to grips with the actual Chinese room argument. That argument rests on the distinction between the formal symbol manipulation that is done by the computer and the mental contents biologically produced by the brain, a distinction I have abbreviated—I hope not misleadingly—as the distinction between syntax and semantics. I will not repeat my answers to all of these objections, but it will help to clarify the issues if I explain the weaknesses of the most widely held objection, argument c—what I call the systems reply. (The brain simulator reply, argument g, is another popular one, but I have already addressed that one in the previous section.)

* * *

The systems reply asserts that of course *you* don't understand Chinese but the whole system—you, the room, the rule book, the bushel baskets full of symbols—does. When I first heard this explanation, I asked one of its proponents, "Do you mean the room understands Chinese?" His answer was yes. It is a daring move, but aside from its implausibility, it will not work on purely logical grounds. The point of the original argument was that symbol shuffling by itself does not give any access to the meanings of the symbols. But this is as much true of the whole room as it is of the person inside. One can see this point by extending the thought experiment. Imagine that I memorize the contents of the baskets and the rule book, and I do all the calculations in my head. You can even imagine that I work out in the open. There is nothing in the "system" that is not in me, and since I don't understand Chinese, neither does the system.

The Churchlands in their companion piece produce a variant of the systems reply by imagining an amusing analogy. Suppose that someone said that light could not be electromagnetic because if you shake a bar magnet in a dark room, the system still will not give off visible light. Now, the Churchlands ask, is not the Chinese room argument just like that? Does it not merely say that if you shake Chinese symbols in a semantically dark room, they will not give off the light of Chinese understanding? But just as later investigation showed that light was entirely constituted by electromagnetic radiation, could not later investigation also show that semantics are entirely constituted of syntax? Is this not a question for further scientific investigation?

Arguments from analogy are notoriously weak, because before one can make the argument work, one has to establish that the two cases are truly analogous. And here I think they are not. The account of light in terms of electromagnetic radiation is a causal story right down to the ground. It is a causal account of the physics of electromagnetic radiation. But the analogy with formal symbols fails because formal symbols have no physical, causal powers. The only power that symbols have, qua symbols, is the power to cause the next step in the program when the machine is running. And there is no question of waiting on further research to reveal the physical, causal properties of 0's and 1's. The only relevant properties of 0's and 1's are abstract computational properties, and they are already well known.

The Churchlands complain that I am "begging the question" when I say that uninterpreted formal symbols are not identical to mental contents. Well, I certainly did not spend much time arguing for it, because I take it as a logical truth. As with any logical truth, one can quickly see that it is true, because one gets inconsistencies if one tries to imagine the converse. So let us try it. Suppose that in the Chinese room some undetectable Chinese thinking really is going on. What exactly is supposed to make the manipulation of the syntactic elements into specifically Chinese thought contents? Well, after all, I am assuming that the programmers were Chinese speakers, programming the system to process Chinese information.

Fine. But now imagine that as I am sitting in the Chinese room shuffling the Chinese symbols, I get bored with just shuffling the—to me—meaningless symbols. So, suppose that I decide to interpret the symbols as standing for moves in a chess game. Which semantics is the system giving off now? Is it giving off a Chinese semantics or a chess semantics, or both simultaneously? Suppose there is a third person looking in through the window, and she decides that the symbol manipulations can all be interpreted as stock-market predictions. And so on. There is no limit to the number of semantic interpretations that can be assigned to the symbols because, to repeat, the symbols are purely formal. They have no intrinsic semantics.

Is there any way to rescue the Churchlands' analogy from incoherence? I said above that formal symbols do not have causal properties. But of course the program will always be implemented in some hardware or another, and the hardware will have specific physical, causal powers. And any real computer will give off various phenomena. My computers, for example, give off heat, and they make a humming noise and sometimes crunching sounds. So is there some logically compelling reason why they could not also give off consciousness? No. Scientifically, the idea is out of the question, but it is not something the Chinese room argument is supposed to refute, and it is not something that an adherent of strong AI would wish to defend, because any such giving off would have to derive from the physical features of the implementing medium. But the basic premise of strong AI is that the physical features of the implementing medium are totally irrelevant. What matters are programs, and programs are purely formal.

The Churchlands' analogy between syntax and electromagnetism, then, is confronted with a dilemma; either the syntax is construed purely formally in terms of its abstract mathematical properties, or it is not. If it is, then the analogy breaks down, because syntax so construed has no physical powers and hence no physical, causal powers. If, on the other hand, one is supposed to think in terms of the physics of the implementing medium, then there is indeed an analogy, but it is not one that is relevant to strong AI.

* * *

Because the points I have been making are rather obvious—syntax is not the same as semantics, brain processes cause mental phenomena—the question arises, How did we get into this mess? How could anyone have supposed that a computer simulation of a mental process must be the real thing? After all, the whole point of models is that they contain only certain features of the modeled domain and leave out the rest. No one expects to get wet in a pool filled with Ping-Pong-ball models of water molecules. So why would anyone think a computer model of thought processes would actually think?

Part of the answer is that people have inherited a residue of behaviorist psychological theories of the past generation. The Turing test enshrines the temptation to think that if something behaves as if it had certain mental processes, then it must actually have those mental processes. And this is part of the behaviorists' mistaken assumption that in order to be scientific, psychology must confine its study to externally observable behavior. Paradoxically, this residual behaviorism is tied to a residual dualism. Nobody thinks that a computer simulation of digestion would actually digest anything, but where cognition is concerned, people are willing to believe in such a miracle because they fail to recognize that the mind is just as much a biological phenomenon as digestion. The mind, they suppose, is something formal and abstract, not a part of the wet and slimy stuff in our heads. The polemical literature in AI usually contains attacks on something the authors call dualism, but what they fail to see is that they themselves display dualism in a strong form, for unless one accepts the idea that the mind is completely independent of the brain or of any other physically specific system, one could not possibly hope to create minds just by designing programs.

Historically, scientific developments in the West that have treated humans as just a part of the ordinary physical, biological order have often been opposed by various rearguard actions. Copernicus and Galileo were opposed because they denied that the earth was the center of the universe; Darwin was opposed because he claimed that humans had descended from the lower animals. It is best to see strong AI as one of the last gasps of this antiscientific tradition, for it denies that there is anything essentially physical and biological about the human mind. The mind according to strong AI is independent of the brain. It is a computer program and as such has no essential connection to any specific hardware.

Many people who have doubts about the psychological significance of AI think that computers might be able to understand Chinese and think about numbers but cannot do the crucially human things, namely—and then follows their favorite human specialty—falling in love, having a sense of humor, feeling the angst of postindustrial society under late capitalism, or whatever. But workers in AI complain—correctly—that this is a case of moving the goalposts. As soon as an AI simulation succeeds, it ceases to be of psychological importance. In this debate both sides fail to see the distinction between simulation and duplication. As far as simulation is concerned, there is no difficulty in programming my computer so that it prints out, "I love you, Suzy"; "Ha ha"; or "I am suffering the angst of postindustrial society under late capitalism." The important point is that simulation is not the same as duplication, and that fact holds as much import for thinking about arithmetic as it does for feeling angst. The point is not that the computer gets only to the 40-yard line and not all the way to the goal line. The computer doesn't even get started. It is not playing that game.

POSTSCRIPT

Will It Be Possible to Build a Computer That Can Think?

Science fiction has played with the idea of "thinking machines" for decades, but is this idea nothing but science fiction? Some scientists do not think so, although they are quick to grant that the technology is not yet nearly ready to produce a convincing example. Still, they are trying, at least in restricted subsets of human intelligence such as game playing. A program called Chinook, for example, is the current world checkers champion (see Ivars Peterson, "Silicon Champions of the Game," *Science News,* August 2, 1997). In February 1996 IBM's "Deep Blue," a chess-playing supercomputer, won and drew games against the human world champion, Garry Kasparov. Although it lost the six-game match, it still demonstrated a skill at something that most people are willing to call "thinking." See Monty Newborn, *Kasparov Versus Deep Blue: Computer Chess Comes of Age* (Springer-Verlag, 1996).

In May 1997 an improved Deep Blue topped its own performance by defeating Kasparov 2–1, with 3 draws, and sent the news media into a frenzy. We are, wrote Charles Krauthammer in *The Weekly Standard,* "creating a new and different form of being. And infinitely more monstrous: creatures sharing our planet who not only imitate and surpass us in logic, who have even achieved consciousness and free will, but are utterly devoid of the kind of feelings and emotions that, literally, humanize human beings. Be afraid." See also Donald Michie, "Slaughter on Seventh Avenue," *New Scientist* (June 7, 1997).

Is chess playing a kind of thinking? When the idea of artificial intelligence was new, workers in the field agreed that it was and set out to achieve it. Even partial success was enough to rouse critics who said that if a machine could do it, it could not be "real" thinking. Deep Blue's complete success, however, seems to have many people *afraid* that chess playing is real thinking and that human primacy in a very fundamental area—in fact, human identity —is now seriously threatened. Moravec, of course, does not feel threatened at all. He develops his ideas at much greater length in his book *Mind Children* (Harvard University Press, 1988) and its successor *Robot, Being: Mere Machine to Transcendent Mind* (Oxford University Press, 1998). In these books, Moravec focuses on the development of motor and sensory apparatus for robots, forecasts the transfer of human minds into immensely capable machines, and speculates on the replacement of biological intelligence by machine intelligence. He also discusses some of the ideas behind the growing field of "artificial life." Also see Steven Levy, *Artificial Life* (Pantheon, 1992).

Moravec's speculations reach far into the future, culminating in a time when humans will exist as streamlined minds residing in "cyberspace," a world simulated within computers, and when the very definition of "reality" must be changed to encompass the new conditions of human life. For an interesting discussion along the same lines, see Frederick Pohl and Hans Moravec, "Souls in Silicon," *Omni* (November 1993).

Not everyone is willing to go as far as Moravec. Searle's essay presented here was paired in the January 1990 issue of *Scientific American* with Paul M. Churchland and Patricia Smith Churchland's "Could a Machine Think?" According to this article, the authors "reject the Turing test as a sufficient condition for conscious intelligence [because it is] very important... that the right sorts of things be going on inside the artificial machine." Unlike Searle, however, the Churchlands believe that true "artificial intelligence, in a nonbiological but massively parallel machine, remains a compelling and discernible prospect."

Searle is by no means alone in objecting to the idea of the mind as a computer program (also known as "strong AI"). Roger Penrose, a renowned physicist and mathematician at the University of Oxford in England, attacks the idea of strong AI vigorously and at length in *The Emperor's New Mind: Concerning Computers, Minds, and the Laws of Physics* (Penguin Books, 1991) and concludes, "Is it not 'obvious' that mere computation cannot evoke pleasure or pain; that it cannot perceive poetry or the beauty of an evening sky or the magic of sounds; that it cannot hope or love or despair; that it cannot have a genuine autonomous purpose?... Perhaps when computations become extraordinarily complicated they can begin to take on the more poetic or subjective qualities that we associate with the term 'mind.' Yet it is hard to avoid an uncomfortable feeling that there must always be something missing from such a picture."

On the Internet . . .

Center for Bioethics: Bioethics Internet Project
The mission of the Center for Bioethics is to advance scholarly and public understanding of ethical, legal, social, and public policy issues in health care.
http://www.med.upenn.edu/~bioethic/

The Foundation for Biomedical Research
The Foundation for Biomedical Research promotes public understanding and support of the ethical use of animals in scientific and medical research.
http://www.fbresearch.org/

The Nature of Wellness
The Nature of Wellness aims to inform the public about the medical and scientific invalidity of animal experimentation and testing and to demonstrate that reliance on animal experimentation and testing is destroying the health care system, the environment, and the economy.
http://home.earthlink.net/~supress/

The Office of Human Radiation Experiments
The Office of Human Radiation Experiments leads the U.S. Department of Energy's efforts to tell the agency's cold war story of radiation research using human subjects.
http://www.ohre.doe.gov/

The Roslin Institute
The Roslin Institute—where Dolly the sheep was cloned—is the major center for research on molecular and quantitative genetics of farm animals and poultry science in the United Kingdom. *http://www.ri.bbsrc.ac.uk/*

PART 5

Ethics

Society's standards of right and wrong have been hammered out over millennia of trial, error, and (sometimes violent) debate. Accordingly, when science and technology offer society new choices to make and new things to do, debates are renewed over whether or not these choices and actions are ethically acceptable. Today there is vigorous debate over such topics as the use of animals in research, performing experiments on human beings, and cloning.

- **Is the Use of Animals in Research Justified?**
- **Is It Ethical to Use Humans as "Experimental Animals"?**
- **Is It Ethically Permissible to Clone Human Beings?**

ISSUE 15

Is the Use of Animals in Research Justified?

YES: Elizabeth Baldwin, from "The Case for Animal Research in Psychology," *Journal of Social Issues* (1993)

NO: Steven Zak, from "Ethics and Animals," *The Atlantic Monthly* (March 1989)

ISSUE SUMMARY

YES: Elizabeth Baldwin, research ethics officer of the American Psychological Association's Science Directorate, argues that animals do not have the same moral rights as humans do, that their use in scientific research is justified by the resulting benefits to both humans and animals, and that their welfare is protected by law.

NO: Research attorney Steven Zak argues that current animal protection laws do not adequately protect animals used in medical and other research and that, for society to be virtuous, it must recognize the rights of animals not to be sacrificed for human needs.

Modern biologists and physicians know a great deal about how the human body works. Some of that knowledge has been gained by studying human cadavers and tissue samples acquired during surgery and through "experiments of nature" (strokes, for example, have taught a great deal about what the various parts of the brain do; extensive injuries from car accidents and wars have also been edifying). Some knowledge of human biology has also been gained from experiments on humans, such as when brain surgery patients agree to let their surgeons stimulate different parts of their brains electrically while the brains are exposed or when cancer patients agree to try experimental treatments.

The key word here is *agree.* Today it is widely accepted that people have the right to consent or not to consent to whatever is done to them in the name of research or treatment. In fact, society has determined that research done on humans without their free and informed consent is a form of scientific misconduct. However, this standard does not apply to animals, experimentation on which has produced the most knowledge of the human body.

Although animals have been used in research for at least the last 2,000 years, during most of that time, physicians who thought they had a workable treatment for some illness commonly tried it on their patients before they had any idea whether or not it worked or was even safe. Many patients, of course, died during these untested treatments. In the mid-nineteenth century, the French physiologist Claude Bernard argued that it was sensible to try such treatments first on animals to avoid some human suffering and death. No one then questioned whether or not human lives were more valuable than animal lives.

Today millions of animals are used in research. Geneticists generally study fruit flies, roundworms, and zebra fish. Physiologists study mammals, mostly mice and rats but also rabbits, cats, dogs, pigs, sheep, goats, monkeys, and chimpanzees. Experimental animals are often kept in confined quarters, cut open, infected with disease organisms, fed unhealthy diets, and injected with assorted chemicals. Sometimes the animals suffer. Sometimes the animals die. And sometimes they are healed, albeit often of diseases or injuries induced by the researchers in the first place.

Not surprisingly, some observers have reacted with extreme sympathy and have called for better treatment of animals used in research. This "animal welfare" movement has, in turn, spawned the more extreme "animal rights" movement, which asserts that animals—especially mammals—have rights as important and as deserving of regard as those of humans. In its most extreme form, this movement insists that animals are persons in every moral sense. Thus, to kill an animal, whether for research, food, or fur, is the moral equivalent of murder.

This attitude has led to important reforms in the treatment of animals and to the development of several alternatives to using animals (see Alan M. Goldberg and John M. Frazier, "Alternatives to Animals in Toxicity Testing," *Scientific American,* August 1989). However, it has also led to hysterical objections to in-class animal dissections, terrorist attacks on laboratories, the destruction of research records, and the theft of research materials (including animals). In 1989 an undersecretary of the Department of Health and Human Services, in attacking the animal rights movement, said, "We must not permit a handful of extremists to deprive millions of the life-sustaining and life-enhancing fruits of biomedical research."

In the following selection, Elizabeth Baldwin argues in the same vein: animals are of immense value with regard to medical, veterinary, and psychological research, but they do not have the same moral rights as humans. Our obligation, she maintains, is not to treat them as persons but to treat them humanely, and there is a sufficient number of laws and regulations to ensure that this is done.

In opposition, Steven Zak, who has written numerous articles on animals with regard to ethics and the law, argues that morality requires society to recognize the right of animals not to be made to suffer at all for the benefit of humans. Therefore, researchers should always find alternative modes of research.

YES

Elizabeth Baldwin

THE CASE FOR ANIMAL RESEARCH IN PSYCHOLOGY

Animal liberationists do not separate out the human animal. A rat is a pig is a dog is a boy.

—Ingrid Newkirk, Director, People for the Ethical Treatment of Animals.

The shock value of this quote has made it a favorite of those defending the use of animals in research. It succinctly states the core belief of many animal rights activists who oppose the use of animals in research. Although some activists work for improved laboratory conditions for research animals, recent surveys suggest that most activists would like to eliminate animal research entirely (Plous, 1991). These activists believe animals have rights equal to humans and therefore should not be used as subjects in laboratory research.

The debate over animal research can be confusing unless one understands the very different goals of animal welfare organizations and animal rights groups. People concerned with animal welfare seek to improve laboratory conditions for research animals and to reduce the number of animals needed. These mainstream goals encompass traditional concerns for the humane treatment of animals, and most researchers share these goals. In contrast, the views of animal rights activists are *not* mainstream, since there are few people who would agree with the above quote from Ingrid Newkirk. Indeed, in a national poll conducted by the National Science Foundation, half the respondents answered the following question affirmatively: "Should scientists be allowed to do research that causes pain and injury to animals like dogs and chimpanzees if it produces new information about human health problems?" (National Science Board, 1991). These findings are particularly impressive given the explicit mention of "pain and injury" to popular animals such as dogs and chimpanzees. My own position is that animals do not have rights in the same sense that humans do, but that people have a responsibility to ensure the humane treatment of animals under their care. Animals have played a pivotal role in improving the human condition, and in return, society should strive to treat them well.

From Elizabeth Baldwin, "The Case for Animal Research in Psychology," *Journal of Social Issues*, vol. 49, no. 1 (1993). References omitted.

BACKGROUND

The modern animal rights movement is intellectual and spiritual heir to the Victorian antivivisection movement in Britain (Sperling, 1988). This 19th-century movement was a powerful force in Britain and arose in part from accelerating changes brought about by science and technology (and the resulting challenges to the prevailing view of humanity's relationship to nature).

The British movement peaked in 1876 with the passage of the Cruelty to Animals Act. This compromise legislation required licenses for conducting animal research, but recognized the societal value of continuing to use animals in research. It was about this time that the scientific community began to organize a defense of animal research. Several challenges to animal research were made in the ensuing 20 years, but in the end, the medical and scientific community were able to successfully protect their interests. The Victorian antivivisection movement, however, did bring about the regulation of research and helped prevent outright abuse (Sperling, 1988).

The beginning of the modern animal rights movement is generally dated to the 1975 publication of *Animal Liberation* by philosopher Peter Singer. Although Singer himself is not an advocate of animal "rights," he provided the groundwork for later arguments that animals have rights—including the right not to be used in research. Most animal rights activists believe animals have a right not to be used for research, food, entertainment, and a variety of other purposes. An inordinate amount of attention is devoted to animal research, however, even though far fewer animals are used for research than for other purposes (Nicoll & Russell, 1990).

There has been a phenomenal growth in the animal rights movement since the publication of Singer's book. People for the Ethical Treatment of Animals (PETA), the leading animal rights organization in the United States, has grown from 18 members in 1981 to more than 250,000 members in 1990. (McCabe, 1990). By any standard, the animal rights movement is a force to be reckoned with.

PHILOSOPHICAL ISSUES

There are two basic philosophies that support the animal rights movement, although activists are often unable to articulate them (Sperling, 1988). These two positions are summarized by Herzog (1990) as the *utilitarian* argument and the *rights* argument.

The utilitarian position is that the greatest good is achieved by maximizing pleasure and happiness, and by minimizing suffering and pain. Although traditionally applied only to humans, Singer argues that animals should be included when considering the greatest good. He states, "No matter what the nature of the being, the principle of equality requires that its suffering be counted equally with the like suffering—insofar as rough comparisons can be made—of any other being" (Singer, 1990, p. 8). Utilitarians would thus argue that animals have an interest equal to that of humans in avoiding pain and suffering, and should therefore not be used in experiments that could cause them harm. Two problems with this philosophy are that (1) it is hard to draw a line between creatures that suffer and creatures that do not, and (2) the argument does not address *qualitative* differences in pain and pleasure across species (Herzog, 1990).

The rights position states that animals possess certain rights based on their inherent value. This philosophy, first developed by Tom Regan (1983), argues that animals have a right not to be used by humans in research (and for many other purposes). Major problems with this position arise in deciding just what rights are and in determining who is entitled to hold them (Herzog, 1990).

While the above positions have been developed relatively recently, the alternative view of animals as qualitatively different from humans has a long history in Judeo-Christian thought. Traditionally, humans were believed to have been created in the image of God and to have dominion over animals. Robb (1988) uses this perspective in arguing that humans are unique by virtue of their capacity for moral choice. Because of this capacity, humans can be held responsible for their choices, and can therefore enter into contractual agreements with binding rights and responsibilities for *both* parties. Robb acknowledges that some animals have human capacities in certain areas, but he argues that this does not make them morally equal to humans or give them rights that take precedence over human needs.

The most persuasive argument for using animals in behavioral research, however, is the untold benefit that accrues to both humans and animals. The benefits of behavioral research with animals have been enumerated by such authors as Miller (1985) and King and Yarbrough (1985), and for most people, these benefits are the reason that they support the continued use of animals in research. This argument—which is basically utilitarian—is the one most often cited by the research community in defense of animal research. In contrast to Singer's utilitarianism, however, animals are not given the same degree of consideration as people.

In conclusion, both sides in the animal rights debate have philosophical underpinnings to support their position, but what often emerges in the rhetoric is not reasoned debate but emotion-laden charges and personal attacks. This is not surprising, given the strong passions aroused in the discussion.

FRAMING THE DEBATE

In the 1980s, activists targeted certain researchers or areas of research that they viewed as vulnerable to attack, and researchers were forced to assume a defensive posture. Unfortunately, activists were right about the vulnerability of individual scientists; little or no institutional defense was mounted against these early attacks. The prevailing attitude was to ignore the activists in hopes that they would go away, and thus attract less attention from the public and the press. This passivity left the early targets of animal rights activists in the position of a man asked, "Why do you beat your wife?" No matter how researchers responded, they sounded defensive and self-serving. It took several years for the research community to realize that animal rights activists were not going away, and that the activists' charges needed to be answered in a systematic and serious manner.

This early failure on the part of the research community to communicate its position effectively left the public with little information beyond what was provided by the animal rights activists. Framing the debate is half the battle, and the research community was left playing catch-up and answering the question, "Why do you abuse your research animals?"

The research community also faced the daunting task of explaining the use of animals in research to a public whose understanding of the scientific method was almost nil. The most difficult misconception to correct was the belief that every research project with animals should produce "useful" results (Orem, 1990). Social scientists who have received Senator William Proxmire's "Golden Fleece Award" are well aware of this line of thinking—a line of thinking that displays a complete misunderstanding of how science works, and ignores the vast amount of basic research that typically precedes each "useful" discovery.

It is difficult for scientific rationales to compete with shocking posters, catchy slogans, and soundbites from the animal rights movement. The most effective response from the scientific community has been to point out innumerable health advances made possible by the use of animals as research models. This approach is something that most people can relate to, since everyone has benefited from these advances.

The early defensive posture of scientists also failed to allay public concerns about the ability of researchers to self-regulate their care and use of research animals. Unlike the participation of humans in research (who are usually able to speak in their own defense and give consent), there seemed to be no one in the system able to "speak" for the animals. Or so people were encouraged to believe by animal rights activists. As discussed below, there are elaborate federal regulations on the use of animals in research, as well as state laws and professional guidelines on the care and use of animals in research.

RESTORING TRUST

Scientists, research institutions, and federal research agencies finally came to realize that the charges being leveled by animal rights activists needed to be publicly—and forcefully—rebutted. Dr. Frederick Goodwin, former Administrator of the Alcohol, Drug Abuse, and Mental Health Administration (ADAMHA), was one of the first federal officials to defend animal research publicly, and point out the difference between animal welfare and animal rights (Booth, 1989). Recently, many more federal officials and respected researchers have publicly spoken on the importance of animal research (Mervis, 1990).

Countering Misinformation

Animal rights literature often uses misleading images to depict animal research—images such as animals grimacing as they are shocked with electricity. These descriptions lead readers to believe animals are routinely subjected to high voltage shocks capable of producing convulsions (e.g., Singer, 1990, pp. 42–45). Such propaganda is far from the truth. In most cases, electric shock (when used at all) is relatively mild—similar to what one might feel from the discharge of static electricity on a cold, dry day. Even this relatively mild use of shock is carefully reviewed by Institutional Animal Care and Use Committees before being approved, and researchers must demonstrate that alternate techniques are not feasible. Stronger shock *is* used in animal research, but it is used to study medical problems such as epilepsy (a convulsive disorder). It is also used to test the effectiveness and side effects of drugs developed to control such disorders. It is not within the scope of this article to refute the myriad charges issued against animal research in general, specific projects, and individual researchers. Suffice it to say that such allegations have been persuasively refuted (Coile & Miller, 1984; Feeney, 1987; Johnson, 1990; McCabe, 1986).

Benefits to Animals

Animal rights activists often fail to appreciate the many benefits to animals that have resulted from animal research. Behavioral research has contributed to improvements in the environments of captive animals, including those used in research (Novak & Petto, 1991). The list of benefits also includes a host of veterinary procedures and the development of vaccines for deadly diseases such as rabies, Lyme disease, and feline leukemia. Research in reproductive biology and captive breeding programs are also the only hope for some animals on the brink of extinction (King et al., 1988).

Regulations and Guidelines

It is clear that many people concerned about the use of animals in research are not aware of the elaborate structure that exists to regulate the care and use of animals in research. This system includes federal regulations under the Animal Welfare Act (U.S. Department of Agriculture, 1989, 1990, 1991), Public Health Service (PHS) policy (Office for Protection from Research Risks, 1986), and state laws that govern the availability of pound animals for research.

The Animal Welfare Act, most recently amended in 1985, is enforced by the USDA's Animal and Plant Health Inspection Service (APHIS). The regulations connected with this law include 127 pages of guidelines governing the use of animals in research. It also includes unannounced inspections of animal research facilities by APHIS inspectors who do nothing but inspect research facilities. Their inspections are conducted to ensure compliance with regulations that include everything from cage size, feeding schedules, and lighting to exercise requirements for dogs and the promotion of psychological well-being among nonhuman primates.

In addition to APHIS inspectors who make unannounced inspections of animal research facilities, there are local Institutional Animal Care and Use Committees (IACUCs) that review each proposed research project using animals. Research proposals must include a justification for the species used and the number of animals required, an assurance that a thorough literature review has been conducted (to prevent unnecessary replication of research), and a consideration of alternatives if available. IACUCs are also responsible for inspecting local animal research facilities to check for continued compliance with state protocols.

Each grant proposal received by a PHS agency (National Institutes of Health, and the Centers for Disease Control) that proposes using animals must contain an assurance that it has been reviewed by an IACUC and been approved. IACUCs must have no less than five members and contain at least one veterinarian, one practicing scientist experienced in research involving animals, one member who is primarily concerned in nonscientific matters (e.g., a lawyer or ethicist), and one member who is not affiliated with the institution in any way and is not an immediate family member of anyone affiliated with the institution (Office for Protection from Research Risks, 1986; USDA, 1989).

Beyond federal animal welfare regulations, PHS policy, and the PHS Guidelines (National Research Council, 1985), there are professional guidelines for the care and use of research animals. Examples include the American Psychological Association's (APA) *Ethical Principles of Psychologists* (1990) and *Guidelines for Ethical Conduct in the Care and Use of Animals* (1993), and the Society for Neuroscience's Handbook (Society for Neuroscience, 1991).

The APA also has a Committee on Animal Research and Ethics (CARE) whose charge includes the responsibility to "review the ethics of animal experimentation and recommend guidelines for the ethical conduct of research, and appropriate care of animals in research." CARE wrote the APA's *Guidelines for Ethical Conduct in the Care and Use of Animals*, and periodically reviews it and makes revisions. These guidelines are widely used by psychologists and other scientists, and have been used in teaching research ethics at the undergraduate and graduate level. The APA's Science Directorate provided support for a conference on psychological well-being of nonhuman primates used in research, and published a volume of proceedings from that conference (Novak & Petto, 1991). The APA also helps promote research on animal welfare by membership in and support for such organizations as the American Association for the Accreditation of Laboratory Animal Care (AAALAC).

AAALAC is the only accrediting body recognized by the PHS, and sets the "gold standard" for animal research facilities. To receive AAALAC accreditation, an institution must go beyond what is required by federal animal welfare regulations and PHS policy. AAALAC accreditation is highly regarded, and those institutions that receive it serve as models for the rest of the research community.

Even with all these safeguards in place, some critics question the ability of the research community to self-regulate its use of animals in research. The system can only be considered self-regulating, however, if one assumes that researchers, institutional officials, members of IACUCs (which must include a member not affiliated with the institution), USDA inspectors, animal care and lab technicians, and veterinarians have identical interests. These are the individuals with the most direct access to the animals used in research, and these are the specialists most knowledgeable about the conditions under which animals are used in research.

In several states, animal rights activists have succeeded in gaining access to IACUC meetings where animal research proposals are discussed. On the whole, however, research institutions have fought—and are still fighting—to keep these meetings closed to the general public. There is a very real fear among researchers that information gleaned from such meetings will be used to harass and target individual researchers. Given the escalating nature of illegal break-ins by such organizations as the Animal Liberation Front, this is a legitimate concern. Indeed, on some campuses "reward posters" offer money to individuals who report the abuse of research animals.

Even though IACUC meetings are generally closed to the public, the elaborate system regulating animal research is by no means a closed one. The most recent animal welfare regulations were finalized after five years of proposals recorded in the *Federal Register;* comments from the public, research institutions, professional associations, animal welfare groups, and animal rights groups; the incorporation of these comments; republication of the revised rules; and so forth. Neither researchers nor animal rights groups were entirely pleased with the final document, but everyone had their say. Although certain elements of the regulatory system rely on researchers, it is hard to imagine a workable system that would fail to use their expertise. The unspoken assumption that researchers cannot be trusted to care for their research animals is not supported by the records of APHIS inspections. Good science demands good laboratory animal care, and it is in a researcher's best interest to ensure that laboratory animals are well cared for.

The Benefits of Behavioral Research With Animals

The use of animals in psychological and behavioral research was an early target of animal rights activists. This research was perceived as a more vulnerable target than biomedical research, which had more direct and easily explained links to specific human health benefits. Psychological and behavioral research also lacked the powerful backing of the medical establishment (Archer, 1986).

There is, of course, a long list of benefits derived from psychological research with animals. These include rehabilitation of persons suffering from stroke, head injury, spinal cord injury, and Alzheimer's disease; improved communication with severely retarded children; methods for the early detection of eye disorders in children (allowing preventive treatment to avoid permanent impairment); control of chronic anxiety without the use of drugs; and improved treatments for alcoholism, obesity, substance abuse, hypertension, chronic migraine headaches, lower back pain, and insomnia (Miller, 1985). Behavioral research with nonhuman primates also permits the investigation of complex behaviors such as social organization, aggression, learning and memory, communication, and growth and development (King et al., 1988).

The nature of psychological and behavioral research makes the development and use of alternatives difficult. It is the behavior of the whole organism, and the interaction among various body systems, that is examined. Computer models may be used, but "research with animals will still be needed to provide basic data for writing computer software, as well as to prove the validity and reliability of computer alternatives" (U.S. Congress, Office of Technology Assessment, 1986). The alternative of using nonliving systems may be possible with epidemiologic data bases for some behavioral research, but chemical and physical systems are not useful for modeling complex behaviors Likewise, in vitro cultures of organs, tissues, and cells do not display the characteristics studied by psychologists.

CONCLUSION

Research psychologists have been asked to eschew emotionalism, and bring logic and reason to the debate over animal research (Bowd, 1990). This is certainly the style most researchers are comfortable with—yet they have also been advised to quit trying to "apply logic and reason in their responses [to animal rights activists]" (Culliton, 1991). Culliton warns that while "animal rights people go for the heart, the biologists go for the head" and are losing the public in the process.

Which path is best? A reasoned approach draws high marks for civility, but will it help scientists in their trench warfare with animal rights activists?

Do animals have rights that preclude their use in laboratory research? I, and the psychologists I help represent, would say no. But researchers do have responsibilities to the animals they use in their research. These responsibilities include ensuring the humane care of their research animals, using the minimum number of animals necessary, and seeing to it that all laboratory assistants are adequately trained and supervised. As stated in the APA's *Ethical Principles*, "Laws and regulations notwithstanding, an animal's immediate protection depends upon the scientist's own conscience" (APA, 1990).

Researchers and others concerned with animal welfare can engage in a useful dialogue as standards of care and use evolve. This dialogue has proven fruitless with animal rights activists, though, since they seem unwilling to compromise or consider other viewpoints. What is the middle ground for a discussion with someone whose goal is the elimination of all research on animals?

The collective decision society has made is that the benefits derived from animal research far outweigh the costs. As public opinion polls indicate, most people are willing to accept these costs but want assurances that animals are humanely cared for. Yes, I'm "speciesist" in the eyes of Ingrid Newkirk—I will never believe my son is a dog is a pig is a rat.

NO

Steven Zak

ETHICS AND ANIMALS

In December of 1986 members of an "animal-liberation" group called True Friends broke into the Sema, Inc., laboratories in Rockville, Maryland, and took four baby chimpanzees from among the facility's 600 primates. The four animals, part of a group of thirty being used in hepatitis research, had been housed individually in "isolettes"—small stainless-steel chambers with sealed glass doors. A videotape produced by True Friends shows other primates that remained behind. Some sit behind glass on wire floors, staring blankly. One rocks endlessly, banging violently against the side of his cage. Another lies dead on his cage's floor.

The "liberation" action attracted widespread media attention to Sema, which is a contractor for the National Institutes of Health [NIH], the federal agency that funds most of the animal research in this country. Subsequently the NIH conducted an investigation into conditions at the lab and concluded that the use of isolettes is justified to prevent the spread of disease among infected animals. For members of True Friends and other animal-rights groups, however, such a scientific justification is irrelevant to what they see as a moral wrong; these activists remain frustrated over conditions at the laboratory. This conflict between the NIH and animal-rights groups mirrors the tension between animal researchers and animal-rights advocates generally. The researchers' position is that their use of animals is necessary to advance human health care and that liberation actions waste precious resources and impede the progress of science and medicine. The animal-rights advocates' position is that animal research is an ethical travesty that justifies extraordinary, and even illegal, measures.

The Sema action is part of a series that numbers some six dozen to date and that began, in 1979, with a raid on the New York University Medical Center, in which members of a group known as the Animal Liberation Front (ALF) took a cat and two guinea pigs. The trend toward civil disobedience is growing. For example, last April members of animal-rights groups demonstrated at research institutions across the country (and in other countries, including Great Britain and Japan), sometimes blocking entrances to them by forming

From Steven Zak, "Ethics and Animals," *The Atlantic Monthly* (March 1989). Copyright © 1989 by Steven Zak. Reprinted by permission.

human chains. In the United States more than 130 activists were arrested, for offenses ranging from blocking a doorway and trespassing to burglary.

To judge by everything from talk-show programs to booming membership enrollment in animal-rights groups (U.S. membership in all groups is estimated at 10 million), the American public is increasingly receptive to the animal-rights position. Even some researchers admit that raids by groups like True Friends and the ALF have exposed egregious conditions in particular labs and have been the catalyst for needed reforms in the law. But many members of animal-rights groups feel that the recent reforms do not go nearly far enough. Through dramatic animal-liberation actions and similar tactics, they hope to force what they fear is a complacent public to confront a difficult philosophical issue: whether animals, who are known to have feelings and psychological lives, ought to be treated as mere instruments of science and other human endeavors....

Animal-rights activists feel acute frustration over a number of issues, including hunting and trapping, the destruction of animals' natural habits, and the raising of animals for food. But for now the ALF considers animal research the most powerful symbol of human dominion over and exploitation of animals, and it devotes most of its energies to that issue. The public has been ambivalent, sometimes cheering the ALF on, at other times denouncing the group as "hooligans." However one chooses to characterize the ALF, it and other groups like it hold an uncompromising "rights view" of ethics toward animals. The rights view distinguishes the animal-protection movement of today from that of the past and is the source of the movement's radicalism.

"THEY ALL HAVE A RIGHT TO LIVE"

Early animal-protection advocates and groups... seldom talked about rights. They condemned cruelty—that is, acts that produce or reveal bad character. In early-nineteenth-century England campaigners against the popular sport of bull-baiting argued that it "fostered every bad and barbarous principle of our nature." Modern activists have abandoned the argument that cruelty is demeaning to human character ("virtue thought") in favor of the idea that the lives of animals have intrinsic value ("rights thought"). Rights thought doesn't necessarily preclude the consideration of virtue, but it mandates that the measure of virtue be the foreseeable consequences to others of one's acts.

"Michele" is thirty-five and works in a bank in the East. She has participated in many of the major ALF actions in the United States. One of the missions involved freeing rats, and she is scornful of the idea that rats aren't worth the effort. "These animals feel pain just like dogs, but abusing them doesn't arouse constituents' ire, so they don't get the same consideration. They all have a right to live their lives. Cuteness should not be a factor."

While most people would agree that animals should not be tortured, there is no consensus about animals' right to live (or, more precisely, their right not to be killed). Even if one can argue, as the British cleric Humphrey Primatt did in 1776, that "pain is pain, whether it be inflicted on man or on beast," it is more difficult to argue that the life of, say, a dog is qualitatively the same as that of a human being. To this, many animal-rights activists would say that every morally relevant characteristic that is lacking in all animals (rationality might be one, according to some ways of defining that term) is also lacking in some "marginal" human beings, such as infants, or the senile, or the severely retarded. Therefore, the activists argue, if marginal human beings have the right to live, it is arbitrary to hold that animals do not. Opponents of this point of view often focus on the differences between animals and "normal" human beings, asserting, for instance, that unlike most human adults, animals do not live by moral rules and therefore are not part of the human "moral community."

The credibility of the animal-rights viewpoint, however, need not stand or fall with the "marginal human beings" argument. Lives don't have to be qualitatively the same to be worthy of equal respect. One's perception that another life has value comes as much from an appreciation of its uniqueness as from the recognition that it has characteristics that are shared by one's own life. (Who would compare the life of a whale to that of a marginal human being?) One can imagine that the lives of various kinds of animals differ radically, even as a result of having dissimilar bodies and environments—that being an octopus feels different from being an orangutan or an oriole. The orangutan cannot be redescribed as the octopus minus, or plus, this or that mental characteristic; conceptually, nothing could be added to or taken from the octopus that would make it the equivalent of the oriole. Likewise, animals are not simply rudimentary human beings, God's false steps, made before He finally got it right with us.

Recognizing differences, however, puts one on tentative moral ground. It is easy to argue that likes ought to be treated alike. Differences bring problems: How do we think about things that are unlike? Against what do we measure and evaluate them? What combinations of likeness and difference lead to what sorts of moral consideration? Such problems may seem unmanageable, and yet in a human context we routinely face ones similar in kind if not quite in degree: our ethics must account for dissimilarities between men and women, citizens and aliens, the autonomous and the helpless, the fully developed and the merely potential, such as children or fetuses. We never solve these problems with finality, but we confront them....

Both advocates and opponents of animal rights also invoke utilitarianism in support of their points of view. Utilitarianism holds that an act or practice is measured by adding up the good and the bad consequences—classically, pleasure and pain—and seeing which come out ahead. There are those who would exclude animals from moral consideration on the grounds that the benefits of exploiting them outweigh the harm. Ironically, though, it was utilitarianism, first formulated by Jeremy Bentham in the eighteenth century, that brought animals squarely into the realm of moral consideration. If an act or practice has good and bad consequences for animals, then these must be entered into the moral arithmetic. And the calculation must be genuinely disinterested. One may not baldly assert that one's own interests count for more. Animal researchers may truly believe that they are impartially weighing all interests when they conclude that human interests overwhelm those of animals. But a skeptical reader will seldom be persuaded that they are in fact doing so....

Even true utilitarianism is incomplete, though, without taking account of rights. For example, suppose a small group of aboriginal tribespeople were captured and bred for experiments that would benefit millions of other people by, say, resulting in more crash-worthy cars. Would the use of such people be morally acceptable? Surely it would not, and that point illustrates an important function of rights thought: to put limits on what can be done to individuals, even for the good of the many. Rights thought dictates that we cannot kill one rights-holder to save another—or even more than one other—whether or not the life of the former is "different" from that of the latter.

Those who seek to justify the exploitation of animals often claim that it comes down to a choice: kill an animal or allow a human being to die. But this claim is misleading, because a choice so posed has already been made. The very act of considering the taking of life X to save life Y reduces X to the status of a mere instrument. Consider the problem in a purely human context. Imagine that if Joe doesn't get a new kidney he will die. Sam, the only known potential donor with a properly matching kidney, himself has only one kidney and has not consented to give it—and his life—up for Joe. Is there really a choice? If the only way to save Joe is to kill Sam, then we would be unable to do so—and no one would say that we chose Sam over Joe. Such a choice would never even be contemplated.

In another kind of situation there *is* a choice. Imagine that Joe and Sam both need a kidney to survive, but we have only one in our kidney bank. It may be that we should give the kidney to Joe, a member of our community, rather than to Sam, who lives in some distant country (though this is far from clear—maybe flipping a coin would be more fair). Sam (or the loser of the coin flip) could not complain that his rights had been violated, because moral claims to some resource—positive claims—must always be dependent on the availability of that resource. But the right not to be treated as if one were a mere resource or instrument—negative, defensive claims—is most fundamentally what it means to say that one has rights. And this is what members of the ALF have in mind when they declare that animals, like human beings, have rights.

Where, one might wonder, should the line be drawn? Must we treat dragonflies the same as dolphins? Surely not. Distinctions must be made, though to judge definitively which animals must be ruled out as holders of rights may be impossible even in principle. In legal or moral discourse we are virtually never able to draw clear lines. This does not mean that drawing a line anywhere, arbitrarily, is as good as drawing one anywhere else.

The line-drawing metaphor, though, implies classifying entities in a binary way: as either above the line, and so entitled to moral consideration, or not. Binary thinking misses nuances of our moral intuition. Entities without rights may still deserve moral consideration on other grounds: one may think that a dragonfly doesn't quite qualify for rights yet believe that it would be wrong to crush one without good reason. And not all entities with rights need be treated in precisely the same way. This is apparent when one compares animals over whom we have assumed custody with wild animals. The former, I think, have rights to our affirmative aid, while the latter have such rights only in certain circumstances. Similar distinctions can be made among human beings, and also between human beings and particular animals. For example, I recently spent $1,000 on medical care for my dog, and I think he had a right to that care, but I have never given such an amount to a needy person on the street. Rights thought, then, implies neither that moral consideration ought to be extended only to the holders of rights nor that all rights-holders must be treated with a rigid equality. It implies only that rights-holders should never be treated as if they, or their kind, didn't matter.

ANIMALS, REFRIGERATORS, AND CAN OPENERS

The question of man's relationship with animals goes back at least to Aristotle, who granted that animals have certain senses—hunger, thirst, a sense of touch—but who held that they lack rationality and therefore as "the lower sort [they] are by nature slaves, and... should be under the rule of a master." Seven centuries later Saint Augustine added the authority of the Church, arguing that "Christ himself [teaches] that to refrain from the killing of animals... is the height of superstition, for there are no common rights between us and the beasts...." Early in the seventeenth century René Descartes argued that, lacking language, animals cannot have thoughts or souls and thus are machines.

One may be inclined to dismiss such beliefs as archaic oddities, but even today some people act as if animals were unfeeling things. I worked in a research lab for several summers during college, and I remember that it was a natural tendency to lose all empathy with one's animal subjects. My supervisor seemed actually to delight in swinging rats around by their tails and flinging them against a concrete wall as a way of stunning the animals before killing them. Rats and rabbits, to those who injected, weighed, and dissected them, were little different from cultures in a petri dish: they were just things to manipulate and observe. Feelings of what may have been moral revulsion were taken for squeamishness, and for most of my lab mates those feelings subsided with time.

The first animal-welfare law in the United States, passed in New York State in 1828, emphasized the protection of animals useful in agriculture. It also promoted human virtue with a ban on "maliciously and cruelly" beating or torturing horses, sheep, or cattle. Today courts still tend to focus on human character, ruling against human beings only for perpetrating the most shocking and senseless abuse of animals....

Most states leave the regulation of medical research to Washington. In 1966 Congress passed the Laboratory Animal Welfare Act, whose stated purpose was not only to provide humane care for animals but also to protect the owners of dogs and cats from theft by proscribing the use of stolen animals. (Note the vocabulary of property law; animals have long been legally classified as property.) Congress then passed the Animal Welfare Act [AWA] of 1970, which expanded the provisions of the 1966 act to include more species of animals and to regulate more people who handle animals. The AWA was further amended in 1976 and in 1985.

The current version of the AWA mandates that research institutions meet certain minimum requirements for the handling and the housing of animals, and requires the "appropriate" use of painkillers. But the act does not regulate research or experimentation itself, and allows researchers to withhold anesthetics or tranquilizers "when scientifically necessary." Further, while the act purports to regulate dealers who buy animals at auctions and other markets to sell to laboratories, it does little to protect those animals....

The 1985 amendments to the AWA were an attempt to improve the treatment of animals in laboratories, to improve enforcement, to encourage the consideration of alternative research methods that use fewer or no animals, and to minimize duplication in experiments. One notable change is that for the first time, research institutions using primates must keep them in environments conducive to their psychological well-being; however, some animal-rights activists have expressed skepticism, since the social and psychological needs of primates are complex, and the primary concern of researchers is not the interests of their animal subjects. Last September [1988] a symposium on the psychological well-being of captive primates was held at Harvard University. Some participants contended that we lack data on the needs of the thirty to forty species of primates now used in laboratories. Others suggested that the benefits of companionship and social life are obvious.

The U.S. Department of Agriculture is responsible for promulgating regulations under the AWA and enforcing the law. Under current USDA regulations the cages of primates need only have floor space equal to three times the area occupied by the animal "when standing on four feet"—in the words of the USDA, which has apparently forgotten that primates have hands. The 1985 amendments required the USDA to publish final revised regulations, including regulations on the well-being of primates, by December of 1986. At this writing the department has yet to comply, and some activists charge that the NIH and the Office of Management and Budget have delayed the publication of the new regulations and attempted to undermine them.

One may believe that virtue thought—which underlies current law—and rights thought should protect animals equally. After all, wouldn't a virtuous person or society respect the interests of animals? But virtue thought allows the law to disregard these interests, because virtue can be measured by at least two yardsticks: by the foreseeable effects of an act on the interests of an animal or by the social utility of the act. The latter standard was applied in a 1983 case in Maryland in which a researcher appealed his conviction for cruelty to animals after he had performed experiments that resulted in monkeys' mutilating their hands. Overturning the conviction, the Maryland Court of Appeals wrote that "there are certain normal human activities to which the infliction of pain to an animal is purely incidental"—thus the actor is not a sadist—and that the state legislature had intended for these activities to be exempt from the law protecting animals.

The law, of course, is not monolithic. Some judges have expressed great sympathy for animals. On the whole, though, the law doesn't recognize animal rights. Under the Uniform Commercial Code, for instance, animals—along with refrigerators and can openers—constitute "goods."

ALTERNATIVES TO US-VERSUS-THEM

Estimates of the number of animals used each year in laboratories in the United States range from 17 million to 100 million: 200,000 dogs, 50,000 cats, 60,000 primates, 1.5 million guinea pigs, hamsters, and rabbits, 200,000 wild animals, thousands of farm animals and birds, and millions of rats and mice. The conditions in general—lack of exercise, isolation from other animals, lengthy confinement in tiny cages—are stressful. Many experiments are painful or produce fear, anxiety, or depression. For instance, in 1987 researchers at the Armed Forces Radiobiology Research Institute reported that nine monkeys were subjected to whole-body irradiation; as a result, within two hours six of the monkeys were vomiting and hypersalivating. In a proposed experiment at the University of Washington pregnant monkeys, kept in isolation, will be infected with the simian AIDS virus; their offspring, infected or not, will be separated from the mothers at birth.

Not all animals in laboratories, of course, are subjects of medical research. In the United States each year some 10 million animals are used in testing products and for other commercial purposes. For instance, the United States Surgical Corporation, in Norwalk, Connecticut, uses hundreds of dogs each year to train salesmen in the use of the company's surgical staple gun. In 1981 and 1982 a group called Friends of Animals brought two lawsuits against United States Surgical to halt these practices. The company successfully argued in court that Friends of Animals lacked "standing" to sue, since no member of the organization had been injured by the practice; after some further legal maneuvering by Friends of Animals both suits were dropped. Last November [1988] a New York City animal-rights advocate was arrested as she planted a bomb outside United States Surgical's headquarters.

In 1987, according to the USDA, 130,373 animals were subjected to pain or distress unrelieved by drugs for "the purpose of research or testing." This figure, which represents nearly seven percent of the 1,969,123 animals reported to the USDA that year as having been "used in experimentation," ignores members of species not protected by the AWA (cold-blooded animals, mice, rats, birds, and farm animals). Moreover, there is reason to believe that the USDA's figures are low. For example, according to the USDA, no primates were subjected to distress in the state of Maryland, the home of Sema, in any year from 1980 to 1987, the last year for which data are available.

Steps seemingly favorable to animals have been taken in recent years. In addition to the passage of the 1985 amendments to the AWA, the Public Health Service [PHS], which includes the NIH, has revised its "Policy on Humane Care and Use of Laboratory Animals," and new legislation has given legal force to much of this policy. Under the revised policy, institutions receiving NIH or other PHS funds for animal research must have an "institutional animal care and use committee" consisting of at least five members, including one nonscientist and one person not affiliated with the institution.

Many activists are pessimistic about these changes, however. They argue that the NIH has suspended funds at noncompliant research institutions only in response to political pressure, and assert that the suspensions are intended as a token gesture, to help the NIH regain lost credibility. They note that Sema, which continues to keep primates in isolation cages (as regulations permit), is an NIH contractor whose principal investigators are NIH employees. As to the makeup of the animal-care committees, animal-rights advocates say that researchers control who is appointed to them. In the words of one activist, "The brethren get to choose."

However one interprets these changes, much remains the same. For example, the AWA authorizes the USDA to confiscate animals from laboratories not in compliance with regulations, but only if the animal "is no longer required... to carry out the research, test or experiment"; the PHS policy mandates pain relief "unless the procedure is justified for scientific reasons." Fundamentally, the underlying attitude that animals may appropriately be used and discarded persists.

If the law is ever to reflect the idea that animals have rights, more-drastic steps—such as extending the protection of the Constitution to animals—must be taken. Constitutional protection for animals is not an outlandish proposition. The late U.S. Supreme Court Justice William O. Douglas wrote once, in a dissenting opinion, that the day should come when "all of the forms of life... will stand before the court—the pileated woodpecker as well as the coyote and bear, the lemmings as well as the trout in the streams."

Suppose, just suppose, that the AWA were replaced by an animal-rights act, which would prohibit the use by human beings of any animals to their detriment. What would be the effect on medical research, education, and product testing? Microorganisms; tissue, organ, and cell cultures; physical and chemical systems that mimic biological functions; computer programs and mathematical models that simulate biological interactions; epidemiologic data bases; and clinical studies have all been used to reduce the number of animals used in experiments, demonstrations, and tests. A 1988 study by the National Research Council, while finding that researchers lack the means to replace all animals in labs, did conclude that current and prospective alternative techniques could reduce the number of animals—particularly mammals—used in research.

Perhaps the report would have been more optimistic if scientists were as zealous about conducting research to find alternatives as they are about animal research. But we should not be misled by discussions of alternatives into thinking that the issue is merely empirical. It is broader than just whether subject A and procedure X can be replaced by surrogates B and Y. We could undergo a shift in world view: instead of imagining that we have a divine mandate to dominate and make use of everything else in the universe, we could have a sense of belonging to the world and of kinship with the other creatures in it. The us-versus-them thinking that weighs animal suffering against human gain could give way to an appreciation that "us" includes "them." That's an alternative too.

Some researchers may insist that scientists should not be constrained in their quest for knowledge, but this is a romantic notion of scientific freedom that never was and should not be. Science is always constrained, by economic and social priorities and by ethics. Sometimes, paradoxically, it is also freed by these constraints, because a barrier in one direction forces it to cut another path, in an area that might have remained unexplored.

Barriers against the exploitation of animals ought to be erected in the law, because law not only enforces morality but defines it. Until the law protects the interests of animals, the animal-rights movement will by definition be radical. And whether or not one approves of breaking the law to remedy its shortcomings, one can expect such activities to continue. "I believe that you should do for others as you would have done for you," one member of the ALF says. "If you were being used in painful experiments, you'd want someone to come to your rescue."

POSTSCRIPT

Is the Use of Animals in Research Justified?

Much debate about the lethal experiments that were conducted on nonconsenting human subjects by the Nazis during World War II, as well as the ensuing trials of the Nazi physicians in Nuremburg, Germany, has established a consensus that no scientist can treat people the way the Nazis did. Informed consent is essential, and research on humans must aim to benefit those same humans.

As these ideas have gained currency, some people have tried to extend them to say that, just as scientists cannot do whatever they wish to humans, they cannot do whatever they wish to animals. Harriet Ritvo, in "Toward a More Peaceable Kingdom," *Technology Review* (April 1992), says that the animal rights movement "challenges the ideology of science itself... forcing experimenters to recognize that they are not necessarily carrying out an independent exercise in the pursuit of truth—that their enterprise, in its intellectual as well as its social and financial dimensions, is circumscribed and defined by the culture of which it is an integral part." The result is a continuing debate, driven by the periodic discovery of researchers who seem quite callous (at least to the layperson's eye) in their treatment of animals (see Kathy Snow Guillermo, *Monkey Business: The Disturbing Case That Launched the Animal Rights Movement,* National Press, 1993) and by the charge that animal rights advocates are misanthropes who just do not understand nature (see Richard Conniff, "Fuzzy-Wuzzy Thinking About Animal Rights," *Audubon,* November 1990).

In the February 1997 issue of *Scientific American,* Andrew N. Rowan presents a debate entitled "The Benefits and Ethics of Animal Research." The opposing articles are Neal D. Barnard and Stephen R. Kaufman, "Animal Research Is Wasteful and Misleading" and Jack H. Botting and Adrian R. Morrison, "Animal Research Is Vital to Medicine." In addition, staff writer Madhusree Mukerjee contributed "Trends in Animal Research." Among books that are pertinent to this issue are Michael P. T. Leahy, *Against Liberation: Putting Animals in Perspective* (Routledge, 1992); Lorenz Otto Lutherer and Margaret Sheffield Simon, *Targeted: The Anatomy of an Animal Rights Attack* (University of Oklahoma Press, 1992); F. Barbara Orlans, *In the Name of Science: Issues in Responsible Animal Experimentation* (Oxford University Press, 1993); Peter Singer, *Animal Liberation,* rev. ed. (Avon Books, 1990); Rod Strand and Patti Strand, *The Hijacking of the Humane Movement* (Doral, 1993); and Deborah Blum, *The Monkey Wars* (Oxford University Press, 1994).

ISSUE 16

Is It Ethical to Use Humans as "Experimental Animals"?

YES: Charles Petit, from "Sunday Interview: A Soldier in the War on AIDS," *San Francisco Chronicle* (January 21, 1996)

NO: Jonathan D. Moreno, from "The Dilemmas of Experimenting on People," *Technology Review* (July 1997)

ISSUE SUMMARY

YES: Science writer Charles Petit interviews an AIDS patient who underwent a highly experimental treatment and considers those who resist human experimentation to be far too cautious.

NO: Bioethicist Jonathan D. Moreno argues that although the requirements of informed consent may pose difficulties for research on human beings, "the simplicity and intuitive force of the ideas articulated in the Nuremberg Code ensure their lasting moral relevance."

Biological and medical experimentation on human beings has a bad reputation. This is partly because of what came to light after World War II: Under the Nazi regime, German researchers had, in the name of science, used prisoners to study amputation, healing, infection, and hypothermia. Many of the subjects died during the experiments. The researchers, however, did not ask their subjects for consent, nor did they supply painkillers or try to put the pieces back together afterward. To them, human prisoners were as disposable as lab rats.

The reputation of psychological experimentation also suffered when—in an effort to learn how people could follow authority, even if it meant committing atrocities against another individual (as the Nazis did during the Holocaust, and as Americans did during the Vietnam War)—social psychologist Stanley Milgram devised an experiment in which subjects were told they were the teachers in an experiment on the effects of punishment on memory. The subject's job was to give the learner electric shocks of increasing intensity each time the learner made an error. However, the "learner" was an actor who purposely gave wrong answers, and the shocks were not real, but the subjects did not know that. Although some of the subjects balked partway through the experiment, an appalling number continued to increase the strength of the shocks, even though the "learner" was in obvious agony. Critics of Milgram's research (and research like it) have objected that the essential role that deception plays in such research means that true informed consent is not possible. Critics also object to the psychological harm that such experiments may cause to subjects by showing them things about themselves that they would rather not know (such as a capacity for cruelty).

American medical researchers have also done some apparently very cruel experiments on human subjects. Consider, for instance, the Tuskegee syphilis project discussed by bioethicist Jonathan D. Moreno in the second of the following selections. In 1932 researchers began to study 400 Tuskegee, Alabama, black men who were infected with syphilis to learn about the progression of the disease. When antibiotics—which can cure syphilis very quickly—became available in the 1940s, the experimenters denied them to their subjects. Studying syphilis was evidently more important to the researchers than saving the lives of their subjects, and it remained so until the media revealed the project in 1972. Public outcry soon resulted in the project's termination and in the 1974 National Research Act, which calls for institutional review boards to approve all federally funded research on human beings.

The radiation experiments performed by the Department of Defense (DoD) are also alarming. In this case, during and after World War II, the DoD exposed terminal patients, retarded children, and others to plutonium without their full knowledge or consent. The DoD felt that the research was so essential that ethical considerations were secondary (see Jonathan D. Moreno, " 'The Only Feasible Means': The Pentagon's Ambivalent Relationship With the Nuremberg Code," *Hastings Center Report,* September–October 1996). After the experiments were revealed, a national ethics commission declared that little damage was done but admitted that the ethical cost was high (see Danielle Gordon, "The Verdict: No Harm, No Foul," *Bulletin of the Atomic Scientists,* January/February 1996). In 1996 President Bill Clinton issued a public apology and offered $4.8 million to a dozen families of subjects of the experiments as reparations.

The existence of the National Research Act implies that human experimentation can be done under suitable circumstances. Those circumstances require informed consent and concern for the subjects' welfare, among other things. To many ethicists, informed consent requires freedom from duress or pressure; it cannot be obtained from prisoners, the retarded, or terminal patients who, in desperate hope of a cure, may be willing to consent to anything. Those desperate patients, however, may have a very different view. Jeff Getty, the subject of the following interview by Charles Petit, sees himself as a soldier in the war against AIDS and, in submitting to a highly controversial experimental treatment, says, "I wanted a chance to charge out of my foxhole and fight."

The other side is apparent in Moreno's essay on the role of informed consent, without duress, in modern medical research. He argues that although the requirements of informed consent may pose difficulties for research on human beings, "the simplicity and intuitive force of the ideas articulated in the Nuremberg Code ensure their lasting moral relevance."

YES

Charles Petit

SUNDAY INTERVIEW: A SOLDIER IN THE WAR ON AIDS

On December 14 [1995] Jeff Getty of Oakland made medical history when a purified mixture of baboon bone marrow cells was infused into his bloodstream in an effort to slow or reverse his advanced AIDS. He hopes the cells from the baboon—an animal naturally resistant to HIV—will migrate to his bone marrow and plant the seeds for a strong immune system able to augment his AIDS-ravaged capacity to fight off infection. The procedure is one of the first attempts to install a nonhuman immune system in a person—two others involving less sophisticated methods have failed—and has inspired extensive debate and controversy. Performed at San Francisco General Hospital, the procedure was approved by the Food and Drug Administration only after protracted hearings. Some critics argue that the experiment's architects, Dr. Suzanne Ildstad of the University of Pittsburgh and Dr. Steven Deeks and Dr. Paul Volberding of the University of California at San Francisco, had not done enough basic research with animals and other methods to justify the risks. Possible hazards include dangerous reactions between the baboon cells and Getty's own tissues, or the incubation in his body of baboon viruses that could infect other people. Some fear that new human diseases are inevitable if animal-to-human transplants become common. Backers of the procedure reply that extreme cases justify risky and unproven medical experiments, and that extensive safeguards are in place to prevent the procedure from creating new diseases. Getty left the hospital after three weeks. He suffered no apparent harm and seems to have recovered from radiation and chemotherapy that temporarily stunned his own immune system to increase chances for the baboon cells to take root. His condition will be monitored closely for months. This week a sample of his bone marrow was to be removed and tested to see if the baboon cells are engrafting. Those test results will take several weeks. This interview was conducted on January 10.

Q: How do you feel?
A: I feel better now than I did before the procedure.

A couple of things have happened that were rather surprising. Since about 1992 I had lost a lot of my sense of smell and taste. We don't know why—possibly because of the radiation or the chemotherapy—but they have returned. I can smell and taste things. And I had some severe asthma since 1992 or so. That has been at least temporarily arrested. I can breathe again.

Given that those two components of my life have been returned and I haven't been feeling sickly and my appetite has been tremendous, I would say life seems to be a little better than before. We have speculated it might be the radiation or the chemotherapy that did this.

I understand I am the first person with late-stage AIDS symptoms to have undergone a pretty effective immune suppression from drugs and radiation and had my immune system come back. This seems to have improved some of my AIDS symptoms, contrary to what several people thought was going to happen. This sort of leads us down another road of research. Everyone involved in this project is kind of pixilated because we are learning so much already. Things we expected were going to happen didn't, and things that weren't expected, did.

Q: This transplant got worldwide attention. What do you have to say to people who wrote you, mostly with good wishes but perhaps some who were not so sympathetic?

A: Well, there were only three (hostile) letters out of hundreds and hundreds of letters of praise and support, mostly from relatives of people who died of AIDS. To those (latter) people, I thank them. I had no idea we were going to touch such a nerve in this country. I was telling a friend today that it seems like a lot of people were feeling hopelessness and despair. There was a pent-up need for somebody to do something, to do anything. If I can say anything to them, it's that your letters and cards, I read them every day. They really got me through the insanity of the confinement. And to the people who sent me the three ugly hate letters, well, it just reminded me that there is an element of evil out there.

Q: What about the procedure itself? Were you frightened? Was there any sensation?

A: The actual procedure took 37 minutes. I was fully conscious and there were several people staring at me in the room. I was quite frightened inside, although people said I didn't show it. I knew there was a chance I could have been killed right at that moment because of what they call an anaphylactic reaction.

I had a funny taste in my mouth almost immediately. And my head felt as if it was rushing, a strange thing, like when I got bonged on the head when I was a kid. I noticed the faster they let the marrow go in the more I got those sensations, so they slowed that process down. The taste? It tasted like blood.

Q: How did your health and feeling progress during the three weeks you were in the hospital?

A: I started out getting beat up by the radiation and the chemotherapy. My face swelled up like a balloon for several days. Then, I thought I was coming out of it and started eating like a horse, thinking I was on my road to recovery.

What I didn't realize at the time was that the lining of my stomach and my intestines was about to die from the radiation. I experienced that wonderful phenomenon at days five and six after the radiation. I had some really bad days in there. I was throwing up and unable to eat and feeling really unsure about what was going to happen next. Then I was specifically told by Lloyd Damon, a doctor and bone marrow specialist at the University of California, not to worry and that was normal.

Once I knew it was supposed to happen I was fine. It took about another five days before I could get to eating well again. From then on it was just boredom and trying to maintain my privacy.

Q: What other treatments have you undergone?

A: You name it. I have had this disease most of my adult life. I started out really aggressive, doing isoprinosine and ribavirin that I was smuggling in from Mexico in pinatas, till I got caught at it. Then I joined a buyers club and got drugs from there.

I did the egg lipid stuff that everybody was doing in 1989. That was weird, drinking egg lipid. I went from there to experimenting with something called thymic hormones. Then I injected myself with Compound Q in 1989, in the buttocks. You are not supposed to do it that way. It is supposed to be a slow infusion over two hours. Kenny (Getty's partner, Ken Klueh) just shot it right in my butt to see what was going to happen. At the time, we didn't know what to do with it. I suffered severely from that. I got necrosis—two large holes that formed where the skin fell in and the muscle dissolved. I ended up on Channel 7 news. The story was, "Please don't do what Jeff Getty did."

Then, in 1991 I started researching what is called allotransfers, or transferring whole white blood cells from a sibling who is not closely HLA (tissue type) matched. I got the University of California to let me try that at Mount Zion Hospital.

My sister donated. At the time I was suffering acute pseudomonas and cryptococcus (infections). I did that three times over the next year. I cleared cryptococcus and the pseudomonas.

Q: In all your public appearances you seem upbeat. Have you had times when you felt like giving up?

A: I think that we all experience that. I don't stay in those black periods very long. A friend of mine, right before he died in 1986, gave me this flag to fly on my sailboat. It says "Don't Give Up the Ship." And that has really become a theme in my life.

The flag theme comes from the War of 1812 where a guy named (John) Lawrence went out to meet the British, and the British just killed everybody. They didn't take any prisoners. Lawrence's dying words were "Don't give up the ship." Then (Oliver Hazard) Perry went out shortly after that on Lake Erie, and he was outnumbered too. Perry said, "Remember what happened to Lawrence," and whipped out this flag that said "Don't Give Up the Ship." They didn't, and they beat the British.

The take-home lesson from that is the same as AIDS. If you think AIDS takes prisoners, you are fooling yourself. That has been my theme all the way through this thing. As long as I keep fighting, I'll stay alive. So far, it has worked.

Q: How did you learn about the bone marrow transplant?

A: I followed Project Inform's Project Immune Restoration, looking over their shoulder all the time. Then Marty (Project Inform director Martin Delaney) told me the details of it. From that point I started a file, which I often do, to track it, and started gathering as many scientific journal articles as I could around the subject of xenogeneic (cross-species) research.

I built a pretty substantial file. It took me about six months. In the summer of '94 I wrote to Dr. Suzanne Ildstad of the University of Pittsburgh and volunteered myself. I said I was a soldier in the front lines of AIDS, someone willing to die for the cause if necessary, and to describe my credentials as someone willing to take chances, I reiterated my treatment history. I was pretty emotional. I said I was sitting in a foxhole, watching shells landing in my friends' foxholes and watching them die one at a time. I wanted a chance to charge out of my foxhole and fight. I have a rather military approach to this disease.

Q: When did you realize you were really a good candidate?

A: In the fall of '94, I finally got hold of the protocol. I realized I qualified. It was almost amazing. The protocol required that I be cytomegalovirus or CMV positive, and cryptococcus negative. Before I did my sister's blood, I was CMV negative, and cryptococcus positive. That turned that right around. It was fate. They wanted me to be CMV positive because baboons are almost all CMV positive.

Q: Some people call you an AIDS warrior. You call yourself an AIDS activist. What does that mean?

A: I do think I have a take-no-prisoners policy in terms of people who are trying to obstruct AIDS research, particularly in the government, and are trying to delay things because they are afraid they are going to lose their jobs. Instead of talking about it, I get out there and take action. We at ACT UP [AIDS Coalition to Unleash Power] Golden Gate believe that action is how you survive and talk is cheap.

Q: Talk about ACT UP Golden Gate.

A: When I first came there I remember thinking, boy, I don't agree with these guys' politics at all. But the first night I spent in a room with those activists, it was like love at first sight. They were just like me. I had finally found my home. The people I started out with, they have all passed away now, but they were great activists.

Now, I would say, I am extremely active working on several projects at the same time. The purpose, our theme, has been to bring forward promising AIDS treatments as fast as possible to gain access to them any way that we can for people who are dying of AIDS. We are trying to do everything we can to save people's lives. We feel that we have done that. We are prolonging lives.

Q: Some researchers felt more basic work should have been done to justify this experiment on a human. Do you appreciate the reasons for the cautions and concerns that led to the long delay?

A: I studied the research very carefully, and I think that (delay) was nonsense. We believe that most of the people who did not want this research to go forward were covering their ass, especially in the government. If you let something like this go through and something goes terribly wrong, there goes your career, your job. But if you stall it until everybody agrees that it is safe, then you'll never get fired for that. Unfortunately, we die waiting.

What I am saying is that the obstacles and the delays that they put in the way of this kind of research are unreasonable. They do not understand the urgency of this disease. And then they will tell me, "Oh, but Jeff, my very good friends have died of AIDS, I understand your pain." That is a standard line that really upsets us. We can almost predict when they are going to say it. We wait for that line. They should know better than to say that.

Q: Some researchers say use of animal tissue for human transplant carries a risk of creating new human diseases. Does this risk worry you?
A: No, not in any way. This particular animal was extremely clean. I would think that in a lot of ways I would be at more risk taking a transfusion from the blood bank, from something like hepatitis C, or from taking blood from my sisters, than from this animal. The tests on this donor far exceeded anything the U.S. blood supply would get. Then there are the other endogenous, possible, hypothetical, unknown viruses that could appear. Well, I am not going to let the fear of hypothetical risks stop me from going forward.
Q: The animal that provided the marrow was killed. What do you say to animal rights groups that want to make use of animals in medical science more difficult or illegal?
A: I think they are making a really big mistake on this one. They are trying to pit animal rights against AIDS, and they are going to lose. We had to sacrifice this one baboon for this experiment for a couple of reasons. I feel really bad about the fact that the baboon had to be sacrificed, and in the future we very well might not have to (kill the baboon donors) if this is ever repeated.

I would sacrifice my own cat if I thought it would help cure this disease, and I was willing to sacrifice my own life. The baboon, by the way, did have a name. It was Raccoon. I feel worse about the hundreds of friends I have lost to AIDS. I have a phone book in which the words "rest in peace" are written on every page. On every page, someone has died.
Q: It appears that the procedure did you no harm. But if it does not help, will it have been worth it?
A: I already feel I have accomplished my goal, which is to do this thing. I don't expect this to work. This is a real long shot. Just to do this, after they said, "No, you can't," is a victory for people with AIDS everywhere in the world.
Q: Being HIV-positive brings an obligation to practice safe sex. Are there any additional limitations on your behavior due to this transplant?
A: They have asked me not to engage in any contact sex at all in which any fluids are exchanged, such as kissing, until we know I have no baboon viruses, and I have agreed to that. There really are not any other restrictions except not to share toothbrushes, razors, water glasses and things like that. It is the same way you would treat someone with hepatitis.
Q: These are anxious times for you. What do you do for enjoyment?
A: I love to take my boat out sailing on the bay. My boat is Mariah, a 30-foot Pearson sloop. For enjoyment I like to go out with my friends, play music, do art work and some stuff like that.
Q: If this doesn't work, is there anything else you can do besides wait?
A: I am always on the prowl for something new. One thing I always tell people with HIV is that when you make your plans, always have a backup plan. So I am looking into the next thing that is out there. For instance, there is something called a thymus transplant, and we are watching that very closely.

People with any stage AIDS, even late-stage AIDS, should remember that as long as you continue to fight and make plans and take action, you may survive. Some of our friends and I have been, at our lowest points, totally wasted with diseases that we were told would kill us for sure. Well, I am still here.

NO

Jonathan D. Moreno

THE DILEMMAS OF EXPERIMENTING ON PEOPLE

Fifty years ago..., the trial of 23 Nazi doctors and medical scientists for performing cruel and inhuman experiments on concentration camp inmates led to the creation of the Nuremberg Code, a milestone in the history of medical ethics. The first line of the code, "The voluntary consent of the human subject is absolutely essential," is generally regarded as the sine qua non for the ethical conduct of research. During the past year, institutions throughout the United States and Europe have been sponsoring events to celebrate the Nuremberg Code as a bulwark of human decency in the pursuit of scientific knowledge.

Although the ideals it embodies are now viewed as unassailable, the code was initially greeted by medical scientists as poorly conceived and unrealistic. For decades, it only sporadically influenced research ethics in policy or in practice; many doctors and scientists resisted applying the principle of informed consent to their own work. The code's uneven influence can be attributed to the extreme circumstances of its origin, the culture of medicine at the time, and the broad phrasing its authors employed. Like so many ethical maxims ("Love thy neighbor as thyself"), the principle of voluntary informed consent seems uncomplicated. Yet 50 years after it was first articulated, we are still struggling to live up to it.

TO DO MORE THAN HAND DOWN JUDGMENTS

The war crimes trial of the Nazi doctors that led to the code was held in Nuremberg, West Germany, from December 1946 to August 1947. Nuremberg was chosen partly for symbolic reasons, for it was there that the Nazi Party held giant, theatrical rallies designed both to impress those faithful to the Reich and to intimidate those who opposed it.

The doctors' trial began a few months after the conclusion of proceedings against two dozen leaders of the Third Reich, including Hermann Goering, Rudolf Hess, and Joachim von Ribbentrop. Although the American forces occupying Germany had not at first planned to conduct an inquest on human

From Jonathan D. Moreno, "The Dilemmas of Experimenting on People," *Technology Review* (July 1997).

experimentation, their decision changed as information emerged about the medical atrocities committed in the concentration camps. The details of what prosecutors called "the medical case" so shocked them that they decided to pursue the matter as a war crime under the charter of the international tribunal.

Medicine had a central place in the Nazi enterprise, for the Nazis believed doctors had a special role in improving the "Volk." Jews, Gypsies, homosexuals, the mentally retarded, and others were singled out as corrupting influences in the German national body, much like bacteria invading the individual. The view that these groups constituted a kind of public health menace implied an instrumental role for the medical profession in the business of "diagnosing" and "treating" the problem.

Although many doctors were involved in the Nazis' racial hygiene policies—and nearly half of German doctors were Nazi party members—those who had access to concentration camp inmates for research purposes had to be well connected with the Nazi political hierarchy. Although we are not accustomed to thinking of the Nazi doctors in the mundane terms of careerism, part of their motivation was typical academic ambition. These scientists wanted to be among the first to make the medical breakthroughs that would advance the military goals of the Third Reich and make them heroes of racial medicine.

The special role given medical science in the Third Reich created an excellent opportunity for a few influential researchers to avail themselves of experimental subjects they could not have obtained under other conditions. The fact that most concentration camp inmates were eventually slated to die helped doctors rationalize their use as research subjects. The urgency of the war effort and the endorsement of the highest state authorities further encouraged these scientists to perform human-subjects research on problems of pressing concern on the battlefield.

One of these was the most efficacious way to thaw Luftwaffe fliers forced to bail out over the frigid waters of the North Sea. To test various thawing techniques Nazi researchers exposed a number of prisoners to freezing conditions and experimented with various methods of reviving them. Other experiments for military purposes included forcing subjects to drink only seawater to determine how long pilots could survive once downed in the ocean and establishing the point at which lungs exploded due to atmospheric pressures, an important issue for fighter pilots seeking to avoid antiaircraft fire. An estimated 100,000 human beings died horrible deaths in the course of experiments at Auschwitz, Buchenwald, Dachau, Sachsenhausen, and other camps.

The brief against the defendants was delivered on December 9, 1946, by chief prosecutor Telford Taylor. In his opening statement, Taylor declared that the men were on trial for "murders, tortures, and other atrocities committed in the name of medical science." But the prosecutors soon discovered that the case raised issues that were more problematic than they had realized—among them the lack of internationally recognized codes or medical ethics by which the behavior of the Nazi doctors could be judged. Nonetheless, nearly eight months later, after harrowing testimony about the experiments by surviving victims, the Nuremberg judges sent seven of the defendants to their deaths and sentenced eight more to lengthy prison terms. (None of those who were imprisoned served a full sentence, and many went on to distinguished careers in postwar Germany.)

The three-judge panel decided that it needed to do more than simply hand down the judgments. The members decided to codify the rules they believed should govern the use of human beings in all medical research. The Nuremberg Code begins: "The voluntary consent of the human subject is absolutely essential. This means that the person involved should... be so situated as to be able to exercise free power of choice, without the intervention of any element of force, deceit, duress, over-reaching, or other ulterior form of constraint or coercion." The code also included provisions requiring that the scientific importance of the question be manifest, that risks to the subjects be kept to a minimum, and that prior experimentation be performed on animals.

Despite its powerful moral influence, the code carried no legal authority. No mechanisms were created to enforce it. In fact, the very circumstances that gave the code its high moral standing—the horrors that surrounded its origins—partly account for its relative lack of influence in the postwar years: ordinary researchers found it hard to believe that the code need be applied to their own work.

EXTRAORDINARY CIRCUMSTANCES

In the early years after Nuremberg, revulsion at the entire death-camp phenomenon was so great that it was difficult for many to see how the Nuremberg Code could apply to normal conditions. As the Yale psychiatrist and law professor Jay Katz put it, the medical community's general attitude was, "It was a good ethics code for barbarians." But not necessarily for everyone else. Confident of their sound motivations and humane instincts, mainstream medical practitioners felt they were inoculated against the evil that had infected the Nazi doctors. That their actions might fall into the same ethical category was difficult for most such practitioners to conceive.

The fact that the Nazi experiments occurred during wartime also made the code seem remote from the conduct of peacetime research. Although the Nuremberg judges utterly rejected national-security concerns as a rationale for the medical experiments, ordinary people may have viewed these crimes less as the actions of culpable individuals than as the result of specific national policies. While governments at war might order such extreme and brutal measures, surely few individuals would do such a thing on their own initiative.

In fact, the principle of informed consent was far from entrenched in American medical practice at the time. Most large-scale research on human subjects, spurred by World War II, was conducted by the military on institutionalized "volunteers"—conscientious objectors, prisoners, people confined to mental institutions, or, in rare cases, military personnel. Researchers preferred institutionalized subjects because it was easier to monitor them.

Although it is clear in retrospect that the Nuremberg judges purposely chose to enshrine the principle of voluntary consent in the broadest possible terms, the very sweep of their language made the code seem poorly crafted. According to historical research conducted by the

President's Advisory Committee on Human Radiation Experiments in 1994, what struck many medical researchers as especially unrealistic was the code's seemingly categorical ban on research with any subjects who could not give voluntary informed consent. This appeared to pose an insurmountable obstacle to progress in key areas of medicine such as pediatrics and psychiatry.

The issue of research on children was an especially thorny one. Some of the most important medical research in history has been aimed at conquering childhood diseases, particularly by developing vaccines. Researchers had to experiment on children in order to ensure the effectiveness of the vaccine on its target population and determine the proper dosage. Some of these experiments were of enormous benefit to their subjects as well as to future generations of children. Nonetheless, it was widely recognized that children are not capable of granting consent. The question confronting researchers, then, was whether the authors of the code intended to abandon this whole field of research, even at the cost of saving children's lives.

The by-now-familiar solution was to establish parental permission as the moral equivalent of the child's consent, on the grounds that parents will act in the best interest of the child. Although obtaining parental consent became standard practice by the late 1950s, it was not required by federal regulations until the late 1970s. This resolution extended to the most vulnerable members of society the principal lesson of Nuremberg.

However, not all of society's vulnerable members were equally well protected. Because there were fewer advocates for the mentally ill than for children, the federal government has never enacted regulations specifically targeting the use of psychiatric patients in research. Research on prisoners—another group whose ability to grant consent is compromised—continued through the mid-1970s, when political pressures from a variety of sources forced the federal government to declare a halt on the grounds that consent cannot be truly voluntary in an inherently coercive environment.

THE DOCTOR KNOWS BEST

Physicians in the United States gradually became willing to concede that the use of unconsenting subjects in experiments that could not benefit them was a dangerous encroachment of science on personal privacy. But therapeutic research—studies involving sick patients who might in some way benefit from experimental treatment—was a different story. Doctors jealously guarded the "therapeutic privilege," the right to withhold information from patients.

Although the field of medicine was changing rapidly, the culture of medical practice still operated largely on an old-fashioned, paternalistic model. According to this view, the trusting, nearly sacred relationship between doctor and patient was based on the premise that the doctor knows best; a good patient was defined as a compliant one. Even if the physician's care involved the use of experimental drugs or devices as part of a scientific study, doctors were reluctant to involve patients in making decisions about their own care.

Nothing symbolized doctors' fears of intrusion into their relationships with patients more than consent forms. In the early 1950s consent forms were used in certain exceptional cases (such as invasive procedures that simply could not be performed without a patient's knowledge and cooperation) but not in the vast bulk of therapeutic research. For the most part, the privacy of the doctor-patient relationship was accepted as serving the public interest.

The philosophy of medical paternalism was codified in the 1964 Helsinki Declaration. Partly in response to the consent problems posed by the Nuremberg Code, the World Medical Association had begun deliberations to formulate its own research code in 1953. The resulting document drew a sharp line between therapeutic and nontherapeutic research; doctors were not required to obtain consent for experimental procedures performed on their patients if this requirement was not "consistent with patient psychology." This might apply, for instance, to terminally ill patients who could become depressed and unwilling to undergo further treatment if informed of their prognoses.

Few doctors involved in research in the 1950s paused to reflect on the contradictions between their own roles as caregivers and as researchers—the so-called double-agent problem. For example, from the 1940s through the early 1960s, scientists performed a series of secret, government-sponsored radiation experiments on patients who were hospitalized, institutionalized, or seeking treatment for other conditions (such as pregnancy), often without obtaining the patients' consent. When records of these experiments recently became public, they provoked widespread outrage. At the time, however, there was no mechanism requiring scientists to obtain consent from their subjects. And even though the researchers must have been aware of the existence of the Nuremberg Code, it is doubtful that they felt its provisions applied to research conducted in the context of the doctor-patient relationship.

AN EMERGING CONSENSUS ON CONSENT

Together these factors worked against implementing the consent requirement of the Nuremberg Code in any formal or consistent fashion in the first decade after it was articulated. But in the early 1960s a wave of medical scandals brought the issue of informed consent to the fore. In 1963, a university team conducted a well-publicized but medically doomed effort to transplant a chimpanzee kidney into a human patient—an experiment conducted partly with federal funds, but without prior animal studies or valid scientific justification. The episode heightened concern about the lack of scientific or government oversight of research using human subjects.

A few years later, Harvard anesthesiology professor Henry Beecher, in an article in the *New England Journal of Medicine,* claimed to have found 22 obvious abuses of human subjects in the recent medical literature. In one of these cases, researchers injected live cancer cells into debilitated patients at the Brooklyn Jewish Chronic Disease Hospital without the subjects' knowledge in order to determine whether their immune systems could mount a defense against the cancer. (The subjects were not harmed by the experiment.)

In 1966, the surgeon general announced a policy to govern human-subjects research supported by Public Health Service grants. It required institutions receiving PHS support to create an institutional review board to oversee research involving human subjects and demanded that researchers obtain consent from their subjects.

But if a single event broke the back of medical paternalism in research it was surely the Tuskegee syphilis study. From the early 1930s to the early 1970s, U.S. Public Health Service doctors had studied more than 400 black men with syphilis in Macon County, Ala. The men were not told they had the disease, nor were they offered treatment —even after the discovery of penicillin made treatment much more effective. When a journalist broke the story, a firestorm of outrage swept the country. The requirement for the "voluntary consent of the human subject" had been systematically abused right here in America, in a study that had begun just around the time the Nazis took power in Germany.

The federal government appointed a commission to investigate the scandal in 1972. Recommendations based on its findings, released in 1978, were incorporated into Department of Health and Human Services regulations in 1981 and extended to all federal agencies conducting or sponsoring human-subjects research in 1991. As a result of the Tuskegee scandal, the requirement for voluntary consent in research became deeply etched in the law and in the minds of many who had not seen the need for vigilance before.

THE CODE TODAY

Although it took decades to gain wide acceptance, the Nuremberg Code exerts a profound influence on the conduct of medical research today. An excellent example is the recent ruling by the Food and Drug Administration (FDA) governing research in the emergency room.

Because standard treatments for some conditions, such as head injuries, are ineffective, doctors want to try experimental drugs and devices that they consider promising, based on laboratory work and animal testing. But many emergency room patients suffering from these conditions are unconscious and relatives may not be present. The inability to obtain consent has slowed the pace of vitally needed research.

The FDA ruling permits researchers to enroll patients with life-threatening conditions in certain studies even if they are unable to grant voluntary informed consent, but only if they have informed the community that such studies are going on and that anyone who is admitted to the emergency department with a serious condition could be assigned as a subject. While media accounts have portrayed the FDA decision as a step back from the principle of informed consent, the fact that the agency had to make an explicit exception shows the tenacity with which we now embrace it. The ruling represents one step in the continuing effort to walk the fine line between medical progress and human rights.

The Nuremberg Code also contributed to a sea change in public attitudes toward research, which in turn has spurred debate over access to experimental treatments. It took decades to develop a consensus within the medical community that even ordinary science conducted by well-meaning researchers can lead to unintended harm. Today a system of carefully designed regulations protects desperate people from medical experiments that might only make them worse. In fact, the protections built into the system are so successful that people have become confident of low risks and high benefits if they are granted access to experimental drugs.

In the 1980s, AIDS activists mounted a series of successful campaigns to broaden access to experimental drugs such as AZT. Similarly, women of reproductive age, who were systematically excluded from many drug trials following the thalidomide tragedy of the early 1960s, have sought to participate equally in research in order to capture the potential benefits of experimental treatments and ensure that drugs or devices are developed to meet their needs. For instance, women excrete medications at different rates than men, and since women are more likely than men to use prescription drugs, advocates have argued that they should be better represented in research studies.

Today, concern about informed consent in medical research is greater than ever. In response to the uproar over the Cold War radiation experiments, President Clinton recently announced that henceforth all secret human research conducted in the name of national security would be subject to the rules of informed consent, and that scientists would reveal the names of the sponsoring agencies to participants. What's more, Clinton's advisory commission on bioethics, appointed last year, will undertake a review of current practices and requirements regarding the use of human subjects. In particular, the commission should address ways to ensure that voluntary consent is meaningful.

For instance, there is evidence that seriously ill patients often overestimate the likelihood that they will benefit from experimental treatments. Early trials of a new drug may be designed to determine its effects on the body, such as the rate at which it is excreted—not whether or not it is likely to cure the patient. In fact, most experimental treatments don't work. In order for consent to be truly voluntary, the commission should require that sick patients receive counseling to ensure they understand the implications of research for themselves.

Another issue confronting the commission is whether states should enable people to declare in advance of a serious illness and loss of capacity their willingness to be a research subject in a study that might help them. At least some people who may be candidates for emergency-room research might be willing to make such an advance declaration. This measure would enable researchers to continue to investigate experimental emergency-room treatments without relying so heavily on unconsented research.

The principle of informed consent continues to pose new dilemmas for medical science. It is not always easy to find the balance between human rights and scientific progress. Yet the simplicity and intuitive force of the ideas articulated in the Nuremberg Code ensure their lasting moral relevance.

POSTSCRIPT

Is It Ethical to Use Humans as "Experimental Animals"?

The February 8, 1996, *San Francisco Chronicle* reported that although Jeff Getty is feeling better, "sophisticated tests were unable to detect any clear signs of the baboon bone marrow infused into his system on December 14. Presumably the foreign cells were destroyed by Getty's own defenses or died without multiplying in his own bone marrow." On the other hand, he *is* feeling better. Perhaps the pretransplant radiation treatment "somehow knocked down his HIV infection." "It's intriguing that there may be unexpected benefits from the process. He's given us a number of interesting avenues to explore," said one of his physicians.

Was the experiment worth doing? Both Getty and his physicians think so, but there are ethicists who find the use of seriously ill patients very troubling, especially when the patients cannot be expected to benefit (Getty's improvement was *not* expected). Indeed, the welfare of the patient is considered paramount in modern medical research, which tests drugs and surgical procedures for efficacy and safety generally by giving them to one group of patients while denying them to another (the control group). Today, when an experimental treatment shows strong signs of being more effective than a control treatment, it is immediately offered to the control group. If the treatment shows signs of causing more harm than no treatment at all, the experiment is halted.

Some critics of human experimentation focus on the reason why the experiments are done. For instance, Arjun Makhijani, in "Energy Enters Guilty Plea," *The Bulletin of the Atomic Scientists* (March–April 1994), argues that the government-sponsored human radiation experiments were so closely linked to military purposes, even when their avowed purposes were something else, they were ethically suspect at best.

Others focus on the risks to patients from such things as animal-borne viruses. In March 1996 the Nuffield Council on Bioethics in London, England, released the report *Animal-to-Human Transplants: The Ethics of Transplantation,* which urges that transplants of animal tissues to humans not be approved until the risks of infection are better understood. The council also has moral reservations about using primates (such as baboons) as tissue sources.

Measures such as the National Research Act have not stopped human experimentation, but they have made such work more difficult. It is therefore worth noting that there are alternative ways of studying processes that hold the potential to damage human health. One is "experiments of nature," in which the subjects of study are victims of human accident and neglect, as well as war.

An unfortunate example came to light at the end of the cold war and the collapse of the Iron Curtain, when it was revealed to the world that the Soviet Union and East Germany ignored precautions against exposing uranium miners and processors to radioactive material but kept careful records of worker exposures and any health effects. Now available to researchers is an archive of data on some 450,000 workers, which German science writer Patricia Kahn has called "the world's biggest data collection on low-level radiation and health—and potentially one of the most valuable for studying the associated cancer risks" (see "A Grisly Archive of Key Cancer Data," *Science*, January 22, 1993). Other data collections covering Soviet nuclear accidents are also now becoming available. As researchers analyze these data, they will surely learn a great deal that they could not learn in other ways (the necessary experiments would never be permitted by institutional review boards).

For more details on the Nazi experiments, see A. Mitscherlich and F. Mielke, *Doctors of Infamy* (Henry Schuman, 1949) and Arthur L. Caplan, ed., *When Medicine Went Mad: Bioethics and the Holocaust* (Humana Press, 1992). The U.S. radiation experiments are also discussed by Charles C. Mann in "Radiation: Balancing the Record," *Science* (January 28, 1994). For an analysis of Stanley Milgram's obedience experiments, see his *Obedience to Authority: An Experimental View* (Harper & Row, 1974). And for more on the ethics of experimentation on human beings, see Caplan's "When Evil Intrudes," *Hastings Center Report* (November–December 1992) and Bernard Barber's "The Ethics of Experimentation With Human Subjects," *Scientific American* (February 1976).

Note that the Nazi experiments were part and parcel of Nazi efforts to improve the "Aryan race." For more on this and a caution on repeating the errors of the past, see Garland E. Allen, "Science Misapplied: The Eugenics Age Revisited," *Technology Review* (August/September 1996).

ISSUE 17

Is It Ethically Permissible to Clone Human Beings?

YES: John A. Robertson, from "A Ban on Cloning and Cloning Research Is Unjustified," Statement at the National Bioethics Advisory Commission Meeting, Washington, D.C. (March 13–14, 1997)

NO: Leon R. Kass, from "The Wisdom of Repugnance," *The New Republic* (June 2, 1997)

ISSUE SUMMARY

YES: John A. Robertson, a medical ethics expert, argues that despite the various objections to cloning, "a ban on all human cloning is both imprudent and unjustified."

NO: Biochemist Leon R. Kass argues that human cloning is "so repulsive to contemplate" that it should be prohibited entirely.

In February 1997 Ian Wilmut and Keith H. S. Campbell of the Roslin Institute in Edinburgh, Scotland, announced that they had cloned a sheep by transferring the gene-containing nucleus from a single cell of an adult sheep's mammary gland into an egg cell whose own nucleus had been removed and discarded. The resulting combination cell then developed into an embryo and eventually a lamb in the same way a normal egg cell does after being fertilized with a sperm cell. That lamb, named Dolly, was a genetic duplicate of the ewe from which the udder cell's nucleus was taken. Similar feats had been accomplished years before with fish and frogs, and mammal embryos had previously been split to produce artificial twins. And in March researchers at the Oregon Regional Primate Research Center announced that they had cloned monkeys by using cells from monkey embryos (not adults). In July the Roslin researchers announced the cloning of lambs from fetal cells—this time cells including human genes. But the reactions of the media, politicians, ethicists, and laypeople have been largely negative. Dr. Donald Bruce, director of the Church of Scotland's Society, Religion and Technology Project, for example, has argued at some length about how "nature is not ours to do exactly what we like with."

Many people seem to agree. In 1994 the U.S. National Advisory Board on Ethics in Reproduction called the whole idea of cloning oneself "bizarre... narcissistic and ethically impoverished." Arthur Caplan, director of the Center for Bioethics at the University of Pennsylvania, wonders, "What is the ethical purpose of even trying?" Conservative columnist George Will asks whether humans are now uniquely endangered since "the great given—a human being is the product of the union of a man and a woman—is no longer a given" and "humanity is supposed to be an endless chain, not a series of mirrors."

Others go further. President Bill Clinton asked the National Bioethics Advisory Commission (see its home page at `http://bioethics.gov/`), chaired by Harold T. Shapiro, president of Princeton University, to investigate the implications of this "stunning" research and to issue a final report by the end of May 1998. In his request, Clinton said, "Any discovery that touches upon human creation is not simply a matter of scientific inquiry. It is a matter of morality and spirituality as well." He also barred the use of U.S. funds to support work on cloning humans. Carl Felbaum, president of the Biotechnology Industry Organization, said in a CNN interview that extending the cloning technique to humans "should be prohibited if necessary by law.... This is not a line we want to cross... not even a line we want to approach."

Yet there are other views. Cloning could serve a great many useful purposes, and further development of the technology could lead to much less alarming procedures, such as growing replacement organs within a patient's body. Some of these benefits were considered when George Washington University researchers, using nonviable embryos, demonstrated that single cells could be removed from human embryos and induced to grow into new embryos. If permitted to develop normally, the cells would grow into genetically identical adults. The resulting adults would be duplicates, but only of each other (like identical twins), not of some preexisting adult. In response to this work, John A. Robertson, in "The Question of Human Cloning," *Hastings Center Report* (March–April 1994), argued that the various objections to cloning were not then sufficient to justify banning or restricting the process.

Did Dolly represent something entirely new? For the very first time, it seemed more than science fiction to say it might soon be possible to duplicate an adult human, not just an embryo. But when Robertson spoke at the National Bioethics Advisory Commission conference held in Washington, D.C., March 13–14, 1997, he repeated his position, saying, "At this early stage in the development of mammalian cloning a ban on all human cloning is both imprudent and unjustified. Enough good uses can be imagined that it would be unwise to ban all cloning and cloning research because of vague and highly speculative fears." The following selection is from that speech.

In the second selection, Leon R. Kass argues that people should trust their initial repugnance about human cloning because it threatens important human values, such as the profundity of sex, the sacredness of the human body, and the value of individuality. Human reproduction must not be debased by turning it into mere willful manufacturing. Kass concludes that human cloning is "so repulsive to contemplate" that it should be prohibited entirely.

YES

John A. Robertson

A BAN ON CLONING AND CLONING RESEARCH IS UNJUSTIFIED

The successful cloning of an adult sheep has startled the public in the speed of its arrival, and in the potential it offers to select and control the genome of offspring. The initial reaction has been hostility and repugnance, and a skepticism that anything but abuse and harm could ensue from human cloning. A more considered response would recognize that there are potential benefits to infertile couples and others from human cloning, and that the harms alleged to flow from cloning are too vague and speculative at this point to justify a ban on cloning or on cloning research.

A crucial point is that it is much too early in the development cycle to make global judgments about human cloning, much less to ban all human cloning research or to declare that anyone who clones another is a criminal. It is still unclear whether the initial successes with cloning sheep and primates will be replicated in those or other nonhuman species. Even if they are, cloning by nuclear transfer may not extend easily to humans, or there may be very little demand for any application, much less the applications that stir the public. Rather than rush to judgment with bans that could deter important research in cloning and related areas, it is important that government officials, advisors, and policymakers proceed carefully and fully assess the issues before determining public policy.

An optimal public policy on human cloning would respect human rights and individual freedom and dignity, including scientific freedom. It would permit cloning to occur where substantial benefit to families or patients would result or important individual freedoms are involved. It would limit or restrict it when tangible harm to others is likely. In assessing harm, deviation from traditional methods of reproduction, including genetic selection of offspring characteristics, is not itself a compelling reason for restriction when tangible harm to others is not present. However, moral or symbolic concerns unrelated to actual harm to persons may appropriately be taken into account when determining the types of research and services to be supported by public funds.

From National Bioethics Advisory Commission. Full Committee Meeting. John A. Robertson. "A Ban on Cloning and Cloning Research Is Unjustified." Statement, March 13–14, 1997. Washington, DC: Government Printing Office, 1997.

A rational assessment of cloning would address (1) how cloning relates to current reproductive and genetic practices; (2) possible beneficial uses, and their relation to prevailing conceptions of procreative liberty; and (3) the harms that cloning could produce.

RELATION TO EXISTING PRACTICES

In significant ways cloning is not qualitatively different from prebirth genetic selection techniques that are now in widespread use. Indeed, cloning appears much less intrusive than the ability to alter and manipulate genes that will follow the development of germline gene therapy that is on the near horizon. Because cloning is situated in a web of other genetic selection practices, there is a danger that in legislating or making policy for cloning alone, practices which now are or will become acceptable will also be restricted.

We now engage in a wide variety of practices to control, influence, or select the genes and characteristics of offspring. Most of these techniques involve carrier and prenatal screening and operate in a negative way by avoiding the conception, implantation, or birth of children with particular characteristics. But there is a large amount of active genetic selection, albeit at the gross level, that occurs in choosing mates or gametes for reproduction, or in deciding which embryos or which fetuses will survive and go to term.[1]

Cloning does differ in some ways from existing selection technology. Because it actively seeks to replicate DNA, it involves positive choice rather than negative deselection, as occurs with most other means of genetic selection. In addition, it selects or replicates the entire genome (except for mitochondria), rather than focus on the presence or absence of particular genes. Yet neither of these differences are qualitatively different from the genetic selection that now occurs in reproductive medicine. If cloning does not lead to tangible harm to others, it should be no less legally available than existing practices are.

Furthermore, cloning is much less radical than the gene alteration technologies on the horizon. Cloning enables a child with the genome of another embryo or person to be born. The genome is taken as it is. Genetic alteration, one the other hand, will change the genome of a person who could have been born with their genome intact.

It would be a serious mistake to make policy for cloning without situating it within the range of genetic selection practices that now occur or are likely to occur as gene therapy is perfected. Sorting out the good and the bad uses of genetic selection across the range of situations in which it arises is a complex matter that requires considerably more than ninety days of study.

POSSIBLE BENEFICIAL USES

Since few scientists and physicians had previously considered the prospect of human cloning, and considerable research remains before human cloning by blastomere separation or nuclear substitution is safe and available, the contributions which cloning might make to treatment of infertility and other diseases are still unclear.

Yet several reasonable grounds for seeking to replicate a human genome are easily imagined. Embryos might be cloned to provide an infertile couple with enough embryos to achieve pregnancy. In that case cloning an embryo could lead to the simultaneous or delayed birth of twins. Temporal separation of the birth of the twins is not necessarily harmful, and may lead to a special form of sibling bonding.[2]

There may be other situations of merit, such as creating embryos from which a child may be able to obtain needed organs or tissue, or creating a twin of a previous child who died. Cloning may also enable a couple seeking an embryo donation to choose more precisely the genome of offspring, thus assuring both that the resulting child has a good genetic start in life and the couple a happy rearing experience. In addition, cloning research is likely to generate insights and knowledge about cellular and genetic development generally, with therapeutic applications beyond treating infertility.

In assessing possible beneficial uses of cloning, a major distinction exists between cloning that occurs in the course of IVF [in vitro fertilization] treatment for infertility and cloning to select the genome of a child for rearing.

Cloning of Embryos in the Course of IVF Treatment

There are several reasons why a couple going through IVF might choose to clone embryos, either by blastomere separation or by nuclear transfer. One would be to obtain enough embryos to achieve pregnancy. Another would be to obtain embryos without going through an additional cycle of hormonal stimulation and egg retrieval. A third would be to have a backup supply of tissue or organs or a replacement child if a tragedy befalls the first.

A cloned embryo may be transferred to the uterus at the same time as its source, thus raising the possibility of intentionally created twins. Or the infertile couple could transfer them at a later time, as might occur if the first cycle failed; if it succeeded and the couple wants a second child; or if it succeeded and the first child died or is in need of tissue or organs. The couple could also donate cloned embryos to other infertile couples who are seeking an embryo donation.

An important point is that requests by couples undergoing IVF to clone by embryo splitting or nuclear transfer may fall within their fundamental freedom to decide whether or not to have offspring.[3] If the ability to clone an embryo and transfer it to a uterus is essential to whether the couple has offspring, then cloning should receive the same protection that other forms of assisted reproduction and genetic selection receive. In that case, dislike or repugnance at how a couple is reproducing will not be a sufficient reason to ban the practice. Unless tangible harm to others is likely to occur, the freedom to use noncoital techniques to reproduce is left to the individuals directly involved. If cloning is essential to a couple's reproduction, it should be similarly treated.

Cloning as a Form of Genetic Selection

In addition to enhancing fertility or obtaining a child for rearing, a reason for cloning would be to produce a child that has a healthy genome. Cloning for this purpose—eugenic cloning—would require that an embryo first be created from existing DNA. In that case an egg would have to be obtained, be denucleated, and the DNA from the source cell removed and placed in it. The resulting cell would then have to be made operable and transferred to a uterus, to enable it to come to term. One or more persons would then have to be prepared to rear the resulting child.

Cloning for genetic selection is also closely related to widely accepted practices that now exist. However, eugenic cloning also has the greatest potential for deviation from those practices, and has generated the most bizarre scenarios and fears. Since only a few forms of eugenic cloning closely relate to current practices, there may be little demand for this form of cloning, even if human cloning technology is perfected.

A situation in which cloning for selection or eugenic purposes is closely related to current practices would arise with a couple who need an embryo donation because both lack viable gametes, but the wife has a functioning uterus and wishes to carry a pregnancy. Ordinarily they would be candidates for an embryo donation. Instead of receiving an embryo left over from another infertile couple's attempts at IVF, which may not be healthy or have been adequately screened for infectious disease or genetic factors, they might well prefer that the embryo they gestate and rear be created from the cell of an adult or child with desirable genes. Or they may request a donation of a cloned embryo from another couple. To proceed with cloning in either case, they would need the consent of the clone source or its parents.

The acceptability of this form of cloning depends first on the acceptability of embryo donation itself.[4] Given a general consensus in favor of embryo donation, a second question concerns whether couples should be free to select the embryos that they are willing to gestate and rear. Since we now allow individuals wide choice over the mates they choose and over the gametes used in assisted reproduction, a strong argument can be made for allowing recipients of embryo donation some choice in the embryos they receive. Accordingly, it is plausible to view cloning existing DNA as a reasonable means of embryo selection to assure that a couple seeking an embryo donation will have a healthy child to rear. Strictly speaking, that couple will not be engaged in genetic reproduction, but they will be involved in having a child whom they will gestate and rear, and thus should be treated equivalently to infertile couples who also provide egg or sperm in forming a family.

The most problematic situation of eugenic cloning would arise if the cloning were not designed to produce a healthy child for rearing by loving parents. Scenarios of abuse and narcissism or excessive power involve such cases. They also illustrate that not all cases of human cloning need be treated similarly, for they are not all equivalent in importance or in their impact on the clone source or on the resulting child. Thus cloning of self as a form of genetic selection might pose different problems and deserve different treatment than cloning embryos in order to treat infertility.[5] Policymakers should distinguish among the differing uses, and prohibit only those which pose a threat of serious harm, not those which serve legitimate needs of infertile couples seeking a healthy child to rear.

If a loving family will rear the child, it is difficult to see why cloning for genetic selection (eugenic cloning) is per se unacceptable. We engage in many forms of genetic selection already, most of which are designed to make sure that a child will be healthy and have good chances in life. Eugenic cloning is but another form of genetic selection, and should not be banned on that ground alone.

HARMS

Given that there are potentially beneficial uses of cloning that fall within current practices in assisted reproduction and genetic selection, a ban on all cloning or on all cloning research can be justified only if cloning always or invariably causes great harm to others. Yet opponents of cloning have been very nonspecific and speculative about harms. The most florid critics imagine power-hungry tycoons or dictators narcissistically cloning themselves, or cloning a race of permanent servants or replicants of limited ability, à la Aldous Huxley's *Brave New World* or Ridley Scott's *Bladerunner.* More moderate opponents talk about the importance of having a unique genome, and how cloning might rob children of a unique identity. At the same time, however, they want to deny the importance of genes alone in creating identity.

A more considered view of the potential harms of human cloning must address three issues. The first issue concerns the rights and status of persons born after cloning by embryo splitting or nuclear transfer. In the most likely cloning scenarios, parents will be seeking a child whom they will love for itself. But even in less benign situations, any resulting child would be a person with all the moral and legal rights of persons, and no more would be the property or subject of the person who commissions or carries out the cloning than any other child.

A second issue concerns whether the child will be harmed because it will have the same DNA as another person, either living or dead. Most negative views of cloning assume that the clone will be exactly identical to the clone source, like the multiple copies of a xeroxed document.

But the child who results from cloning will not be the same person as the clone source, even if the two share many physical characteristics, for its rearing environment and experiences will be different.[6] Indeed, religious commentators have noted that such a child will have its own soul. Given the importance of nurture in making us who we are, the danger that the person cloned will be a mere copy or replica is highly fanciful.

A key issue in assessing harm to the child who results from cloning (whether by nuclear transfer or embryo splitting) is that that child would not have existed but for the cloning procedure at issue. Prior to the cloning, it did not exist. It came into being only as a result of cloning. In a crucial sense it has not been harmed because it has no other way to be born but with the DNA chosen for it. Nor can it said to be harmed because its life itself is so full of suffering or confused identity that any existence as a clone is less preferable than nonexistence.

Of course, it might be preferable if parents had had a child whose DNA has not been copied from another. Yet that option will usually not be present in many of the situations involving cloning which couples face. It will be either the clone or no child at all. A policy requiring no child at all would interfere with their procreative liberty. It could not be justified as protecting the child with the DNA of another, for such a child would exist only if the cloning occurred.

A third issue is the need to recognize that the great discomfort with cloning —and calls for its prohibition—may be rooted in the discomfort felt by the deliberate and intentional choice of another's genome that cloning represents. This discomfort arises regardless of whether harm to offspring or families can be shown. The very idea of selecting a child's DNA appears too instrumental or manipulative, and risks treating children as means rather than ends. To prevent such an attitude toward children, some persons would ban all cloning and all cloning research.

There are two problems with this view of harm. One is that it is subjective and personal, reflecting a view about our relationship to offspring that is not universally or even necessarily widely shared. Depending on the needs and purposes which cloning serves, people will vary in their perceptions of whether it instrumentalizes or commodifies children.

A second problem with this view is that it paints with too broad a brush. If taken seriously, this view would condemn all deliberate decisions to have children, whether by current assisted reproductive and genetic screening practices or by coital conception. For infertile couples and couples at risk of offspring with severe genetic disease, the children they seek are truly children of choice and could be said to serve selfish ends. Indeed, the same can be said of most cases of having children. There is no basis for singling out cloning as the most egregious form of instrumentalization of offspring, if it is one at all.

In sum, it is difficult to show actual harm to offspring, families, or society from the cloning scenarios most likely to occur. There may be harm or offense to particular notions of how conception and children should be chosen and born. But such purely moral or symbolic concerns are not a sufficient basis for overcoming procreative choice or banning beneficial uses, even though they may appropriately enter into federal research funding policy.

CONCLUSION

At this early stage in the development of mammalian cloning a ban on all human cloning is both imprudent and unjustified. Enough good uses can be imagined that it would be unwise to ban all cloning and cloning research because of vague and highly speculative fears. Nor need all cases of cloning be treated the same, for they differ in their intent and effects on the clone source and resulting individual. As with other technological innovations, science fiction should not drive science policy.

NOTES

1. JA Robertson, "Genetic Selection of Offspring Characteristics," 76 *Boston University Law Review* 421–482 (1996).
2. JA Robertson, "The Question of Human Cloning," 24 *Hastings Center Report* 6–14 (1994).
3. JA Robertson, *Children of Choice: Freedom and the New Reproductive Technologies* 32–42 (Princeton University Press, 1994).
4. JA Robertson, "Ethical and Legal Issues in Human Embryo Donation," 64 *Fertility and Sterility* 885–895 (1994).
5. Some cases of self-cloning might appear to fall within an infertile couple's attempts to have a child. Further analysis of self-cloning is needed to determine the circumstances, if any, in which it might be acceptable.
6. As Thomas Murray has pointed out, a clone of Mel Gibson might look like Mel Gibson, but he will not be Mel Gibson. Testimony Before U.S. House of Representatives, March 5, 1994.

NO

Leon R. Kass

THE WISDOM OF REPUGNANCE

Our habit of delighting in news of scientific and technological breakthroughs has been sorely challenged by the birth announcement of a sheep named Dolly. Though Dolly shares with previous sheep the "softest clothing, woolly, bright," William Blake's question, "Little Lamb, who made thee?" has for her a radically different answer: Dolly was, quite literally, made. She is the work not of nature or nature's God but of man, an Englishman, Ian Wilmut, and his fellow scientists. What's more, Dolly came into being not only asexually —ironically, just like "He [who] calls Himself a Lamb"—but also as the genetically identical copy (and the perfect incarnation of the form or blueprint) of a mature ewe, of whom she is a clone. This long-awaited yet not quite expected success in cloning a mammal raised immediately the prospect—and the specter—of cloning human beings: "I a child and Thou a lamb," despite our differences, have always been equal candidates for creative making, only now, by means of cloning, we may both spring from the hand of man playing at being God.

After an initial flurry of expert comment and public consternation, with opinion polls showing overwhelming opposition to cloning human beings, President Clinton ordered a ban on all federal support for human cloning research (even though none was being supported) and charged the National Bioethics Advisory Commission to report in ninety days on the ethics of human cloning research. The commission (an eighteen-member panel, evenly balanced between scientists and non-scientists, appointed by the president and reporting to the National Science and Technology Council) invited testimony from scientists, religious thinkers and bioethicists, as well as from the general public. It is now deliberating about what it should recommend, both as a matter of ethics and as a matter of public policy.

Congress is awaiting the commission's report, and is poised to act. Bills to prohibit the use of federal funds for human cloning research have been introduced in the House of Representatives and the Senate; and another bill, in the House, would make it illegal "for any person to use a human somatic cell for the process of producing a human clone." A fateful decision is at hand. To clone or not to clone a human being is no longer an academic question.

... [S]ome cautions are in order and some possible misconceptions need correcting. For a start, cloning is not Xeroxing. As has been reassuringly reiterated, the clone of Mel Gibson, though his genetic double, would enter the world hairless, toothless and peeing in his diapers, just like any other human infant. Moreover, the success rate, at least at first, will probably not be very high: the British transferred 277 adult nuclei into enucleated sheep eggs, and implanted twenty-nine clonal embryos, but they achieved the birth of only one live lamb clone. For this reason, among others, it is unlikely that, at least for now, the practice would be very popular, and there is no immediate worry of mass-scale production of multicopies. The need of repeated surgery to obtain eggs and, more crucially, of numerous borrowed wombs for implantation will surely limit use, as will the expense; besides, almost everyone who is able will doubtless prefer nature's sexier way of conceiving.

Still, for the tens of thousands of people already sustaining over 200 assisted-reproduction clinics in the United States and already availing themselves of in vitro fertilization, intracytoplasmic sperm injection and other techniques of assisted reproduction, cloning would be an option with virtually no added fuss (especially when the success rate improves)....

In anticipation of human cloning, apologists and proponents have already made clear possible uses of the perfected technology, ranging from the sentimental and compassionate to the grandiose. They include: providing a child for an infertile couple; "replacing" a beloved spouse or child who is dying or has died; avoiding the risk of genetic disease; permitting reproduction for homosexual men and lesbians who want nothing sexual to do with the opposite sex; securing a genetically identical source of organs or tissues perfectly suitable for transplantation; getting a child with a genotype of one's own choosing, not excluding oneself; replicating individuals of great genius, talent or beauty—having a child who really could "be like Mike"; and creating large sets of genetically identical humans suitable for research on, for instance, the question of nature versus nurture, or for special missions in peace and war (not excluding espionage), in which using identical humans would be an advantage. Most people who envision the cloning of human beings, of course, want none of these scenarios. That they cannot say why is not surprising. What is surprising, and welcome, is that, in our cynical age, they are saying anything at all.

THE WISDOM OF REPUGNANCE

"Offensive." "Grotesque." "Revolting." "Repugnant." "Repulsive." These are the words most commonly heard regarding the prospect of human cloning. Such reactions come both from the man or woman in the street and from the intellectuals, from believers and atheists, from humanists and scientists. Even Dolly's creator has said he "would find it offensive" to clone a human being.

People are repelled by many aspects of human cloning. They recoil from the prospect of mass production of human beings, with large clones of look-alikes, compromised in their individuality; the idea of father-son or mother-daughter twins; the bizarre prospects of a woman giving birth to and rearing a genetic copy of herself, her spouse or even her deceased father or mother; the grotesqueness of conceiving a child as an exact replacement for another who has died; the utilitarian creation of embryonic genetic duplicates of oneself, to be frozen away or created when necessary, in case of need for homologous tissues or organs for transplantation; the narcissism of those who would clone themselves and the arrogance of others who think they know who deserves to be cloned or which genotype any child-to-be should be thrilled to receive; the Frankensteinian hubris to create human life and increasingly to control its destiny; man playing God. Almost no one finds any of the suggested reasons for human cloning compelling; almost everyone anticipates its possible misuses and abuses. Moreover, many people feel oppressed by the sense that there is probably nothing we can do to prevent it from happening. This makes the prospect all the more revolting.

* * *

Revulsion is not an argument; and some of yesterday's repugnances are today calmly accepted—though, one must add, not always for the better. In crucial cases, however, repugnance is the emotional expression of deep wisdom, beyond reason's power fully to articulate it. Can anyone really give an argument fully adequate to the horror which is father-daughter incest (even with consent), or having sex with animals, or mutilating a corpse, or eating human flesh, or even just (just!) raping or murdering another human being? Would anybody's failure to give full rational justification for his or her revulsion at these practices make that revulsion ethically suspect? Not at all. On the contrary, we are suspicious of those who think that they can rationalize away our horror, say, by trying to explain the enormity of incest with arguments only about the genetic risks of inbreeding.

The repugnance at human cloning belongs in this category. We are repelled by the prospect of cloning human beings not because of the strangeness or novelty of the undertaking, but because we intuit and feel, immediately and without argument, the violation of things that we rightfully hold dear. Repugnance, here as elsewhere, revolts against the excesses of human willfulness, warning us not to transgress what is unspeakably profound....

* * *

Typically, cloning is discussed in one or more of three familiar contexts, which one might call the technological, the liberal and the meliorist. Under the first, cloning will be seen as an extension of existing techniques for assisting reproduction and determining the genetic makeup of children. Like them, cloning is to be regarded as a neutral technique, with no inherent meaning or goodness, but subject to multiple uses, some good, some bad. The morality of cloning thus depends absolutely on the goodness or badness of the motives and intentions of the cloners....

The liberal (or libertarian or liberationist) perspective sets cloning in the context of rights, freedoms and personal empowerment. Cloning is just a new option for exercising an individual's right to reproduce or to have the kind of child that he or she wants. Alternatively, cloning enhances our liberation (especially women's liberation) from the confines of nature, the vagaries of change, or the necessity for sexual mating. Indeed, it liberates women from the need for men altogether....

The meliorist perspective embraces valetudinarians and also eugenicists....

These people see in cloning a new prospect for improving human beings—minimally, by ensuring the perpetuation of healthy individuals by avoiding the risks of genetic disease inherent in the lottery of sex, and maximally, by producing "optimum babies," preserving outstanding genetic material, and (with the help of soon-to-come techniques for precise genetic engineering) enhancing inborn human capacities on many fronts. Here the morality of cloning as a means is justified solely by the excellence of the end....

* * *

These three approaches, all quintessentially American and all perfectly fine in their places, are sorely wanting as approaches to human procreation. It is, to say the least, grossly distorting to view the wondrous mysteries of birth, renewal and individuality, and the deep meaning of parent-child relations, largely through the lens of our reductive science and its potent technologies. Similarly, considering reproduction (and the intimate relations of family life!) primarily under the political-legal, adversarial and individualistic notion of rights can only undermine the private yet fundamentally social, cooperative and duty-laden character of child-bearing, child-rearing and their bond to the covenant of marriage....

The technical, liberal and meliorist approaches all ignore the deeper anthropological, social and, indeed, ontological meanings of bringing forth new life. To this more fitting and profound point of view, cloning shows itself to be a major alteration, indeed, a major violation, of our given nature as embodied, gendered and engendering beings—and of the social relations built on this natural ground. Once this perspective is recognized, the ethical judgment on cloning can no longer be reduced to a matter of motives and intentions, rights and freedoms, benefits and harms, or even means and ends. It must be regarded primarily as a matter of meaning: Is cloning a fulfillment of human begetting and belonging? Or is cloning rather, as I contend, their pollution and perversion? To pollution and perversion, the fitting response can only be horror and revulsion; and conversely, generalized horror and revulsion are prima facie evidence of foulness and violation. The burden of moral argument must fall entirely on those who want to declare the widespread repugnances of humankind to be mere timidity or superstition.

Yet repugnance need not stand naked before the bar of reason. The wisdom of our horror at human cloning can be partially articulated, even if this is finally one of those instances about which the heart has its reasons that reason cannot entirely know....

THE PERVERSITIES OF CLONING

First, an important if formal objection: any attempt to clone a human being would constitute an unethical experiment upon the resulting child-to-be. As... animal experiments... indicate, there are grave risks of mishaps and deformities. Moreover, because of what cloning means, one cannot presume a future cloned child's consent to be a clone, even a healthy one. Thus, ethically speaking, we cannot even get to know whether or not human cloning is feasible.

I understand, of course, the philosophical difficulty of trying to compare a life with defects against nonexistence. Several bioethicists, proud of their philosophical cleverness, use this conundrum to embarrass claims that one can injure a child in its conception, precisely because it is only thanks to that complained-of conception that the child is alive to complain. But common sense tells us that we have no reason to fear such philosophisms. For we surely know that people can harm and even maim children in the very act of conceiving them, say, by paternal transmission of the AIDS virus, maternal transmission of heroin dependence or, arguably, even by bringing them into being as bastards or with no capacity or willingness to look after them properly. And we believe that to do this intentionally, or even negligently, is inexcusable and clearly unethical....

* * *

Cloning creates serious issues of identity and individuality. The cloned person may experience concerns about his distinctive identity not only because he will be in genotype and appearance identical to another human being, but, in this case, because he may also be twin to the person who is his "father" or "mother" —if one can still call them that. What would be the psychic burdens of being the "child" or "parent" of your twin? The cloned individual, moreover, will be saddled with a genotype that has already lived. He will not be fully a surprise to the world. People are likely always to compare his performances in life with that of his alter ego. True, his nurture and his circumstance in life will be different; genotype is not exactly destiny. Still, one must also expect parental and other efforts to shape this new life after the original—or at least to view the child with the original version always firmly in mind....

Since the birth of Dolly, there has been a fair amount of doublespeak on this matter of genetic identity. Experts have rushed in to reassure the public that the clone would in no way be the same person, or have any confusions about his or her identity: as previously noted, they are pleased to point out that the clone of Mel Gibson would not be Mel Gibson. Fair enough. But one is shortchanging the truth by emphasizing the additional importance of the intrauterine environment, rearing and social setting: genotype obviously matters plenty. That, after all, is the only reason to clone, whether human beings or sheep. The odds that clones of Wilt Chamberlain will play in the NBA are, I submit, infinitely greater than they are for clones of Robert Reich....

Genetic distinctiveness not only symbolizes the uniqueness of each human life and the independence of its parents that each human child rightfully attains. It can also be an important support for living a worthy and dignified life. Such arguments apply with great force to any large-scale replication of human individuals. But they are sufficient, in my view, to rebut even the first attempts to clone a human being. One must never forget that these are human beings upon whom our eugenic or merely playful fantasies are to be enacted.

Troubled psychic identity (distinctiveness), based on all-too-evident genetic identity (sameness), will be made much worse by the utter confusion of social identity and kinship ties....

Social identity and social ties of relationship and responsibility are widely connected to, and supported by, biological kinship. Social taboos on incest (and adultery) everywhere serve to keep clear who is related to whom (and especially which child belongs to which parents), as well as to avoid confounding the social identity of parent-and-child (or brother-

and-sister) with the social identity of lovers, spouses and co-parents. True, social identity is altered by adoption (but as a matter of the best interest of already living children: we do not deliberately produce children for adoption). True, artificial insemination and in vitro fertilization with donor sperm, or whole embryo donation, are in some way forms of "prenatal adoption"—a not altogether unproblematic practice. Even here, though, there is in each case (as in all sexual reproduction) a known male source of sperm and a known single female source of egg—a genetic father and a genetic mother—should anyone care to know (as adopted children often do) who is genetically related to whom.

In the case of cloning, however, there is but one "parent." The usually sad situation of the "single-parent child" is here deliberately planned, and with a vengeance. In the case of self-cloning, the "offspring" is, in addition, one's twin; and so the dreaded result of incest—to be parent to one's sibling—is here brought about deliberately, albeit without any act of coitus. Moreover, all other relationships will be confounded....

* * *

Human cloning would also represent a giant step toward turning begetting into making, procreation into manufacture (literally, something "handmade"), a process already begun with in vitro fertilization and genetic testing of embryos. With cloning, not only is the process in hand, but the total genetic blueprint of the cloned individual is selected and determined by the human artisans.... In clonal reproduction, ... and in the more advanced forms of manufacture to which it leads, we give existence to a being not by what we are but by what we intend and design. As with any product of our making, no matter how excellent, the artificer stands above it, not as an equal but as a superior, transcending it by his will and creative prowess. Scientists who clone animals make it perfectly clear that they are engaged in instrumental making; the animals are, from the start, designed as means to serve rational human purposes. In human cloning, scientists and prospective "parents" would be adopting the same technocratic mentality to human children: human children would be their artifacts.

Such an arrangement is profoundly dehumanizing, no matter how good the product. Mass-scale cloning of the same individual makes the point vividly; but the violation of human equality, freedom and dignity are present even in a single planned clone....

* * *

Finally, and perhaps most important, the practice of human cloning by nuclear transfer—like other anticipated forms of genetic engineering of the next generation—would enshrine and aggravate a profound and mischievous misunderstanding of the meaning of having children and of the parent-child relationship. When a couple now chooses to procreate, the partners are saying yes to the emergence of new life in its novelty, saying yes not only to having a child but also, tacitly, to having whatever child this child turns out to be. In accepting our finitude and opening ourselves to our replacement, we are tacitly confessing the limits of our control. In this ubiquitous way of nature, embracing the future by procreating means precisely that we are relinquishing our grip, in the very activity of taking up our own share in what we hope will be the immortality of human life and the human species. This means that our children are not *our* children: they are not our property, not our possessions. Neither are they supposed to live our lives for us, or anyone else's life but their own. To be sure, we seek to guide them on their way, imparting to them not just life but nurturing, love, and a way of life; to be sure, they bear our hopes that they will live fine and flourishing lives, enabling us in small measure to transcend our own limitations. Still, their genetic distinctiveness and independence are the natural foreshadowing of the deep truth that they have their own and never-before-enacted life to live. They are sprung from a past, but they take an uncharted course into the future....

MEETING SOME OBJECTIONS

The defenders of cloning, of course, are not wittingly friends of despotism. Indeed, they regard themselves mainly as friends of freedom: the freedom of individuals to reproduce, the freedom of scientists and inventors to discover and devise and to foster "progress" in genetic knowledge and technique. They want large-scale cloning only for animals, but they wish to preserve cloning as a human option for exercising our "right to reproduction"—our right to have children, and children with "desirable genes." As law professor John Robertson points out, under our "right to reproduce" we already practice early forms of unnatural, artificial and extramarital reproduction, and we already practice early forms of eugenic choice. For this reason, he argues, cloning is no big deal.

We have here a perfect example of the logic of the slippery slope, and the slippery way in which it already works in this area. Only a few years ago, slippery slope arguments were used to oppose artificial insemination and in vitro fertilization using unrelated sperm donors. Principles used to justify these practices, it was said, will be used to justify more artificial and more eugenic practices, including cloning. Not so, the defenders retorted, since we can make the necessary distinctions. And now, without even a gesture at making the necessary distinctions, the continuity of practice is held by itself to be justificatory.

The principle of reproductive freedom as currently enunciated by the proponents of cloning logically embraces the ethical acceptability of sliding down the entire rest of the slope—to producing children ectogenetically from sperm to term (should it become feasible) and to producing children whose entire genetic makeup will be the product of parental eugenic planning and choice. If reproductive freedom means the right to have a child of one's own choosing, by whatever means, it knows and accepts no limits.

But, far from being legitimated by a "right to reproduce," the emergence of techniques of assisted reproduction and genetic engineering should compel us to reconsider the meaning and limits of such a putative right. In truth, a "right to reproduce" has always been a peculiar and problematic notion. Rights generally belong to individuals, but this is a right which (before cloning) no one can exercise alone. Does the right then inhere only in couples? Only in married couples? Is it a (woman's) right to carry or deliver or a right (of one or more parents) to nurture and rear? Is it a right to have your own biological child? Is it a right only to attempt reproduction, or a right also to succeed? Is it a right to acquire the baby of one's choice? ...

BAN THE CLONING OF HUMANS

What, then, should we do? We should declare that human cloning is unethical in itself and dangerous in its likely consequences. In so doing, we shall have the backing of the overwhelming majority of our fellow Americans, and of the human race, and (I believe) of most practicing scientists. Next, we should do all that we can to prevent the cloning of human beings. We should do this by means of an international legal ban if possible, and by a unilateral national ban, at a minimum. Scientists may secretly undertake to violate such a law, but they will be deterred by not being able to stand up proudly to claim the credit for their technological bravado and success. Such a ban on clonal baby-making, moreover, will not harm the progress of basic genetic science and technology. On the contrary, it will reassure the public that scientists are happy to proceed without violating the deep ethical norms and intuitions of the human community....

I appreciate the potentially great gains in scientific knowledge and medical treatment available from embryo research, especially with cloned embryos. At the same time, I have serious reservations about creating human embryos for the sole purpose of experimentation. There is something deeply repugnant and fundamentally transgressive about such a utilitarian treatment of prospective human life. This total, shameless exploitation is worse, in my opinion, than the "mere" destruction of nascent life. But I see no added objections, as a matter of principle, to creating and using *cloned* early embryos for research purposes, beyond the objections that I might raise to doing so with embryos produced sexually.

And yet, as a matter of policy and prudence, any opponent of the manufacture of cloned humans must, I think, in the end oppose also the creating of cloned human embryos.... We should allow all cloning research on animals to go forward, but the only safe trench that we can dig across the slippery slope, I suspect, is to insist on the inviolable distinction between animal and human cloning.

Some readers, and certainly most scientists, will not accept such prudent restraints, since they desire the benefits of research. They will prefer, even in fear and trembling, to allow human embryo cloning research to go forward.

Very well. Let us test them. If the scientists want to be taken seriously on ethical grounds, they must at the very least agree that embryonic research may proceed if and only if it is preceded by an absolute and effective ban on all attempts to implant into a uterus a cloned human embryo (cloned from an adult) to produce a living child. Absolutely no permission for the former without the latter.

The National Bioethics Advisory Commission's recommendations regarding this matter should be watched with the greatest care. Yielding to the wishes of the scientists, the commission will almost surely recommend that cloning human embryos for research be permitted. To allay public concern, it will likely also call for a temporary moratorium—not a legislative ban—on implanting cloned embryos to make a child, at least until such time as cloning techniques will have been perfected and rendered "safe" (precisely through the permitted research with cloned embryos). But the call for a moratorium rather than a legal ban would be a moral and a practical failure. Morally, this ethics commission would (at best) be waffling on the main ethical question, by refusing to declare the production of human clones unethical (or ethical). Practically, a moratorium on implantation cannot provide even the minimum protection needed to prevent the production of cloned humans.

Opponents of cloning need therefore to be vigilant. Indeed, no one should be willing even to consider a recommendation to allow the embryo research to proceed unless it is accompanied by a call for *prohibiting* implantation and until steps are taken to make such a prohibition effective.

* * *

Technically, the National Bioethics Advisory Commission can advise the president only on federal policy, especially federal funding policy. But given the seriousness of the matter at hand, and the grave public concern that goes beyond federal funding, the commission should take a broader view. (If it doesn't, Congress surely will.) . . .

The proposal for such a legislative ban is without American precedent, at least in technological matters, though the British and others have banned cloning of human beings, and we ourselves ban incest, polygamy and other forms of "reproductive freedom." Needless to say, working out the details of such a ban, especially a global one, would be tricky, what with the need to develop appropriate sanctions for violators. Perhaps such a ban will prove ineffective; perhaps it will eventually be shown to have been a mistake. But it would at least place the burden of practical proof where it belongs: on the proponents of this horror, requiring them to show very clearly what great social or medical good can be had only by the cloning of human beings. . . .

The president's call for a moratorium on human cloning has given us an important opportunity. In a truly unprecedented way, we can strike a blow for the human control of the technological project, for wisdom, prudence and human dignity. The prospect of human cloning, so repulsive to contemplate, is the occasion for deciding whether we shall be slaves of unregulated progress, and ultimately its artifacts, or whether we shall remain free human beings who guide our technique toward the enhancement of human dignity.

POSTSCRIPT

Is It Ethically Permissible to Clone Human Beings?

Have humans already been cloned? In 1978 writer David Rorvik claimed to document the deed in *In His Image: The Cloning of a Man* (Lippincott), in which he describes "Max," a millionaire, who had hired a scientist to set up a lab somewhere in Asia, tap local women for eggs, refine the necessary techniques, and produce Max, Jr. Rorvik's claims provoked controversy reminiscent of that surrounding the current issue, but in the end no one believed him.

Now, however, the technique has been shown to work in sheep and monkeys. It seems very much on the verge of possibility for humans, and the debate over whether or not that possibility is desirable is vigorous. For the moment, the debate has been settled by the report of the National Bioethics Advisory Commission, which commission chair Harold T. Shapiro summarized in the July 11, 1997, issue of *Science*. He said, "[The commission] made every effort to consult widely with ethicists, theologians, scientists, scientific societies, physicians, and others in initiating an analysis of the many scientific, legal, religious, ethical, and moral dimensions of the issue [including] potential risks and benefits of using this technique to create children and a review of the potential constitutional challenges that might be raised if new legislation were to restrict [its use]."

Speaking to the commission, Ruth Macklin, of the Albert Einstein College of Medicine, said, "It is absurd to maintain that the proposition 'cloning is morally wrong' is self-evident.... If I cannot point to any great benefits likely to result from cloning, neither do I foresee any probable great harms, provided that a structure of regulation and oversight is in place. If objectors to cloning can identify no greater harm than a supposed affront to the dignity of the human species, that is a flimsy basis on which to erect barriers to scientific research and its applications."

Nathan Myhrvold, chief technology officer at Microsoft, takes a different tack in opposing bans on cloning in "Human Clones: Why Not? Opposition to Cloning Isn't Just Luddism—It's Racism," *Slate* (March 13, 1970). In it, he argues, "Calls for a ban on cloning amount to discrimination against people based on another genetic trait—the fact that somebody already has an identical DNA sequence."

The cloning of humans may thus be put off for years, but work with animals will continue. Indeed, in August 1997 researchers at ABS Global, a biotechnology company in Wisconsin, announced that they had cloned calves from the cells of month-old embryos and introduced Gene, a six-month-old Holstein, to photographers.

CONTRIBUTORS TO THIS VOLUME

EDITOR

THOMAS A. EASTON is a professor of life sciences at Thomas College in Waterville, Maine, where he has been teaching since 1983. He received a B.A. in biology from Colby College in 1966 and a Ph.D. in theoretical biology from the University of Chicago in 1971. He has also taught at Unity College, Husson College, and the University of Maine. He is a prolific writer, and his articles on scientific and futuristic issues have appeared in the scholarly journals *Experimental Neurology* and *American Scientist*, as well as in such popular magazines as *Astronomy*, *Consumer Reports*, and *Robotics Age*. He is also the science columnist for the online magazine *Tomorrowsf* (`http://www.tomorrowsf.com/`). His publications include *Focus on Human Biology*, 2d ed. (HarperCollins, 1995), coauthored with Carl E. Rischer, and *Careers in Science*, 3rd ed. (National Textbook, 1996). Dr. Easton is also a well-known writer and critic of science fiction.

STAFF

David Dean List Manager
David Brackley Developmental Editor
Juliana Poggio Associate Developmental Editor
Rose Gleich Administrative Assistant
Brenda S. Filley Production Manager
Juliana Arbo Typesetting Supervisor
Diane Barker Proofreader
Lara Johnson Graphics
Richard Tietjen Publishing Systems Manager

AUTHORS

ELIZABETH BALDWIN is a research ethics officer for the American Psychological Association's Science Directorate. Her work involves a broad range of research ethics issues, including those relating to the use of animals in research. Prior to her position at the American Psychological Association, she worked at the Congressional Research Service in the Division of Science Policy. She holds a B.A. in biology, an M.S. in entomology, and an M.A. in science, technology, and public policy.

WILFRED BECKERMAN is a fellow of Balliol College at Oxford University in Oxford, England, where he has been teaching since 1975. He has also been a professor of economy at the University of London and chair of the Department of Political Economy at the University College in London. He is coauthor, with Stephen Clark, of *Poverty and Social Security in Britain Since 1961* (Oxford University Press, 1982). He received a Ph.D. from Trinity College at Cambridge University in 1950.

SVEN BIRKERTS is the author of three books of criticism, including *American Energies: Essays on Fiction* (William Morrow, 1992). He has won the National Book Critics Circle Citation of Excellence in Reviewing, a P.E.N. Speilvogel/Diamondstein Special Citation for *The Electric Life: Essays on Modern Poetry* (William Morrow, 1989), and Lila Wallace–Reader's Digest Foundation and Guggenheim fellowships. His essays and reviews have appeared in the *New York Times Book Review,* the *Atlantic Monthly, Harper's Magazine,* and the *New Republic.*

PAUL BRODEUR is an author and a staff writer for the *New Yorker* magazine. He has published books on asbestos, ozone depletion, and the electromagnetic field–cancer link, including *Currents of Death: Power Lines, Computer Terminals, and the Attempt to Cover Up Their Threat to Your Health* (Simon & Schuster, 1989). He has won the National Magazine Award, the Sidney Hillman Foundation Award, and the American Association for the Advancement of Science Award, and the United Nations Environment Program has named him to its Global 500 Roll of Honor.

LESTER R. BROWN is president of the Worldwatch Institute. He is the author or coauthor of dozens of books, including *Tough Choices: Facing the Challenge of Food Scarcity* (W. W. Norton, 1996) and *Full House: Reassessing the Earth's Population Carrying Capacity,* with Hal Kane (W. W. Norton, 1994).

DANIEL CALLAHAN, a philosopher, is cofounder and president of the Hastings Center in Briarcliff Manor, New York, where he is also director of International Programs. He is the author or editor of over 31 publications, including *Ethics in Hard Times* (Plenum Press, 1981), coauthored with Arthur L. Caplan; *Setting Limits: Medical Goals in an Aging Society* (Simon & Schuster, 1987); and *The Troubled Dream of Life: In Search of Peaceful Death* (Simon & Schuster, 1993). He received a Ph.D. in philosophy from Harvard University.

EDWARD W. CAMPION is deputy editor of the *New England Journal of Medicine* and an assistant professor of medicine at Harvard Medical School and Massachusetts General Hospital.

MARY H. COOPER is a staff writer for the Congressional Quarterly's *CQ Researcher,* a weekly magazine providing in-depth analysis of current issues. She is the author of *The Business of Drugs* (Congressional Quarterly, 1988).

RICHARD DAWKINS is the Charles Simonyi Professor of the Public Understanding of Science at Oxford University and the recipient of the American Humanist Association's 1996 Humanist of the Year Award.

DANIEL C. DENNETT is the Distinguished Arts and Sciences Professor at Tufts University, where he is also director of the Center for Cognitive Studies. He is the author of *Brainstorms: Philosophical Essays on Mind and Psychology* (MIT Press, 1980), *Elbow Room: The Varieties of Will Worth Wanting* (MIT Press, 1984), and *Consciousness Explained* (Little, Brown, 1991).

A. K. DEWDNEY is a computer scientist at the University of Western Ontario in London, Ontario, Canada. For several years, he wrote *Scientific American*'s "Computer Recreations" and "Mathematical Recreations" columns.

FRANK DRAKE is a professor of astronomy and astrophysics at the University of California, Santa Cruz, where he has also served as dean of natural sciences. He is president of the SETI Institute and former president of the Astronomical Society of the Pacific, which is one of the world's leading astronomical organizations.

ANNE H. EHRLICH is a senior research associate in biological sciences at Stanford University in Stanford, California. She is coauthor, with Paul R. Ehrlich, of *Betrayal of Science and Reason: How Anti-Environmental Rhetoric Threatens Our Future* (Island Press, 1996) and coeditor, with John Birks, of *Hidden Dangers: Environmental Consequences of Preparing for War* (Sierra Club Books, 1991).

PAUL R. EHRLICH is the Bing Professor of Population Studies and a professor of biological sciences at Stanford University in Stanford, California. His many publications include *The Population Bomb* (Ballantine Books, 1971) and *Healing the Planet: Strategies for Resolving the Environmental Crisis* (Addison-Wesley, 1991), coauthored with Anne H. Ehrlich.

ROSS GELBSPAN was an editor and reporter at the *Philadelphia Bulletin,* the *Washington Post,* and the *Boston Globe* over a 30-year period. In 1984 he was corecipient of the Pulitzer Prize for public-service reporting.

DAVID H. GUSTON is an assistant professor of public policy at the Eagleton Institute of Policies at Rutgers–The State University of New Jersey. In 1990–1991 he served on the staff of the Panel on Scientific Responsibility and the Conduct of Research at the National Academy of Sciences. He is coeditor, with Kenneth Keniston, of *The Fragile Contract: University Science and the Federal Government* (MIT Press, 1994).

JACK HITT is a contributing editor to *Harper's Magazine.*

JAMES P. HOGAN is a writer of science fiction novels. Before moving from the United Kingdom to the United States in 1977, he was a systems design engineer.

JAMES HUGHES, a sociologist, is assistant director of research in the MacLean Center for Clinical Medical Ethics, Department of Medicine, at the University of Chicago and the editor of *Doctor-Patient Studies.* His special interests include health care reform and the social construction of personhood at the intersection of medical and environmental ethics. He received his doctorate in sociology from the University of Chicago in 1994.

LEON R. KASS is the Addie Clark Harding Professor in the College and the Committee on Social Thought at the University of Chicago and an adjunct scholar at the American Enterprise Institute. A trained physician and biochemist, he is the author of *Toward a More Natural Science: Biology and Human Affairs* (Free Press, 1985).

KENNETH KENISTON is the Andrew W. Mellon Professor of Human Development in the Program in Science, Technology, and Society at the Massachusetts Institute of Technology. He is coeditor, with David H. Guston, of *The Fragile Contract: University Science and the Federal Government* (MIT Press, 1994).

ANDREW KIMBRELL is policy director of the Foundation on Economic Trends in Washington, D.C., which was founded in 1977 to disseminate information through lectures and the distribution of educational materials on issues such as the environment, religion, genetics, and engineering in order to effect social change.

JESSE MALKIN writes about economic issues for *Investor's Business Daily.* He is a Rhodes scholar, and he holds B.A. degrees from Oxford University and Oberlin College.

JOHN S. MAYO is president emeritus of Lucent Technologies Bell Laboratories, formerly AT&T Bell Laboratories.

JOHN MERCHANT is president of RPU Technology in Needham, Massachusetts. He is a former senior staff engineer at Loral Infrared and Imaging Systems in Lexington, Massachusetts.

HANS MORAVEC is a principal research scientist in the Robotics Institute at Carnegie Mellon University in Pittsburgh, Pennsylvania, and director of the university's Mobile Robot Laboratory. He received a Ph.D. from Stanford University in 1980 for his design of a TV-equipped, computer-controlled robot that could negotiate cluttered obstacle courses. His publications include *Mind Children: The Future of Robot and Human Intelligence* (Harvard University Press, 1988).

JONATHAN D. MORENO teaches bioethics at the State University of New York Health Science Center in Brooklyn and directs a human research ethics project at the University of Pennsylvania's Center for Bioethics. He was a senior staff member of the President's Advisory Committee on Human Radiation Experiments, and he is the author of *Deciding Together: Bioethics and Moral Consensus* (Oxford University Press, 1995).

NATIONAL ACADEMY OF SCIENCES is a private, nonprofit, self-perpetuating society of distinguished scholars engaged in scientific and engineering research. The academy is dedicated to the furtherance of science and technology and to their use for the general welfare. Upon the authority of the charter granted to it by Congress in 1863, the academy has a mandate that requires it to advise the federal government on scientific and technical matters. Dr. Bruce Alberts is president of the National Academy of Sciences.

CHARLES PETIT is a science writer with the *San Francisco Chronicle.*

JOHN A. ROBERTSON is the Vinson and Elkins Chair at the University of Texas School of Law. He is the author of *Children of Choice: Freedom and the New Reproductive Technologies* (Princeton University Press, 1996) and *The Rights of the Critically Ill: The Basic ACLU Guide to the Rights of Critically Ill and Dying Patients* (Ballinger Publishing, 1983), as well as numerous law review articles on the subject of reproductive technology. An expert on medical ethics in the courtroom, he has served on numerous ethics committees, including the President's Commission for the Study of Ethical Problems in Medicine and Biomedical and Behavioral Research.

JOHN R. SEARLE is a professor of philosophy at the University of California, Berkeley.

JOHN SHANAHAN is vice president and counsel of the Alexis de Tocqueville Institution in Arlington, Virginia.

ANDREW L. SHAPIRO is a contributing editor to *The Nation* and a fellow of the Twentieth Century Fund.

DAVA SOBEL is a science and medicine writer for several newspapers and magazines, including *Harvard Magazine, Omni, Good Housekeeping,* and the *New York Times Book Review.* She is a former science reporter for the *New York Times,* and she is the author of *The Incredible Planets: New Views of the Solar Family* (Reader's Digest Association, 1992) and *Longitude* (Walker, 1995).

WEN STEPHENSON is editor of the *Atlantic Monthly*'s online edition.

WORLD BANK comprises five organizations: the International Bank for Reconstruction and Development (IBRD), the International Development Association (IDA), the International Finance Corporation (IFC), the Multilateral Investment Guarantee Agency (MIGA), and the International Centre for the Settlement of Investment Disputes (ICSID). The IBRD, frequently called the "World Bank," aims to reduce poverty and improve living standards by promoting sustainable growth and investments in people. The bank provides loans, technical assistance, and policy guidance to help its developing-country members achieve this objective.

STEVEN ZAK is an attorney in Los Angeles, California. He received a B.A. in psychology from Michigan State University in 1971, an M.S. from the Wayne State University School of Medicine in 1975, and a J.D. from the University of Southern California Law School in 1984. He has written about animals with regard to ethics and the law for numerous publications, including the *Los Angeles Times,* the *New York Times,* and the *Chicago Tribune.*

ROBERT ZUBRIN, a former senior engineer at Lockheed Martin Astronautics, is executive chairman of the National Space Society and founder of Pioneer Astronautics in Lakewood, Colorado. He is coeditor, with Stanley Schmidt, of *Islands in the Sky: Bold New Ideas for Colonizing Space* (John Wiley, 1996).

INDEX

abortion, 211, 229
"aggressive mimicry," 160–161
agriculture: controversy over future food supply and, 68–85; global warming and, 100; ozone depletion and, 114
AIDS, experimental treatments for, 320–324, 331
Algeny (Rifkin), 219
American Association for the Accreditation of Laboratory Animal Care (AAALAC), 305
American Psychological Association (APA), 305, 307
Anderson, French, 214, 215–218
animals, controversy over use of, in research, 300–316
Antarctica, ozone hole over, 112–128
antivivisection movement, 301
Arctic, ozone hole over, 112–128
artificial intelligence, 183, 185; controversy over, 274–293

Baldwin, Elizabeth, on use of animals in research, 300–307
batteries, universal robots and, 279
Beckerman, Wilfred, on global warming, 98–107
Benedick, Richard Elliot, 116–117
Birkerts, Sven: on computers and literacy, 254–261; reaction to views of, 262–270
bone marrow transplantation (BMT), 216; between human and baboon to treat AIDS, 320–324
Brave New World (Huxley), 226
Brodeur, Paul, 141; on electromagnetic fields, 132–139
bromine monoxide, 112, 113
Brown, David, 132, 133, 134, 135
Brown, George, 5, 8
Brown, Lester R., on food supply for future generations, 77–85
brownlash, controversy over environmental regulations and, 157–164
"bubble boy" syndrome, genetic engineering therapy for, 215–218

Callahan, Daniel, on science as a faith, 26–34
Campion, Edward W., on electromagnetic fields, 140–143
cancer, 214, 216; electromagnetic fields and, 132–143; skin, and ultraviolet radiation, 114, 121, 126–127
carbon dioxide, controversy over global warming and, 90–108
Carnegie Commission on Science, Technology, and Government, 4, 12
catalysis argument, ozone and, 124
Chargoff, Erwin, 210, 211
Chernobyl, nuclear disaster at, 7
Clean Water Act (CWA), 149
China, agricultural productivity in, 80
Chinese room argument, artificial intelligence and, 284–291
chlorofluorocarbons (CFCs), 94, 96; controversy over ozone depletion and, 112–128, 155
Churchland, Patricia Smith, 286, 287, 290, 291
Churchland, Paul M., 286, 287, 290, 291
climate, controversy over global warming and, 90–107
Cline, William, 101, 103, 104–105
Clinton, Bill, 5
Clipper Chip, 248
cloning, controversy over, of humans, 336–350
Cocconi, Giuseppe, 194, 196
cold war, effect of, on R&D, 7, 11, 15, 18
computer models, use of, in medical research, 306, 315
computers: controversy over artificial intelligence and, 274–293; controversy over, as hazardous to literacy, 254–270; controversy over information revolution and, 236–249
connectionist model, of artificial intelligence, 287
conservationists, 32
Cooper, Mary H., on ozone depletion, 112–120
cooperative research and development agreements (CRADAs), 20
Copernicus, Nicholas, 53–54, 292
creationism: controversy over replacing theory of evolution with, 44–62; controversy over science as a faith and, 38
Cruelty to Animals Act, of Great Britain, 301
cybernetics, 275, 276
cystic fibrosis, 211, 212

Darwin, Charles, 45, 52–53, 54–57
Dawkins, Richard, on science as a faith, 35–39
democracy, science and, 8–10
Dennett, Daniel C., on replacing theory of evolution with creationism, 52–62
Dewdney, A. K., on search for extraterrestrial intelligence, 200–206
discrimination, genetic engineering and, 224, 227–228
diversity, genetic engineering and, 224, 226–227
dualism, artificial intelligence and, 292
Drake, Frank: on search for extraterrestrial intelligence, 192–199; reaction to views of, 200–201, 204
Ducks Unlimited, 154
dumb robots, 277

education, computers and, 238, 239, 241
Ehrlich, Anne H., on environmental regulations, 157–164
Ehrlich, Paul R., on environmental regulations, 157–164
electromagnetic fields, controversy over health risks of, 132–143
Ellsaesser, Hugh, 123, 125–126, 127
embryo donation, 339
embryos, 212
environmental issues, 17, 32; controversy over global warming and, 90–107; controversy over ozone depletion and, 112–128
environmental regulations, controversy over, 148–164
Equal Employment Opportunity Commission (EEOC), 214
eugenic cloning, 338–340
eugenics, controversy over genetic engineering and, 210–230
evolution, 35–36, 275; controversy over replacing, with creationism, 44–62
expert systems, 280
extraterrestrial life, controversy over search for, 192–206

faith, controversy over science as, 26–39
FEED, 264–265

fertilizer, agricultural productivity and, 81
food, controversy over future supply of, 68–85
fossil fuel, controversy over global warming and, 90–107
fundamentalism, genetic engineering and, 233–234
fuzzy logic, 280

Galileo (Galelei), 292
Gelbspan, Ross, on global warming, 90–97
General Circulation Model, of climate, 94
genetic engineering, controversy over, 210–230
geothermal energy, 104
germline genetic engineering, controversy over, 210–230
Geschwind, Sandy, 132, 133, 134, 135
Getty, Jeff, interview with, 320–324
Gingrich, Newt, 177
glaciers, global warming and, 92
global warming, controversy over, 90–107
Glover, Jonathan, 219, 220
Gore, Al, 98
Gould, Stephen Jay, 46
greenhouse effect, controversy over global warming and, 90–107
Guidelines for Ethical Conduct in the Care and Use of Animals (American Psychological Association), 305
Guilford, CT, controversy over electromagnetic fields in, 132–136, 139
Guston, David H., on federal government involvement in decision making about science, 4–13

halons, 112, 116, 119
Helvarg, David, 160
Hitt, Jack, on replacing theory of evolution with creationism, 44–51
Hoffman, Dave, 117–118, 120
Hogan, James P., on ozone depletion, 121–128
Hughes, James, on genetic engineering, 219–230
Human Genome Project, 216
humans: controversy over cloning of, 336–350; controversy over use of, in medical research, 320–331
hydrochlorofluorocarbons (HCFCs), 119
hydrogen, as energy source, 96, 104

Ildstad, Suzanne, 320, 323
imagery, universal robots and, 278–288
immune system: genetic engineering and, 215–218; ultraviolet radiation and, 114
in vitro cultures, use of, in medical research, 306, 315
in vitro fertilization (IVF) treatment, cloning and, 338
infectious diseases, spread of, and global warming, 92, 98
information revolution, 187; controversy over, 236–249
Institutional Animal Care and Use Committees (IACUCs), 303, 304, 305
intellectual property rights, 19
"intelligent design" (ID) theory, 46
Intergovernmental Panel on Climate Change (IPCC), 90, 91, 95, 98, 101, 105
International Rice Research Institute (IRRI), 78
irrigation, agricultural productivity and, 74

James, William, 57
Japan, agricultural productivity in, 79
Johnson, Phillip, 45–46

Kass, Leon R., on cloning human beings, 342–350
Keniston, Kenneth, on federal government involvement in decision making about science, 4–13
Kimbrell, Andrew, on genetic engineering, 210–218
Kurylo, Michael, 112–113

Laboratory Animal Welfare Act, 312
Leyner, Mark, 257
literacy, controversy over computers as hazardous to, 254–270

Malkin, Jesse, on global warming, 98–107
Malthus, Thomas, 72
Manhattan Project, 7
manned missions, controversy over, in space exploration, 170–188
Mars, and controversy over manned missions in space exploration, 170–188
Mars Direct, 171–177
Mayo, John S., on information revolution, 236–243
McCaffrey, Larry, 257
McGarrity, Gerald J., 214, 217
medical research: controversy over use of animals in, 300–316; controversy over use of humans in, 320–331
melanoma, 211; ultraviolet radiation and, 114, 127
Mendel, Gregor, 54
Merchant, John, on manned missions in space exploration, 179–188
merit review, competitive, and federal science research funding, 21–22
META-SETI (Megachannel ExtraTerrestrial Assay), 195, 196, 197
Michaels, Pat, 93–94
Molina, Mario, 115, 116, 123, 125
Montecito, CA, controversy over electromagnetic fields in, 138–139
Montreal Protocol, 113, 116, 117
Moravec, Hans, on artificial intelligence, 274–282
Moreno, Jonathan D., on use of humans in medical research, 325–331
Morrison, Philip, 194, 195, 196
murine leukemia virus (MULV), 216

National Academy of Sciences (NAS), 4, 8, 12, 127, 210; on federal government involvement in decision making about science, 14–22
National Aeronautics and Space Administration (NASA), 112, 115, 120, 153, 171; controversy over search for extraterrestrial intelligence and, 192–206
National Oceanic and Atmospheric Administration (NOAA), 92, 115, 118
National Science and Technology Council, 6, 11
National Science Foundation (NSF), 4, 5, 19, 300
neural nets, 280
Newkirk, Ingrid, 300, 307
Noah's Flood, creationism and, 45, 47
nuclear energy, 104
nuclear weapons, 113, 226
Nuremberg Code, use of humans in medical research and, 325–331

Office of Management and Budget (OMB), 5, 6, 313
ozone depletion, 94; controversy over, 112–128, 155

Paglia, Camille, 257
patents, on forms of life, 27
Pease, Robert, 122–123, 124
pesticides, 123

Petit, Charles, on use of humans in medical research, 320–324
photodissociation, 122, 123
photosynthesis, 114
photovoltaic cells, 104
polar vortex, 116, 125
preservationists, 32
privacy: genetic engineering and, 222; information revolution and, 244–249
"progressive creationists," 45
Project Ozma, 192, 194, 196, 200, 201
Project Sentinel, 194–195
Prolog logical inference computer language, 280
property rights, environmental regulations and, 150–151, 153–154
psychology, animal research in, 300–307
Public Health Service (PHS), 304, 305, 314
punctuated equilibrium, 46

R&D (research and development), and controversy over federal government involvement in decision making about science, 4–22
Remote Projection Units (RPUs), 180–188
research: controversy over federal government involvement in decision making about scientific, 4–22; controversy over use of animals in, 300–316; controversy over use of humans in, 320–331
Resource Conservation and Recovery Act (RCRA), 149
retroviruses, 216
Rifkin, Jeremy, 210, 213, 214–215, 219, 223, 224
rights argument, use of animals in medical research and, 301–302, 309, 311
Robertson, John A., on cloning human beings, 336–341
robots, controversy over artificial intelligence and, 274–293
Rogers, Stanfield, 213, 215
Rowan County, NC, controversy over electromagnetic fields in, 136–138
Rowland, Sherwood, 115, 116, 123, 125
rule utilitarianism, genetic engineering and, 221

Sagan, Carl, 177
Santayana, George, 57–58
schistosomiasis, 98
Science Advisory Committee, 6
sea level, global warming and, 91, 94, 101
search for extraterrestrial intelligence (SETI), controversy over, 192–206
Searle, John R., on artificial intelligence, 283–293
Shanahan, John, on environmental regulations, 148–156
Shapiro, Andrew L., on information revolution, 244–249
sickle-cell anemia, 211, 212
Singer, Peter, 301, 302
Singer, S. Fred, 93, 94, 126
skin cancer, ultraviolet radiation and, 114, 121, 126–127
Small Business Innovation Research, 20
small businesses, environmental regulations and, 149–150
Sobel, Dava, on search for extraterrestrial intelligence, 192–199
Solomon, Susan, 117–118, 120
space program, 5, 7, 123; controversy over manned missions in space exploration and 170–188
Stephenson, Wen, on computers and literacy, 262–270

T lymphocytes, 215
Tarter, Jill, 194, 195
"theistic evolutionist," 45
Turing, Alan M., 283, 284, 292
Tuskegee syphilis study, 330

ultraviolet (UV) radiation, ozone depletion and, 112–128
UNESCO Declaration on the Protection of the Human Genome, 221
United Nations Enviornmental Program (UNEP), 114, 117, 119–120
United States, agricultural productivity in, 79, 80
United States Department of Agriculture (USDA), 304, 305, 306, 313, 314, 315
universal robots, 274–282
Upper Atmosphere Research Satellite (UARS), 120
utilitarianism: animal rights and, 301–302, 310–311; genetic engineering and, 221

virtual reality, 238, 239, 241
volcanoes, ozone depletion and, 113, 117–119, 120, 124

Wald, George, 210, 211
Whitmore, John, 44, 45, 47
Wise, Kurt, 48–50
wise-use movement, 160–161
World Bank, 78; on food supply for future generations, 68–76

xenogenic treatment, for AIDS, 320–324

yellow fever, 92
young-earther creationists, 45, 47, 48

Zak, Steven, on use of animals in research, 308–316
Zich, Robert, 257–258
zooplankton, decline in, 91–92, 113–114
Zubrin, Robert, on manned missions in space exploration, 170–178